TRANSPOSITION

SYMPOSIA OF THE
SOCIETY FOR GENERAL MICROBIOLOGY*

*Published by the Cambridge University Press, except for the first Symposium, which was published by Blackwell's Scientific Publications Limited.

TRANSPOSITION

EDITED BY
A. J. KINGSMAN, K. F. CHATER
AND S. M. KINGSMAN

FORTY-THIRD SYMPOSIUM OF
THE SOCIETY FOR
GENERAL MICROBIOLOGY
HELD AT
THE UNIVERSITY OF WARWICK
APRIL 1988

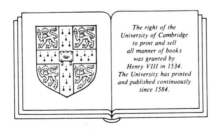

The right of the
University of Cambridge
to print and sell
all manner of books
was granted by
Henry VIII in 1534.
The University has printed
and published continuously
since 1584.

Published for the Society for General Microbiology

CAMBRIDGE UNIVERSITY PRESS
CAMBRIDGE
NEW YORK   NEW ROCHELLE
MELBOURNE   SYDNEY

Published by the Press Syndicate of the University of Cambridge
The Pitt Building, Trumpington Street, Cambridge CB2 1RP
32 East 57th Street, New York, NY 10022, USA
10 Stamford Road, Oakleigh, Melbourne 3166, Australia

First published 1988

Printed in Great Britain at The Bath Press, Avon

British Library cataloguing in publication data
Transposition. – (symposium of the Society
  for General Microbiology ; 43rd).
  1. Micro-organism – Physiology
  2. Deoxyribonucleic acid.   3. Translocation
  (Genetics).
  I. Kingsman, A.J.   II. Chater, K.F.
  III. Kingsman, S.M.   IV. Series
  576'.11   QR92.D/

Library of Congress cataloguing in publication data
Society for General Microbiology. Symposium
  (43rd : 1988 : University of Warwick)
  Transposition : Forty-third Symposium of the Society
for General Microbiology, held at the University of
Warwick, April 1988.
  Includes index.
  1. Microbial genetics – Congresses.   2. Translocation
(Genetics) – Congresses.   I. Kingsman, A.J.   II. Chater,
K.F.   III. Kingsman, S.M.   IV. Title.
  QH434.S63   1988   576'.139   87–36749

ISBN 0 521 35464 1

CONTRIBUTORS

ADAMS, S. E. Department of Molecular Biology, British Biotechnology Ltd, Brook House, Watlington Road, Cowley, Oxford OX4 5LY, UK

ADZUMA, K. Laboratory of Molecular Biology, National Institute of Diabetes, and Digestive and Kidney Diseases, National Institutes of Health, Bethesda, MA 20892, USA

BERG, D. E. Departments of Microbiology and Immunology, and of Genetics, Washington University School of Medicine, St Louis, MO 63110, USA

BIBB, M. J. John Innes Institute and AFRC Institute of Plant Science Research, Colney Lane, Norwich NR4 7UH, UK

CHATER, K. F. John Innes Institute and AFRC Institute of Plant Science Research, Colney Lane, Norwich NR4 7UH, UK

CLEWELL, D. B. Departments of Oral Biology and Microbiology/Immunology, Schools of Dentistry and Medicine, and The Dental Research Institute, The University of Michigan, Ann Arbor, MI 48109, USA

CRAIGIE, R. Laboratory of Molecular Biology, National Institute of Diabetes, and Digestive and Kidney Diseases, National Institutes of Health, Bethesda, MA 20892, USA

DATTA, N. 9 Dukes Avenue, London W4 2AA, UK

DODSON, K. W. Department of Microbiology and Immunology, Washington University School of Medicine, St Louis, MO 63110, USA

FALVEY, E. E. Department of Molecular Biophysics and Biochemistry, Yale University, 333 Cedar St., New Haven, CT 06510, USA

FINNEGAN, D. J. Department of Molecular Biology, University of Edinburgh, King's Buildings, Mayfield Road, Edinburgh EH9 3JR, UK

FLANNAGAN, S. E. Departments of Oral Biology and Microbiology/Immunology, Schools of Dentistry and Medicine, and The Dental Research Institute, The University of Michigan, Ann Arbor, MI 48109, USA

FULTON, S. M. Department of Biochemistry, South Parks Road, Oxford OX1 3QU, UK

GAWRON-BURKE, C. Departments of Oral Biology and Microbiology/Immunology, Schools of Dentistry and Medicine, and The Dental Research Institute, The University of Michigan, Ann Arbor, MI 48109, USA

GRINDLEY, N. D. F. Department of Molecular Biophysics and Biochemistry, Yale University, 333 Cedar St., New Haven, CT 06510, USA

HAAS, R. Max-Planck Institut für Biologie, Infektgenetik, Spemannstrasse 34, D-74 Tübingen, FRG

HARTL, D. L. Department of Genetics, Washington University School of Medicine, St Louis, MO 63110, USA

HATFULL, G. F. Department of Molecular Biophysics and Biochemistry, Yale University, 333 Cedar St., New Haven, CT 06510, USA

HENDERSON, D. J. John Innes Institute and AFRC Institute of Plant Science Research, Colney Lane, Norwich NR4 7UH, UK

HOPWOOD, D. A. John Innes Institute and AFRC Institute of Plant Science Research, Colney Lane, Norwich NR4 7UH, UK

JONES, J. M. Departments of Oral Biology and Microbiology/Immunology, Schools of Dentistry and.Medicine, and The Dental Research Institute, The University of Michigan, Ann Arbor, MI 48109, USA

KAZIC, T. Department of Microbiology and Immunology, Washington University School of Medicine, St Louis, MO 63110, USA

KINGSMAN, A. J. Department of Biochemistry, South Parks Road, Oxford OX1 3QU, UK

KINGSMAN, S. M. Department of Biochemistry, South Parks Road, Oxford OX1 3QU, UK

KUFF, E. L. Laboratory of Biochemistry, National Cancer Institute, National Institutes of Health, Bethesda, MA 20892, USA

LODGE, J. K. Department of Microbiology and Immunology, Washington University School of Medicine, St Louis, MO 63110, USA

LUEDERS, K. K. Laboratory of Biochemistry, National Cancer Institute, National Institutes of Health, Bethesda, MA 20892, USA

MALIM, M. H. Department of Biochemistry, South Parks Road, Oxford OX1 3QU, UK

MEYER, T. F. Max-Planck Institut für Biologie, Infektgenetik, Spemannstrasse 34, D-74 Tübingen, FRG

MIZUUCHI, K. Laboratory of Molecular Biology, National Institute of Diabetes, and Digestive and Kidney Diseases, National Institutes of Health, Bethesda, MA 20892, USA

MIZUUCHI, M. Laboratory of Molecular Biology, National Institute of Diabetes, and Digestive and Kidney Diseases, National Institutes of Health, Bethesda, MA 20892, USA

MURPHY, E. The Public Health Research Institute, 455 First Avenue, New York, NY 10016, USA

PAYS, E. Department of Molecular Biology, Free University of Brussels, 67 rue des Chevaux, B1640 Rhode St Genèse, Belgium

PHADNIS, S. H. Department of Microbiology and Immunology, Washington University School of Medicine, St Louis, MO 63110, USA

PUTTE, P. VAN DE Laboratory of Molecular Genetics, University of Leiden, PO Box 9505, 2300 RA Leiden, The Netherlands

RATHJEN, P. D. Department of Biochemistry, South Parks Road, Oxford OX1 3QU, UK

RIMPHANITCHAYAKIT, V. Department of Molecular Biophysics and Biochemistry, Yale University, 333 Cedar St., New Haven, CT 06510, USA

RIO, D. C. Whitehead Institute for Biomedical Research, Nine Cambridge Center, Cambridge, MA 02142, USA and Department of Biology, Massachusetts Institute of Technology, Cambridge, MA 02139, USA

SAEDLER, H. Max-Planck Institut für Zuchtungsforschung, 5000 Köln, FRG

SALVO, J. J. Department of Molecular Biophysics and Biochemistry, Yale University, 333 Cedar St., New Haven, CT 06510, USA

SAWYER, S. A. Department of Genetics, Washington University School of Medicine, St Louis, MO 63110, USA, and Department of Mathematics, Washington University, St Louis, MO 63110, USA

SCHELL, J. Max-Planck Institut für Züchtungsforschung 5000 Köln, FRG

SCHWARZ-SOMMER, Z. Max-Planck Institut für Züchtungsforschung, 5000 Köln, FRG

SENGHAS, E. Departments of Oral Biology and Microbiology/Immunology, Schools of Dentistry and Medicine, and The Dental Research Institute, The University of Michigan, Ann Arbor, MI 48109, USA

WILSON, W. Department of Biochemistry, South Parks Road, Oxford OX1 3QU, UK

YAMAMOTO, M. Departments of Oral Biology and Microbiology/Immunology, Schools of Dentistry and Medicine, and The Dental Research Institute, The University of Michigan, Ann Arbor, MI 48109, USA

CONTENTS

EDITORS' PREFACE

The arrangement of genetic information in prokaryotic and eukaryotic genomes is not fixed. Recombination can result in inversions, deletions and translocations of large segments of DNA. In addition, there are a number of discrete genetic elements which can move to new locations in the genome. These movable elements are called TRANSPOSONS. They can move to random sites in the genome and, as a consequence, they can cause mutations by disrupting functional genes. Many transposons are present in multiple copies and so provide movable targets for homologous recombination resulting in genomic rearrangements. In some cases the transposon moves to a specific site. The process of discrete genetic elements moving to new sites in the genome is called *transposition*. The aim of this book is to discuss the properties of transposons, and the mechanisms and consequences of transposition in a range of different prokaryotic and eukaryotic organisms.

This subject is appropriate for discussion by the Society for General Microbiology not only because transposition is widespread amongst prokaryotic and eukaryotic microbes but also because some of the transposons in non-microbial eukaryotic cells may have their origins in the microbial world. Some of these transposons, such as the *Drosophila* P elements, show structural similarities to bacterial transposons, and others, such as the yeast Ty element, are characteristically viral. We have been fortunate in assembling a group of distinguished experts in the area of transposition to join us in this Symposium. Our aim has been to describe some transposons in detail but, whenever possible, to illustrate general principles. We hope that, by bringing together information about diverse systems in one Symposium, we may discover new relationships between prokaryotic and eukaryotic systems and identify conserved features in the regulation and mechanisms of transposition.

Leaving aside the pioneering studies by Barbara McLintock of transposition in maize, much of the molecular analysis of transposition stems from the discovery of transposons in Gram-negative bacteria (the historical background of these studies is established in Naomi Datta's introduction), and recent rapid developments in our knowledge of transposons, in Gram-positive bacteria and in eukaryotes, both of which are given extensive coverage in this book, have widened our view of the general nature of these elements. The effects of transposition vary widely. In some cases, the host organism bene-fits, with dire consequences for human populations: antibiotic resis-tance is stably introduced into bacterial populations by transposition, and antigenic variation mediated by site-specific transposition/re-combination in trypanosomes and bacteria allows them to escape host defences. In other cases transposition is random and can result in mutations in the host; this occurs to some extent in yeast and plants and an extreme example is the highly mutagenic hybrid dysge-nesis which results from P and I element transposition in some strains of *Drosophila*. Movement of DNA may also occur between genomes: mouse retro-elements may be transported between genomes after copackaging into retroviral particles and the T-DNA of Ti plasmids moves between kingdoms from bacterial to plant genomes.

The mechanism of transposition is becoming well understood in a number of systems. There has been tremendous progress in the understanding of transposition via site-specific recombination in bac-teriophage Mu and Th elements. A quite different mechanism has been established for the yeast transposon Ty where an RNA interme-diate is involved. In many cases the transposition process appears to be controlled by genetic background. In the case of P elements and some plant transposons, many of the elements themselves are defective through deletion, but they can be activated to transpose when appropriate genetic crosses introduce an intact element. In other cases there appear to be cellular factors which control expres-sion and transposition. For example, the expression of mouse retro-transposons and *Drosophila* P elements differs in germline and somatic cells. Some of the prokaryotic transposons regulate transpo-sition via transposon-encoded gene products, and yeast Ty elements may limit transposition by sequestering intermediates in virus-like particles.

Transposons can also be exploited experimentally. In this book we have examples of the construction of mutagenesis and gene transfer vectors and of novel approaches to vaccine production through the analysis of transposons and transposition.

We believe that this book will prove valuable to both genetics and microbiology specialists, from advanced undergraduates to mature researchers, and for those involved in practical aspects of medical microbiology in which transposition and DNA rearrange-ments are increasingly seen to be of major importance. We thank the various contributors for their excellent and conscientiously pre-pared contributions and for their receptiveness towards editorial suggestions, and the Society for General Microbiology (personified in this case by Crawford Dow and Caroline Alderson) for providing the scaffolding on which this volume was built.

November 1987

A. J. Kingsman
K. F. Chater
S. M. Kingsman

INTRODUCTION

NAOMI DATTA

9 Dukes Avenue, London W4 2AA

The notion of transposable genes, or specific transposable DNA sequences, has profound effects in all aspects of biology. It must make biologists think 'Whatever is going on?' and 'How does this concept affect our own discipline and its accepted tenets?'. The notion has a very short history. Only the latest generation of biologists have met it during their formal scientific education; their seniors cannot afford to ignore it or they will be left behind as knowledge and ideas progress at the present great rate.

Mutations in *Escherichia coli* resulting from the insertion of phage Mu were reported over 20 years ago (Taylor, 1963); soon after, the insertion sequences (IS), causing strongly polar mutations in *E.coli* were discovered (Jordan, Saedler & Starlinger, 1968). In the early 1970s, transmissible, plasmid-determined antibiotic resistance was being studied; in particular, we were classifying resistance plasmids by compatibility (or incompatibility) in *E.coli* K12. Recombinations between unrelated plasmids, in which one acquired resistance genes from another, occurred unexpectedly frequently. It was R. W. Hedges, aware of the findings on both Mu and ISs, who first suggested that the TEM β-lactamase gene, commonly found in plasmids of many different incompatibility (Inc) groups, was carried on a specific transposable DNA sequence. With A. E. Jacob, he found evidence to confirm the idea and coined the term *transposon* (Hedges & Jacob, 1974); that first described is Tn*1*. In parenthesis, it should be pointed out that the work on antibiotic resistance was of the kind smiled upon by grant-giving bodies, because of its obvious medical significance. But without knowledge derived from more academic research on the genetics of bacteria, the implications of these observations might not have been understood.

Thereafter, many antibiotic-resistance transposons were identified. How wonderfully effective they are in disseminating resistance genes! For example, transposons indistinguishable from Tn*1*, or the almost identical Tn*3*, are found in a great variety of plasmids in many bacterial genera, all over the world. They are responsible for the relatively recent emergence of penicillin resistance in *Haemophilus influenzae* and *Neisseria gonorrhoeae*. Tetracycline resistance

in Gram-negative bacteria was already common in the 1950s, before
resistance plasmids were discovered, as evidenced by surveys published in Europe and America. Very likely Tn*10* was already widely
distributed. Trimethoprim, a synthetic drug, not an antibiotic, was
not used in medicine until 1968 (later in USA) and in 1974, English
surveys showed trimethoprim-resistant bacteria to be uncommon,
both in hospitals and in the community. The trimethoprim-resistant
transposon, Tn7, was identifed in 1976 and within a few years it
appeared in plasmids of many Inc groups, in bacteria isolated in
far-distant places, and in the chromosomes of wild strains.

The penicillins, tetracyclines and trimethoprim, as well as other
antibiotics, are manufactured and consumed on an enormous scale,
exerting strong selection for the spread of resistance genes. It might
be expected that the selection would also affect the vectors of those
genes, leading to higher incidences of plasmids in bacteria and of
transposase functions. However, the frequency of plasmids (without
resistances) in bacteria isolated in the years before antibiotics were
used in medicine, was found to be comparable to the present frequency (Datta & Hughes, 1983; Hartl & Sawyer, this Symposium),
nor was there any evidence to suggest that IS*1* has increased in
frequency as a result of its selection, in Tn*9*, by chloramphenicol
usage (D. L. Hartl, personal communication). It would be of interest
to look in pre-antibiotic bacteria for other transposase genes, e.g.
those of the Tn*3* family, of Tn7 and of IS*10*.

There are many examples of transposing elements in bacteria.
Although unifying hypotheses on their mechanisms are useful, these
elements do not follow one set of rules. In general, they have inverted
repeat sequences at their outer ends and, in general, they generate
duplications of host DNA at their insertion sites; but transposing
sequences without these properties exist. Since their discovery, much
has been learnt about the effects of transposons on host genomes.
Some questions on the mechanisms of transposition, and its control,
have been answered, for some examples. The answers have been
found by most ingenious and imaginative experiments, using techniques, both *in vivo* and *in vitro*, developed at a breathtaking rate.
Each element apparently encodes, at least, an enzyme, *transposase*,
whose exact functions are still not clear, but which recognises, and
acts upon, the end sequences of that same element. Besides this,
each controls its transposing ability by strategies that seem as
ingenious as those of the experimenters analysing them (e.g.
Kleckner, 1986). Much genetic information is packed into transposon

DNA. Various molecules, protein or RNA, inhibit transcription or translation of transposase genes. Secondary structures in mRNA play a part in control, as do host functions. Transposition of Tn7, whose natural history was referred to above, involves a whole set of proteins, apparently unrelated to those of other known elements (Smith & Jones, 1984, 1986: Rogers *et al.* 1986).

In the brief review, above, of the discovery of transposition, all eukaryotes are omitted. Barbara McClintock's work on transposable elements in maize was well before any bacterial transposon was known. Molecular analysis of the elements in maize had to wait for the development of *in vitro* DNA technology in bacteria. With that technology, the genetics of higher plants and animals becomes accessible to new understanding in a most exciting way.

Although the known history of transposons is so short, there is no reason to suppose them to be a new phenomenon. Speculation on their role in the evolution of both pro- and eukaryotes is of great interest and it will be considered in the Symposium.

REFERENCES

DATTA, N. & HUGHES, V. M. (1983). Plasmids of the same Inc groups in enterobacteria before and after the medical use of antibiotics. *Nature*, **306**, 616–17.

HEDGES, R. W. & JACOB, A. E. (1974). Transposition of ampicillin resistance from RP4 to other replicons. *Molecular and General Genetics*, **132**, 31–40.

JORDAN, E., SAEDLER, H. & STARLINGER, P. (1968). 0° and strong polar mutations in the *gal* operons are insertions. *Molecular and General Genetics*, **102**, 353–60.

KLECKNER, N. (1986). Mechanism and regulation of Tn*10* and IS*10* transposition. *Regulation of Gene Expression – 25 Years on*. 39th Symposium of the Society for General Microbiology, ed. I. R. Booth & C. F. Higgins, pp. 221–37. Cambridge: Cambridge University Press.

ROGERS, M., EKATERINAKI, N., NIMMO, E. & SHERRATT, D. (1986). Analysis of Tn*7* transposition. *Molecular and General Genetics*, **205**, 550–56.

SMITH, G. M. & JONES, P. (1984). Effects of deletions in transposon 7 on its frequency of transposition. *Journal of Bacteriology*, **157**, 962–4.

SMITH, G. M. & JONES, P. (1986). Tn*7* transposition: a multigene process. Identification of a regulatory gene product. *Nucleic Acid Research*, 7915–27.

TAYLOR, A. (1963). Bacteriophage-induced mutations in *Escherichia coli*. *Proceedings of the National Academy of Sciences, USA*, **50**, 1043–51.

PROKARYOTIC SYSTEMS

GENOME FLUX IN *STREPTOMYCES COELICOLOR* AND OTHER STREPTOMYCETES AND ITS POSSIBLE RELEVANCE TO THE EVOLUTION OF MOBILE ANTIBIOTIC RESISTANCE DETERMINANTS

K. F. CHATER, D. J. HENDERSON, M. J. BIBB AND D. A. HOPWOOD

John Innes Institute and AFRC Institute of Plant Science Research, Colney Lane, Norwich NR4 7UH, UK

INTRODUCTION

Modern methods of genetic analysis lead to the rapid detection of transposable elements and DNA rearrangements in intensively studied groups of bacteria. In the case of *Streptomyces* there has recently been a considerable increase in the availability and use of molecular genetic techniques for the study of various species (Hopwood *et al.*, 1985; Hunter, 1985), and several natural phenomena leading to the creation of new junctions between DNA sequences have inevitably been discovered. A comprehensive survey of the literature on this topic has recently been prepared by Hütter & Eckhardt (1988). In the major part of this chapter we focus on DNA rearrangements in the most studied strain, *S. coelicolor* A3(2), but with constant reference to related phenomena present (and sometimes known in more detail) in other streptomycetes. By this means we aim to build up an overview of the elements contributing to genome flux in this distinctive group of organisms. Among the plasmids, phages, IS elements and transposons, and specific amplification and deletion events that are reviewed, there are several phenomena that show some degree of novelty when considered against knowledge derived mainly from *E. coli*, and we have little doubt that their detailed molecular analysis, still at an early stage, will be rewarding.

It is not surprising to encounter novelty in streptomycetes because they differ from the other well-studied Gram-positive bacteria such as *Bacillus, Staphylococcus* and *Streptococcus* in at least two ways that have an important bearing on their genetic phenomenology.

First they possess DNA very rich in guanine and cytosine (more than 70 mole % [G + C]), providing what would normally be an unfavourable target for many kinds of transposition events: bacterial transposable elements often preferentially insert into [A + T]-rich DNA (Calos & Miller, 1980). Secondly they are multicellular, mycelial organisms with a complex life cycle. Spores germinate to form a coherent, branching vegetative mycelium that colonises and solubilises insoluble organic matter in soil, and then aerial hyphae grow on the substrate mycelium and subsequently metamorphose into chains of spores. The selection pressures associated with this life-cycle may be expected to differ from those encountered by non-differentiating bacteria, and perhaps to give rise to different genetic phenomena.

Their complex morphology is one of two particularly compelling reasons for studying streptomycetes. The other is their importance as producers of antibiotics, and it is from this aspect that interest in their genetic stability or instability mainly stems: antibiotic production is often a markedly unstable phenotype, with economically disadvantageous consequences. In addition, it has been proposed (Walker & Walker, 1970; Benveniste & Davies, 1973) that the pool of resistance genes present in antibiotic-producing streptomycetes may be the origin of resistance genes residing on mobile elements in clinical isolates of bacteria. In a second, highly speculative section of this chapter we discuss the extent to which current knowledge of *Streptomyces* antibiotic resistance genes, gene expression and DNA mobility supports this view.

AN OVERVIEW OF DNA REARRANGEMENTS AND EXTRACHROMOSOMAL DNA ELEMENTS KNOWN IN *STREPTOMYCES COELICOLOR* A3(2)

As shown in Fig. 1, the total genome of the wild type *S. coelicolor* A3(2) contains several disparate DNA species that are physically separable from the main chromosome (which, at $5–6 \times 10^3$ kb, is perhaps 1.5 to 2 times larger than the *E. coli* chromosome: Genthner, Hook & Strohl, 1985; Gladek & Zakrzewska, 1984), and is organised in such a way (probably as an actual circular molecule) as to give rise to a circular linkage map (Hopwood, 1967). These extrachromosomal elements are all absent from the closely related species *S. lividans* 66 (which may well contain other such elements).

SCP1, a giant linear plasmid

The largest of the extrachromosomal elements is the plasmid SCP1, originally defined genetically by its fertility properties and its specification of production of, and resistance to, the antibiotic methylenomycin A. Recently, however, orthogonal field-alternation gel electrophoresis has revealed SCP1 as an apparently linear molecule of some 340 kb (Kinashi *et al.*, 1987; T. Garbe & T. Meitinger, personal communication). (Relatively small linear plasmids were earlier found in several other streptomycetes (reviewed by Hütter & Eckhardt, 1988), but giant molecules like SCP1 have now been found in several species (Kinashi *et al.*, 1987).) There is strong genetic evidence that SCP1 can recombine with the *S. coelicolor* chromosome, at various locations, either integrating or giving SCP1-primes (for a review, see Chater & Hopwood, 1984). The most studied result of such events is the formation of a so-called NF strain in which SCP1 is integrated at the '9 o'clock' region of the chromosome, giving rise to high-frequency chromosome donation in crosses with SCP1⁻ strains (Hopwood *et al.*, 1973). Circumstantial evidence suggesting that an insertion sequence IS466, present in SCP1 and in the chromosome, is involved in the integration event is discussed in a later section.

Covalently closed circular (CCC) DNA species

Conventional CCC DNA preparations from strain A3(2) give two kinds of molecule. One, SCP2, is a low copy-number plasmid of 30 kb (Schrempf *et al.*, 1975), which acts as a sex factor (Bibb *et al.*, 1977), even though there is no evidence of physical interaction or homology with the chromosome. SCP2 does, however, have a short region of homology with SCP1 (M. J. Bibb, unpublished). The second CCC DNA species is a '2.6 kb minicircle', present at only one copy per 10–20 chromosomes. This DNA appears to originate from one or both of two linear copies found integrated in the *S. coelicolor* chromosome (Lydiate *et al.*, 1986: see below).

SLP1 and SLP4

There are at least two other plasmid-like elements in *S. coelicolor*, which are detected when *S. lividans* exconjugants from *S. coelicolor* A3(2) × *S. lividans* 66 crosses are analysed. These are SLP1 (Bibb *et al.*, 1981) and SLP4 (Hopwood *et al.*, 1983). SLP4 has not been characterised physically (and is not shown in Fig. 1), but SLP1 is a 17 kb sequence present as a single chromosomally integrated copy

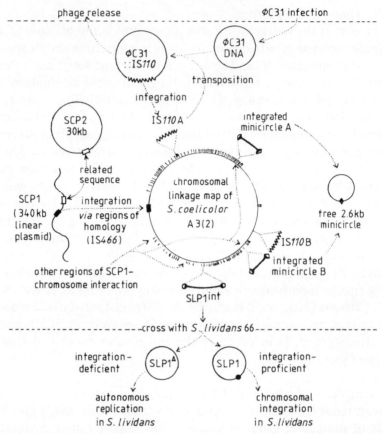

Fig. 1. Physically defined genetic elements and their interactions in *Streptomyces coelicolor* A3(2). The circumference of the central circle is marked with radial striations that indicate the map locations of known genes on the *S. coelicolor* linkage map. The positions on this map of IS*110*A, IS*110*B and the integrated minicircle sequence B are not known *accurately* in relation to most nearby markers and the positions of marked sequences in SCP1 are placed arbitrarily. At least one more copy of IS*110* is present in the genome. It is not known whether all IS*110* copies are active, though all can act as sites for integration of φC31 :: IS*110* derivatives. These and other phenomena are discussed in more detail in the text.

in *S. coelicolor*. The nature of the interactions between SLP1 and the host chromosome are discussed below, along with evidence that such plasmidogenic sequences are widespread in *Streptomyces*.

The insertion sequence IS*110*

One other mobile genetic element has been detected in the *S. coelicolor* A3(2) chromosome. This element, IS*110*, is a 1.55 kb sequence present at 3–4 copies (Chater *et al.*, 1985). Two copies were mapped to the chromosome, at widely separated sites. IS*110* can transpose into a preferred site in the DNA of ϕC31, a wide host range *Streptomyces* temperate phage used as a cloning vector (Chater, 1986). IS*110* has been sequenced, and it too is discussed in more detail below.

Deletions and amplifications

S. coelicolor also displays genetic instability. The most studied case of this involves high frequency mutation to give chloramphenicol-sensitive derivatives with even more obvious instability; they frequently lose large segments of (presumptively) chromosomal DNA including an arginine biosynthetic gene, and sequences adjacent to the deletion become tandemly amplified to a moderate degree (Flett *et al.*, 1987). Such deletions, coupled with more extreme amplification of DNA, sometimes to copy numbers so high that the amplified DNA constitutes up to 30% of the total genome, have also been studied in considerable detail in several other species (see below and, for recent reviews, Cullum *et al.*, 1987; Dyson *et al.*, 1987; Schrempf *et al.*, 1987; Hütter & Eckhardt, 1988).

The Pgl phenotype

A different kind of instability, the physical basis of which is undetermined, involves so-called 'phage growth limitation' (the Pgl$^+$ phenotype: Chinenova *et al.*, 1982) which prevents ϕC31 (but not various other phages) forming plaques on *S. coelicolor* A3(2). The Pgl$^+$ parent strain gives rise to Pgl$^-$ derivatives at a frequency of 10^{-2} to 10^{-3} per spore, and Pgl$^-$ derivatives revert at a similar rate (Chinenova *et al.*, 1982). It remains to be determined whether these oscillations are determined by DNA rearrangements (see the articles by Meyer & Haas and van de Putte in this volume) or by some epigenetic means (for example, DNA methylation: Holliday & Pugh, 1975). The Pgl$^+$ phenotype is interesting not only for its instability, but also for its mechanism: a Pgl$^+$ strain can be lysed or lysogenised by ϕC31 only if the phage was previously grown on a Pgl$^-$ mutant or a different streptomycete. This unusual reversal of a classical restriction-modification pattern may provide an example of the effect of multicellularity on genetic interactions in *Streptomyces*, since it provides an absolute barrier to subsequent ϕC31 propagation in a mycelium even though the hyphal compartment initially infected may be sacrificed (Chinenova *et al.*, 1982).

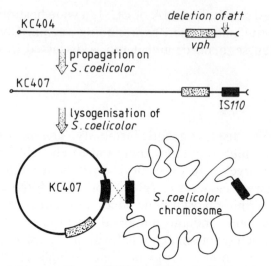

Fig. 2. Detection of transposition of IS*110* into DNA of a φC31 derivative. KC404 is an *attP*-site-deleted derivative of the temperate phage φC31, and contains a cloned *vph* gene (conferring viomycin resistance). It cannot normally transduce *S. coelicolor* to give viomycin-resistant lysogens because of the absence of *attP*. If, however, it acquires a copy of IS*110* by transposition (as in KC407), it can subsequently integrate into a chromosomal copy of IS*110* by homologous recombination, thereby giving a viomycin-resistant lysogen (Chater *et al.*, 1985).

INSERTION SEQUENCE-LIKE ELEMENTS OCCUR IN *STREPTOMYCES*

IS110 and other IS-like elements acquired by temperate phages

The organisation of IS110

At least four independent occurrences of transposition of IS*110* from the *S. coelicolor* chromosome into a favoured site in φC31 derivatives have been observed (Fig. 2; Chater *et al.*, 1985; R. J. Neal, personal communication). This region of the phage has been sequenced with and without the inserted element (Bruton & Chater, 1987). The element had inserted into a run of seven C residues, and the resulting junctions were in runs of 11 and 15 C residues (making it impossible to determine whether a short duplication of the target site had occurred). There is perceptible homology between the sequences flanking the insertion site and sequences at the ends of IS*110* (Fig. 3) which may perhaps account for the choice of insertion site. IS*110* itself resembles 'classical' insertion sequences in a number of ways: it contains short terminal inverted repeats (though these are not quite terminal and, at 10/15 bases matched, are very

(a)

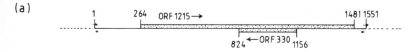

(b)

(c)

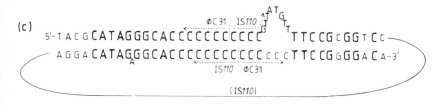

(d)

Fig. 3. Organisation of IS*110* and features of sequences at the junctions between IS*110* and its target site in a φC31 derivative (from Bruton & Chater, 1987). (a) Coding capacity of IS*110*, based on its DNA sequence. No other substantial potential open reading frames were apparent. (b) Sequence of the 'unoccupied' target site for IS*110* insertion in φC31 DNA. The region around the target site is directly (and imperfectly) repeated in nearby DNA, as indicated by the lines above and below the sequence. Bases identical in the two sequences are given in large letters. (c) Sequences at the left and right junctions of IS*110* with φC31 DNA show substantial homology with each other (shown by large letters). (d) An imperfect inverted repeat is found close to the ends of IS*110* (potentially base-pairing sequences are given in large letters).

imperfect), and one major open reading frame (ORF 1215) with a possible smaller one (ORF 330) within its anti-sense strand (Fig. 3). The potential proteins specified by both genes would be basic, like those predicted for genes in other IS-elements (Kleckner, 1981). DNA related, but not identical, to IS*110* was detectable by Southern blot analysis in a significant minority of other streptomycetes (Chater *et al.*, 1985).

An IS110-related element from S. clavuligerus

No hybridisation of IS*110* was found to DNA from the taxonomically distant *S. clavuligerus*, but it now appears that a related (but interestingly different) element is present there. This element (still unnamed) was discovered by chance as a 1.4 kb insertion into the multicopy plasmid pIJ702 after propagation in *S. clavuligerus* (an organism important for its production of the β-lactams clavulanic acid and cephamycin). Southern blotting revealed one hybridising region in the *S. clavuligerus* genome. The sequence of the relevant region of pIJ702 with and without the element was determined (Leskiw *et al.*, 1986 and personal communication). The junctions with pIJ702 show no sequence duplication at the target site, and there is no suggestion of inverted repeats at the ends of the element. Most of its length is occupied by an open reading frame, which, while very different in DNA sequence from the IS*110* ORF 1215, would encode a markedly similar protein (about 35% of residues in a 375 amino acid overlap would be identical, with only three single-residue gaps in the sequence alignment: C. J. Bruton, B. Leskiw & K. F. Chater, unpublished). On the assumption that this protein mediates transposition, the differences between the ends of the *S. clavuligerus* element and those of IS*110* are surprising. We will return to this in a later section on the 2.6 kb minicircle.

Insertion sequences in other temperate phages
The presence of an IS-like element (IS*281*) has been reported in another phage, φC43, which is homoimmune with φC31 and contains DNA very closely related to that of φC31 along most of its length (Lomovskaya *et al.*, 1980). IS*281* is present in the φC43 prophage found in the natural lysogen *S. lividans* 803, and as several other copies in the 803 genome (Sladkova, 1986). (At least one copy of an IS*281*-related sequence is also present in *S. coelicolor* A3(2) and *S. lividans* 66.) IS*281* is located close to the *att* site of φC43 prophage. Most, if not all, of the φC43 phages released from *S. lividans* 803 contain deletions adjacent to or including IS*281* sequences (probably because the intact φC43 genome plus IS*281* would be too large for stable packaging into phage particles) and usually the nearby *attP* site is deleted. Further evidence that IS*281*, like transposable elements in general, causes deletions was obtained by studying a φC43 phage carrying an entire copy of IS*281*: further deletions from within or next to the element occured rather often (10^{-2}: Sladkova *et al.*, 1984).

A third example of acquisition of a host DNA sequence is provided by phage SH10, which is unrelated to ϕC31. Here, a 1.2 kb insertion has repeatedly been observed in the same site, close to a 0.78 kb deletion found in certain clear plaque deletion mutants. The element appears to re-establish the ability of the phage to lysogenise. Sequences related to it were found in *S. chrysomallus* 40341 and *S. lividans* 66 (S. Klaus, personal communication).

It is notable from these examples that in each case the element in question has provided a potential means for prophage integration, in a phage mutant that has lost its original capacity to integrate, as a result of an insertion near to the site of the mutation. The reasons for such site selectivity are not clear, but it appears that this region of the phages may be subject to relatively frequent loss and acquisition of DNA with functions related to transposition. The possible relevance of these observations to the evolution of the integration mechanisms of temperate phages is returned to below in our consideration of the 2.6 kb minicircle.

Directed searches for IS elements in streptomyces

Spontaneous mutations

The IS elements described above were all discovered by chance. However, some systematic attempts to find IS elements have been made. One approach has been to look for spontaneous insertion mutations into genes in which there is a positive selection for forward mutations. Three of these systems have been studied in *S. coelicolor* and are of potentially wide applicability. Mutants in genes for galactokinase, glucose kinase and sulphate metabolism (cysteine biosynthesis) can be selected by resistance to the respective substrate analogues 2-deoxygalactose (Brawner *et al.*, 1985; Kendall *et al.*, 1987), 2-deoxyglucose (Hodgson, 1982) and selenate (Lydiate *et al.*, 1987). The DNA regions containing these genes have been cloned and used as probes in Southern blots of DNA from collections of spontaneous mutants. In a survey of *galK* mutants of *S. coelicolor* and *S. lividans*, U. Ali-Dunkrah, K. Kendall & J. Cullum (personal communication) found that a few percent of *S. lividans* mutants have rearrangements which are not simple deletions. Among *glk* mutants (Fisher *et al.*, 1987) and selenate resistant mutants (Lydiate *et al.*, 1987), deletions accounted for a significant fraction of the mutations and there was no clear evidence of insertion mutations.

A fourth system, specific to *S. albus* G, involves the inactivation of the *Sal*I restriction and modification (RM) genes to give the R⁻M⁻ phenotype. Here, selection for the R⁻ phenotype is possible because of the availability of a mutant temperature-sensitive for modification (Chater & Wilde, 1980), such that survivors at 37 °C are almost all R⁻ mutants. R⁻ mutants are abundant (10^{-3} to 10^{-4} among spores produced at 28 °C) and almost all are also M⁻. Using the cloned genes for the *Sal*I RM system (Rodicio & Chater, 1987) as a probe in Southern blots of the DNA of one such mutant and of other R⁻M⁻ mutants obtained by more laborious screening, it was found that all the R⁻M⁻ mutants (but none of a small collection of R⁻M⁺ mutants) contained a similar insertion of about 1 kb in apparently the same position, close to or in the region between the restriction and modification genes. This element has now been cloned (Rodicio & Chater, 1987).

SCP1 integration may involve IS466

Another approach to the discovery of IS elements is to examine regions involved in plasmid-chromosome interactions, which usually involve IS elements in Gram-negative bacteria (see Iida *et al.*, 1983, for a review). In certain high-fertility donor strains of *S. coelicolor*, the plasmid SCP1 becomes linked to chromosomal DNA, either by integration, as in NF strains (Hopwood *et al.*, 1973) or in the formation of SCP1-primes (Vivian & Hopwood, 1973; Hopwood & Wright, 1976). Following the discovery by Hodgson & Chater (1981) that NF formation was accompanied by inactivation of the chromosomally located *dagA* gene (for agarase), Kendall & Cullum (1984) cloned the region containing *dagA* and later used it to probe strains with autonomous or integrated copies of SCP1 (Kendall & Cullum, 1986). They found that a short segment of DNA located close to *dagA* is homologous to a segment of SCP1, and that in NF strains this segment is deleted, along with *dagA* itself. The *dagA* linked element, of 2–2.5 kb, has recently been shown to transpose (K. Miyashita & J. Cullum, personal communication; it has been designated IS466), strengthening the likelihood that *dagA* may be part of a transposon (Hodgson & Chater, 1981).

In summary, there is now abundant evidence of IS-like elements in various *Streptomyces* species. Study of them is still at a very early stage, but enough of them show unusual features to encourage more detailed analysis. It is not yet clear whether any of these elements form (or can be persuaded to form) parts of compound transposons.

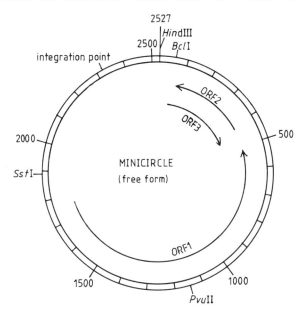

Fig. 4. Features of the 2.6 kb minicircle deduced from its DNA sequence (D. J. Henderson, unpublished results). Insertion of DNA at the *Hin*dIII, *Bc*II or *Ss*II sites does not interfere with the ability of the minicircle to integrate into the *S. lividans* genome, although the effects of these insertions on the formation of free minicircle are unknown (Lydiate *et al.*, 1986; D. J. Henderson, unpublished results).

THE 2.6 kb MINICIRCLE, A TRANSPOSABLE ELEMENT WITH A CIRCULAR INTERMEDIATE

Insertion specificity and organisation of the minicircle

The low-abundance 2.6 kb minicircle DNA of *S. coelicolor* can mediate the stable integration of DNA cloned into it, into the chromosome of *S. lividans* and many other streptomycetes. This was shown with *in vitro* constructions: the minicircle joined to DNA from an *attP*-deleted φC31 derivative (Lydiate *et al.*, 1986), and pBR327::minicircle structures (Lydiate *et al.*, 1987). In each case, *Streptomyces* selective markers were part of the construct. Most, but not all, of such insertion events in *S. lividans* used a preferred integration site in the chromosome, and a particular sequence of the minicircle was always involved in forming the new junctions.

The complete sequence of the minicircle has been determined (D. J. Henderson, unpublished results). It contains a large open reading frame (ORF1) through half its length, and two small open reading frames (ORF2 and ORF3), occupying opposite strands of the same DNA region (Fig. 4). The minicircle integration point is

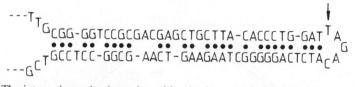

Fig. 5. The integration point (arrow) used by the free minicircle of *S. coelicolor* A3(2) to integrate into the chromosome is flanked by an imperfect inverted repeat sequence. The figure shows one strand of the minicircle duplex DNA drawn to emphasise the bases (marked with dots) that contribute to the inverted repeat (D. J. Henderson, unpublished results).

located in the long non-coding region (Fig. 4). The integration point is flanked by an imperfect inverted repeat (Fig. 5). Preliminary DNA sequence data on the minicircle–chromosome junction points suggest that no duplication occurs at the target site during insertion, and that there is no obvious homology between the minicircle integration site and the preferred chromosomal target (D. J. Henderson, unpublished results). In this latter respect, the minicircle differs from transposons such as Tn*916* (Clewell *et al.*, this volume) and Tn*554* (Murphy, this volume) which also seem to use unusual transposition modes. Remarkably, the predicted product of the large ORF shows 23% identity to that of ORF 1215 of IS*110* (over 415 amino acid residues, with only nine gaps in the sequence alignment), and a comparable degree of similarity to that of the ORF of the *S. clavuligerus* IS element (D. J. Henderson, C. J. Bruton, B. Leskiw & K. F. Chater, unpublished results) despite marked differences in their DNA sequences. No perceptible amino acid sequence relatedness has been detected, however, between these proteins and those deduced from the DNA sequences of any *E. coli* IS elements. Thus, it is possible that related polypeptides are involved in transposition processes involving three apparently quite different modes of integration, as judged by the sequences of insertion junction points and by the general organisation of the three elements concerned.

Speculation on the evolution of minicircle DNA and temperate phages

Although functional analysis of the minicircle is yet to be carried out, it is attractive to speculate that ORF2 and ORF3 have roles in site-specific recombination events involving the insertion site. Then, the organisation of the minicircle would resemble that of prophage ϕC43 (see above) in that an ORF related to those of insertion sequences (i.e. ORF1) would be closely linked to a site-specific

recombination system. This suggests a plausible route for the evolutionary origin of minicircular DNA; a prophage-chromosome *att* site
junction becomes flanked by direct repeats of an IS, so that a single
homologous recombination event between these repeats causes loop-
ing-out of a minicircle-like element. A variety of interactions of the
new element with other phages could lead to their acquisition of
a new *att* region, providing a specialised version of the modular
evolution of temperate phages discussed by Campbell & Botstein
(1983). An involvement of minicircles in on-going lateral transmission of *att* regions would not, of course, exclude the possibility
that minicircles may (also?) have evolved independently of temperate phages and then provided a *primary* source of phage integration
systems.

PLASMIDS EXHIBITING SITE-SPECIFIC INTERACTIONS WITH THE CHROMOSOME ARE WIDESPREAD IN *STREPTOMYCES*

Interaction of SLP1 with the host chromosome

The 2.6 kb minicircle, with its extremely low copy number and small
size, seems unlikely to be a replicon (although this possibility is
not excluded). However, many streptomycetes harbour chromosomal segments which, like the minicircle, appear capable of generating
extrachromosomal circular forms, but undoubtedly are replicons.
(The term 'excision' is not used since it is not clear whether any
literal cutting-out from the chromosome occurs, rather than some
kind of replicative mechanism, during generation of the extrachromosomal elements.) A paradigm of such elements is SLP1 from
S. coelicolor, whose properties as a plasmid are reviewed elsewhere
(Hopwood *et al.*, 1986; Omer & Cohen, 1987). Here we focus on
the site-specific interactions between SLP1 and the host chromosome, which have been studied mainly in *S. lividans*, at least partly
because no SLP1-cured derivative of *S. coelicolor* has yet been isolated. When full-length SLP1 derivatives that carry the *attP* site used
in integration are introduced into *S. lividans*, they integrate efficiently into a site, *attB*, corresponding to the *S. coelicolor* site. (The
majority of SLP1 plasmids transferred to *S. lividans* from *S. coelicolor* do not contain the whole 17 kb element, and lack sequences
required for integration, although some, such as SLP1.2 (Fig. 6),
retain the *attP* site; these plasmids do not integrate and they remain

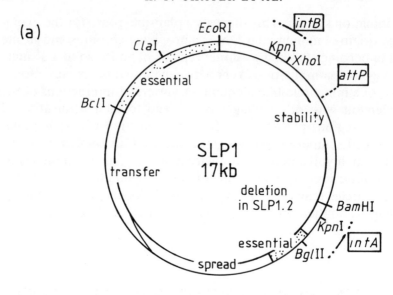

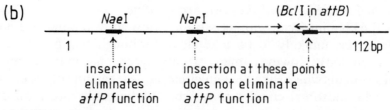

Fig. 6. Regions of SLP1 concerned with integration. (a) Map of the intact circular form of SLP1 presumed sometimes to occur transiently during mating between *S. coelicolor* A3(2) and *S. lividans* 66 (Hopwood *et al.*, 1985; Omer & Cohen, 1986; C. A. Omer, personal communication). The regions marked *intA* and *intB* are required for site-specific integration involving the *attB* site (see text). SLP1.2 is a naturally occurring derivative of SLP1 that lacks *intA* and replicates autonomously in *S. lividans* rather than integrating (Bibb *et al.*, 1981). (b) The 112 bp region of the SLP1 *attP* site homologous with the *S. lividans* 66 *attB* site. An imperfect inverted repeat is indicated by broken arrows.

autonomous: Bibb *et al.*, 1981; Omer & Cohen, 1984, 1986.) Integration requires, in addition to *attP* and the chromosomal *attB* site, two regions of SLP1: one (*intA*) in the '5 o'clock' region, and the other (*intB*) in the '1 o'clock' region (Fig. 6, C. A. Omer, D. Stein & S. N. Cohen, personal communication). The modes of action of *intA* and *intB* are not established. The *attP* and *attB* regions and the left and right junctions of SLP1 with the *S. lividans* chromosome (*attL* and *attR*) have been sequenced (Omer & Cohen, 1986), revealing that integration takes place by recombination within a common sequence (111 out of 112 bp are identical). The recombination event is virtually always in the 94 bp of common sequence to the left of the single mismatch: an observation made possible by a *Bcl*I restriction fragment length polymorphism caused by the mismatch (Omer & Cohen, 1986). Recent experiments have shown that *in vitro* insertion of DNA at the *Nae*I site at bp 16–21 of the common region inactivates the *attB*, and probably the *attP*, site, whereas the *Nar*I site (bp 46–51) and *BCl*I site (bp 90–95) can be disrupted without significantly affecting *attB* function (Fig. 6b). An extensive inverted repeat (bp 67–111: Fig. 6b), which drew the initial attention of Omer & Cohen (1986), does not seem after all to be centrally implicated in *att* function (C. A. Omer, S. Lee & S. N. Cohen, personal communication).

Site-specific recombination of other SLP1-like elements

If, as now seems likely, the functionally important part of SLP1 *attP* resides in its leftmost 50 bp or so, then it comes to resemble more closely what has been observed in analogous plasmidogenic sequences from rather distantly related actinomycetes. In both pMEA1 of the rifamycin producer *Nocardia mediterranei* (Madon *et al.*, 1987) and pSE101 of the erythromycin producer *Saccharopolyspora erythraea* (formerly *Streptomyces erythraeus*) (Katz *et al.*, 1987; L. Katz, personal communication), plasmid integration involves recombination within a 47 bp sequence shared by the relevant *attP* and *attB* sites. Recent data also indicate that a sequence of 49 bp is in common between pSAM2 (present as both integrated and autonomous copies in the spiramycin producer *S. ambofaciens*: Pernodet *et al.*, 1984) and the preferred site in *S. lividans* DNA for pSAM2 integration (F. Boccard, J.-L. Pernodet & M. Guerineau, personal communication). The details of the equivalent sites in other plasmids (catalogued by Hopwood *et al.*, 1986, and Hütter & Eckhardt, 1988) that have arisen from larger replicons (usually assumed to be the chromosome) are undetermined. However, it now seems acceptable to make the generalisation that there is a class of element, widespread among actinomycetes, whose direct interaction with the host genome involves site-specific recombination between homologous DNA regions of about 50 bp. This mechanism of recombination does not appear to be closely related to that involved in the formation and integration of the 2.6 kb minicircle.

HIGH FREQUENCY DELETION AND AMPLIFICATION EVENTS THAT OCCUR IN *STREPTOMYCES* GENOMES, AND THEIR OCCASIONAL ASSOCIATION WITH REPEATED DNA SEQUENCES

Deletions and amplifications in S. coelicolor and S. lividans

Instability of chloramphenicol resistance

More than a decade ago, several laboratories independently discovered that *S. coelicolor* A3(2) gives rise to chloramphenicol-sensitive (Cml^S) derivatives at a relatively high frequency (10^{-2} to 10^{-3} per spore: reviewed by Chater & Hopwood, 1984). The genetic and physiological basis has been studied in parallel in *S. coelicolor* and its close relative *S. lividans* 66 (for reviews see Cullum *et al.*, 1987; Schrempf *et al.*, 1987; Dyson *et al.*, 1987; Flett & Cullum, 1987). There are suggestions of a location of the Cml^R determinant near the 3 o'clock 'silent' region of the linkage maps of *S. coelicolor* (Fedorenko *et al.*, 1988) and *S. lividans* (Dyson *et al.*, 1987). It has been proposed that the Cml^R determinant is transposable (Sermonti *et al.*, 1978), but there are still no firm data on which to base this suggestion (Chater & Hopwood, 1984). In recent experiments by Flett & Cullum (1987), a Cml super-resistant mutant of *S. coelicolor* was found to contain 20- to 50-fold amplification of some restriction fragments which, when cloned and used as a probe in Southern blots, turned out to be deleted, along with at least 40 kb of flanking DNA, from Cml^S mutants of both *S. coelicolor* and *S. lividans*. (It is not yet clear how these observations are related to the reversibility of the Cml^R–Cml^S change observed by Freeman *et al.*, 1977, or to the finding of Danilenko *et al.*, 1986, that Cml^S derivatives are deamplified for a DNA segment that is amplified in the wild-type Cml^R *S. lividans* parent: see Flett & Cullum, 1987, for a discussion.) Remarkably, these initial deletions are often accompanied by further genetic instability: many (but not all) Cml^S mutants give rise spontaneously at high frequency (1–7% in *S. coelicolor*: Flett *et al.*, 1987; 10–25% in *S. lividans*: Schrempf *et al.*, 1987 and Cullum *et al.*, 1987) to arginine auxotrophs in which the *argG* gene is deleted (Ishihara *et al.*, 1985) and a characteristic adjacent segment is amplified (to a low extent in *S. coelicolor*: Flett *et al.*, 1987; and more highly in *S. lividans*: Schrempf *et al.*, 1987; Cullum *et al.*, 1987; Dyson *et al.*, 1987).

Other examples of genetic instability and DNA amplification

In both *S. coelicolor* and *S. lividans*, instability of resistance to other antibiotics is also observed (Danilenko *et al.*, 1977; Schrempf *et al.*, 1987) and at least in the case of tetracycline sensitivity in *S. lividans*, DNA instability results, leading to DNA amplification and to effects on primary metabolism (Schrempf *et al.*, 1987). These events in *S. coelicolor* and *S. lividans* are examples of a phenomenon widespread in *Streptomyces* (reviewed by Hütter & Eckhardt, 1988; Cullum *et al.*, 1987 and Schrempf *et al.*, 1987), in which unselected deletions, often removing genes for recognisable phenotypic characteristics such as antibiotic production and resistance, melanin production and morphological differentiation as well as *argG*, are adjacent to amplified DNA sequences.

In most examples the amplified DNA has no known function. There are, however, a few cases where amplified DNA contains genes controlling a recognisable phenotype: in *S. lividans*, a 93 kb amplifying DNA sequence contains a mercury resistance determinant (J. Altenbuchner, personal communication); in *S. tendae*, selection for increased production of an α-amylase inhibitor resulted in a strain in which the relevant structural gene has become associated with a highly amplified 37 kb sequence (Koller, 1986); in a rubradirin-producing strain of *S. achromogenes*, a spectinomycin resistance gene is part of an 8 kb segment capable of massive amplification to give high-level spectinomycin resistance detected in appropriate selective conditions (Hornemann *et al.*, 1987); an *S. rimosus* strain contains an amplifiable kanamycin resistance determinant (Potekhin & Danilenko, 1985); and an *S. antibioticus* strain contains an amplifying fragment that influences oleandomycin resistance and production (Orlova & Danilenko, 1983).

Directly repeated DNA sequences sometimes flank amplifiable DNA

In several cases, the unamplified copy of an amplifiable DNA sequence is flanked by directly repeated sequences of a size comparable with that of IS elements. Examples are the 1 kb sequences flanking the *S. lividans* and *S. coelicolor* sequences amplified in *argG* mutants (Altenbuchner & Cullum, 1985; Dyson *et al.*, 1987; Cullum *et al.*, 1987); the 2.2 kb sequences flanking an amplifiable sequence in a tylosin-producing *S. fradiae* strain (Fishman *et al.*, 1985); and the 0.8 kb sequences flanking the amplifiable spectinomycin-resistance gene of *S. achromogenes* var *rubradiris* (Hornemann *et al.*, 1987). (Directly repeated sequences of such length are not involved

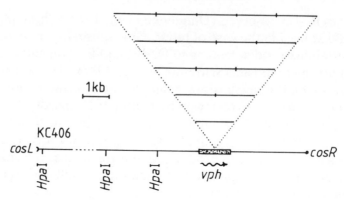

Fig. 7. A transposed amplified DNA sequence of *S. coelicolor* A3(2) and *S. lividans* 66 detected in a φC31 phage derivative, KC406. The wavy arrow represents *vph* mRNA, and *cosL* and *cosR* are the cohesive ends of KC406 DNA. KC406 was obtained by S. G. Foster (unpublished) as a plaque morphology mutant of a highly under-packaged phage, KC404 (Chater, 1986). It contains up to five tandemly arranged copies of a 1.4 kb fragment of *S. lividans* DNA inserted in the *vph* gene, With five copies of the element, the phage is about 5% over-packaged.

in some other amplification events (reviewed by Hütter & Eckhardt, 1988) and are therefore not a *sine qua non* for all amplification.) The organisation of the direct repeats is reminiscent of compound transposons such as Tn*9* and Tn*204* (Calos & Miller, 1980), but at present none of the *Streptomyces* sequences has been shown to transpose, and a preliminary report (Cullum *et al.*, 1987) states that the DNA sequence of the 1 kb elements that flank the sequences amplified in *S. lividans* Cml^S Arg$^-$ mutants does not suggest close similarity to IS elements, at least with respect to the organisation of the ends of the elements.

Transposition and amplification of a segment of S. lividans DNA

Foster (1983) isolated a spontaneously occurring derivative (KC406) of KC404, a φC31-derived cloning vector, into which a 1.5 kb segment of *S. lividans* DNA had transposed. Remarkably, this segment underwent amplification *in situ*, up to the limits imposed by the packaging constraints of the phage (Fig. 7). It was also able accurately to excise from its insertion site, regenerating the viomycin resistance phenotype specified by the *vph* gene into which the insertion had initially occurred (I. Tobek & K. F. Chater, unpublished results). In Southern blotting of *S. lividans* and *S. coelicolor* DNA using KC406 DNA as probe, both species gave similar complex patterns of strong and weak bands, suggesting that *part* of the inserted DNA is a dispersed repeated sequence (I. Tobek & K. F. Chater, unpublished results). This example linking transposition and amplification supports the notion that situations in which repeated DNA sequences flank unique DNA have probably originated by a transposition event.

Amplifying DNA sequences as selfish DNA

The mechanisms of these deletion and amplification phenomena in streptomycetes are still obscure, though two groups have proposed models (Dyson *et al.*, 1987; Young & Cullum, 1987). Even less clear is the advantage of the events. One possibility is to invoke 'selfish' DNA (Orgel & Crick, 1980): if genetic transformation occurs in nature, associated with DNA released from cells lysed as part of normal colony development (see final section below), then amplification of particular DNA segments in response to stress would present them with a survival advantage (particularly if the amplified DNA included sequences capable of homology-independent integration).

OCCURRENCE OF A TN3-LIKE TRANSPOSON IN *STREPTOMYCES FRADIAE*

Discovery of Tn4556

Although IS-like elements occur in streptomycetes, clearcut evidence of compound transposons such as Tn5 or Tn10 (Berg *et al.*, this volume) is lacking, and there remains (at least) this missing link in the chain proposed to carry antibiotic resistance from streptomycetes to clinically important bacteria. Recently, however, it has been discovered that a neomycin-producing strain of *S. fradiae* contains a 6.8 kb Tn3-like (class II) transposon (Chung, 1987). The element (Tn4556) was detected not by virtue of any resistance phenotype but because its spontaneous insertion stabilised an otherwise highly unstable plasmid based on a prophage origin of replication. Thus, selection for maintenance of this plasmid allowed Chung (1987) to obtain many independent derivatives containing Tn4556, and in each case insertion was at a different position in the originally unstable replicon.

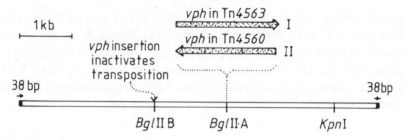

Fig. 8. Tn*4556*, a class II transposon originating from a neomycin-producing strain of *S. fradiae* (Chung, 1987). Insertion of *vph* in *Bgl*II site B inactivates transposition; insertion of *vph* in *Bgl*II site A does not significantly influence transposition. In orientation I, *vph* expression appears to depend on transcription from host promoters adjacent to the site of insertion.

Genetically marked derivatives of Tn4556

Fortuitously, one of the insertions resulted in a derivative in which the element was immediately flanked by *Bam*HI sites, allowing it to be cloned into a plasmid, pMT660, originally and presciently developed for transposon delivery by Birch & Cullum (1985). pMT660 is a point mutant derivative of the *tsr* (thiostrepton-resistance gene)-containing vector pIJ702 (Katz *et al.*, 1983) which is unable to replicate at elevated temperatures. A viomycin resistance gene (*vph*) was cloned into pMT660::Tn*4556* at either of the two *Bgl*II sites present in the transposon (Fig. 8). When the various pMT660::Tn*4556*::*vph* derivatives were introduced into *S. lividans* and the transformants incubated at 39 °C, which is non-permissive for pMT660 replication, viomycin-resistant thiostrepton-sensitive (Vio^{R}Thio^S) survivors were obtained if derivatives carrying *vph* in one of the *Bgl*II sites were used. Derivatives with *vph* in the other *Bgl*II site gave no Vio^R Thio^S survivors in this test. In many, if not all, of the surviving colonies Tn*4556*::*vph* had transposed to various locations in the *S. lividans* genome. Thus, pMT660:: Tn*4556*::*vph* derivatives are now potentially usable for conventional transposon mutagenesis in *Streptomyces*, making a valuable contribution to the growing number of techniques available for genetic manipulation. Interestingly, the *vph* gene inserted into Tn*4556* lacks its own promoter (Bibb *et al.*, 1985) and requires readthrough transcription for its expression. This is apparently provided by the transposon in one orientation of *vph* (i.e. in Tn*4560*) but not in the other (i.e. in Tn*4563*); for Tn*4563*, transcriptional activation of *vph* seems to require readthrough from a transcription unit into which the transposon has inserted (Chung, 1987).

Tn4556 is related to Tn3

Molecular analysis of Tn*4556* has only just begun, but already the ends of the element have been found to be 38 bp inverted repeats of each other, their sequence resembling those of transposons of the Tn*3* family (S. T. Chung, personal communication). It appears that members of the Tn*3* family are the most ubiquitous transposons in prokaryotes, with differently marked versions being found in *E. coli* (e.g. Tn*3*, ampicillin resistance), *Pseudomonas* (e.g. Tn*501*, mercury resistance) and Gram-positive bacteria such as *Staphylococcus* and *Streptococcus* (e.g. Tn*551* and Tn*917*, erythromycin resistance) (Murphy, this volume). (Note that pseudomonads contain relatively high, and staphylococci and streptococci contain very low, proportions of G + C in their DNA.) Where tested, all these elements have proven capable of transposition in *E. coli*. It may well be that this family is a major agency by which genetic information has been disseminated 'horizontally' between bacterial groups.

STREPTOMYCES AND THE EVOLUTION OF TRANSPOSABLE ANTIBIOTIC RESISTANCE

In the preceding section, current information on elements contributing to DNA flux in *Streptomyces* has been summarised. This information is also relevant to the question of more widespread genome flux involving *Streptomyces*, which is raised by the often-quoted hypothesis that many antibiotic resistance genes found on the R-factors of enteric bacteria originated from *Streptomyces* or other antibiotic-producing actinomycetes (Walker & Walker, 1970; Benveniste & Davies, 1973). In the discussion that follows, we have attempted to evaluate the impact of recent results on this hypothesis. Relevant factors include the comparison of genes for biochemically similar resistance mechanisms from streptomycetes and other bacteria; the availability of mechanisms for transposition of *Streptomyces* resistance genes; the extent to which these genes could be transcribed and translated in other bacteria; and the opportunities for DNA transfer from streptomycetes into other bacteria. It is also noted here that some bacteria other than actinomycetes produce antibiotics, sometimes of chemical classes typical of actinomycetes: for example, aminoglycoside production, and resistance through a phosphotransferase enzyme, have been described in a strain of *Bacillus circulans* (Courvalin *et al.*, 1977; Herbert *et al.*, 1986).

Relationships between Streptomyces resistance enzymes and genes, and those associated with transposable resistance in clinical isolates of bacteria

The original formulation of the hypothesis of Walker & Walker (1970) and Benveniste & Davies (1973) was stimulated by biochemical evidence that antibiotic-inactivation mechanisms (now extended to resistance mechanisms such as exclusion, as in tetracycline resistance (Ohnuki *et al.*, 1985) and target site modification, as in erythromycin resistance (Uchiyama & Weisblum, 1985)) are often similar in streptomycetes and clinical isolates of bacteria.

Recently it has become possible to make the first comparisons of relevant resistance genes at the DNA and (hence) protein sequence levels. Although nucleotide sequence homology could not be detected (not surprisingly in view of the much higher overall G + C content of *Streptomyces* DNA compared with that of most of the organisms from which the other genes were isolated), similar amino acid sequences were found in streptomycete and R-factor determinants conferring resistance to aminoglycoside (Thompson & Gray, 1983; Herbert *et al.*, 1986; Distler *et al.*, 1987) and macrolide (Uchiyama & Weisblum, 1985; Odelson *et al.*, 1986) antibiotics. These amino acid sequence similarities are often used to support the notion of horizontal gene transfer (e.g. Trieu-Cuot *et al.*, 1987). In Table 1 attempts have been made to quantitate these similarities and to compare them to those observed for streptomycete and non-streptomycete proteins of analogous function but which are not involved in antibiotic resistance. These latter comparisons are between genes and species where there would be little reason to predict either the occurrence of lateral gene transfer or unusual structural constraints on the proteins compared. The streptomycete amino acid sequences were optimally aligned with their potential homologues; the degree of amino acid identity was then determined, and its statistical significance estimated. For every alignment of resistance proteins, the degree of amino acid sequence identity and its general distribution throughout the length of the proteins would tend to rule out the possibility of convergent evolution and strongly suggests that each of the streptomycete resistance determinants shares a common ancestor (i.e. is strictly homologous) with its counterpart(s).

However, comparable degrees of similarity are also observed for proteins not involved in antibiotic resistance, e.g. α-amylase and the enzymes of the galactose operon. Thus, the observed high level of amino acid sequence similarity between the streptomycete and R-factor resistance proteins in Table 1 provides no indication of their acquisition by horizontal gene transfer.

One might tend, therefore, to look for evidence at the nucleotide sequence level. However, the high degree of nucleotide sequence divergence that would have had to occur between the high G + C streptomycete genes and their R-factor counterparts will undoubtedly complicate such comparisons and it is not clear that current methods of analysis can cater for such differences. Even DNA hybridisation may not be indicative of a recent common evolutionary history. For example, the α-amylase genes of *S. limosus* and *Drosophila melanogaster* show regions of approximately 60 bp with greater than 80% homology, which ought to be detectable, albeit at a low level of stringency, by Southern analysis (given the high G + C content of both α-amylase genes and the degree of amino acid sequence homology (Long *et al.*, 1987), such nucleotide sequence similarity between two presumably distantly related genes could have been predicted).

With this cautionary note in mind there exists, nevertheless, nucleotide sequence homology that may be indicative of horizontal gene transfer between genera. Within the actinomycetes, strong hybridisation and closely similar restriction maps have been observed between cloned aminoglycoside resistance determinants from the aminoglycoside-producers *Micromonospora purpurea* and *S. tenebrarius* (Skeggs *et al.*, 1987). More remarkably Suarez *et al.* (1987, and personal communication) found that a fosfomycin-resistance gene cloned from a fosfomycin-producing strain of *S. fradiae* hybridises strongly, and shares at least four restriction sites, with a transposable fosfomycin-resistance determinant originally discovered in clinical isolates of *Serratia marcescens*. Preliminary DNA sequence analysis confirms the marked similarity (J. E. Suarez, personal communication). This strongly suggests a close evolutionary relationship between the two genes, and future detailed DNA sequence comparisons will be of great interest.

If *Streptomyces* genes were transferred to other bacteria, could they be expressed?

Current evidence suggests that only a minority of *Streptomyces* promoters are recognised in *E. coli* (Bibb & Cohen, 1982; Jaurin &

Table 1. *Comparison between the amino acid sequences deduced for analogous genes from streptomycetes and other species.*

Streptomycete	Non-streptomycete	Identity I/O	Identity %	Gaps/100	NAS	Z	References
Antibiotic-resistance determinants:							
Aminoglycoside phospho-transferase							
S. fradiae (268, 73)	*Escherichia coli* Tn5	99/241	41.1	4.9	441	37.8	1, 2
	Escherichia coli Tn903	93/238	38.6	4.4	440	47.8	1, 3
	Bacillus circulans	91/242	37.6	4.4	388	36.9	1, 4
	Streptococcus faecalis	87/238	36.6	4.5	346	26.2	1, 5
Streptomycin phospho-transferase							
S. griseus (307, 74)	*Escherichia coli* Tn5	73/250	29.2	10.2	326	12.6	6, 7
23S rRNA methylase							
S. erythraeus (370, 72)	*Arthrobacter* sp.	184/318	57.9	3.1	460	69.4	8, 9
	Bacteroides fragilis	67/233	28.8	4.6	302	20.5	8, 10
	Streptococcus sanguis	44/163	27.0	2.3	423	17.2	8, 11
	Bacillus licheniformis	61/252	24.2	3.1	389	24.2	8, 12
	Staphylococcus aureus	55/228	24.1	4.3	378	20.6	8, 13
Chloramphenicol acetyl-transferase							
S. acrimycini (219, 71)	*Escherichia coli* Tn9	84/214	39.3	2.3	602	53.0	14, 15
	Bacillus pumilis	69/209	33.0	1.9	713	50.4	14, 16
	Staphylococcus aureus	65/209	31.1	1.4	654	34.4	14, 17
Beta-lactamase							
S. albus G (314, 68)	*Bacillus cereus*	98/244	40.2	3.0	519	63.8	18, 19
	Escherichia coli pBR322	98/246	39.8	2.4	460	31.8	18, 20
	Bacillus licheniformis	113/294	38.4	2.7	464	41.8	18, 21
	Staphylococcus aureus	84/259	37.4	3.4	399	38.1	18, 22

Streptomycete		Non-streptomycete		Identity		Gaps/100	NAS	Z	References
				I/O	%				
Non-antibiotic-resistance determinants:									
Galactokinase									
S. lividans	(395, 75)	Escherichia coli	(382, 53)	112/346	32.4	3.7	294	44.5	23, 24
UDP-4-epimerase									
S. lividans	(319, 71)	Escherichia coli	(338, 54)	119/309	38.6	4.1	371	47.9	23, 25
Galactose-1-phosphate uridyltransferase									
S. lividans	(317, 70)	Escherichia coli	(347, 56)	76/204	37.3	4.3	421	33.9	23, 25
Alpha-amylase									
S. limosus	(566, 71)	Drosophila melanogaster	(494, 63)	199/420	47.2	5.1	466	55.9	26, 27
		Mus musculus	(508, 44)	177/410	43.2	8.6	456	57.9	26, 28
Extracellular protease									
S. griseus	(185, 70)	Myxobacter 495	(198, ?)	58/152	38.2	5.2	371	13.7	29, 30

Table 1. The length of each protein (in amino acids) and the mol % G + C of its coding sequence are given after each species name. Identity was calculated from aligned regions only; unmatched positions and unaligned termini were not included; I/O indicates the number of identities in the region of overlap aligned by the SEQHP program (Kanehisa, 1982) and is also expressed as a percentage. Gaps/100 were determined by dividing the number of gaps in the aligned region by the average number of residues in the aligned segments and multiplying by 100. Normalised aligned scores (NAS) were calculated by dividing the alignment scores produced by the SEQHP program by the number of aligned residues in a given comparison and multiplying by 100. An estimate of the statistical significance of each alignment was made using the program SEQDP (Kanehisa, 1981). Each alignment score was compared with the mean score of 30 attempted alignments of randomised versions of the two sequences; the difference, Z, is expressed in standard deviation units. Z values of 3–6 and > 6 are taken as possibly and probably representative of an authentic relationship respectively (Doolittle, 1982; Lipman & Pearson, 1985). References: 1, Thompson & Gray, 1983; 2, Beck et al., 1982; 3, Oka et al., 1981; 4, Herbert et al., 1986; 5, Trieu-Cout & Courvalin, 1983; 6, Distler et al., 1987; 7, Mazodier et al., 1985; 8, Uchiyama & Weisblum, 1985; 9, Roberts et al., 1985; 10, Rasmussen et al., 1986; 11, Horinouchi et al., 1983; 12, Gryczan et al., 1984; 13, Horinouchi & Weisblum, 1982; 14, I. A. Murray & W. V. Shaw, personal communication; 15, Alton & Vapnek, 1979; 16, Harwood et al., 1983; 17, Bruckner & Matzura, 1985; 18, Dehottay et al., 1987; 19, Sloma & Gross, 1983; 20, Sutcliffe, 1978; 21, Neugebauer et al., 1981; 22, Chan, 1986; 23, C. W. Adams, J. A. Fornwald, F. J. Schmidt, M. Rosenberg & M. E. Brawner, personal communication; 24, Debouck et al., 1985; 25, Lemarie & Muller-Hill, 1986; 26, Long et al., 1987; 27, Boer & Hickey, 1986; 28, Hagenbuechle et al., 1980; 29, Henderson et al., 1987; 30, Olson et al., 1970.

Cohen, 1985; Deng *et al.*, 1986; Buttner & Brown, 1987). This partly reflects the greater heterogeneity of *Streptomyces* RNA polymerase (Westpheling *et al.*, 1985) but is probably also a consequence of the relative [G + C]-richness of regions of *Streptomyces* promoters that are not involved in specific contacts with RNA polymerase (Bibb *et al.*, 1983; Westpheling *et al.*, 1985). Certain *Streptomyces* resistance genes have been expressed in *E. coli* by readthrough from vector promoters (*vph* from *S. vinaceus*: Kieser *et al.*, 1982; *aph* from *S. fradiae*: Rodgers *et al.*, 1982), and what little evidence there is does not indicate that there are translational barriers to *Streptomyces* gene expression in *E. coli*. It may well be anticipated that transcription and translation of heterologous genes will be more limited in Gram-positive genera such as *Bacillus, Staphylococcus* and *Streptococcus*, which contain [A + T]-rich DNA and which appear to be very fastidious about promoter structure (I. Smith, unpublished data, quoted in Westpheling *et al.*, 1985) and ribosome-binding site recognition (McLaughlin *et al.*, 1981). A suitable transcriptional fusion of a *Streptomyces* resistance gene to a promoter capable of expression in a new host might conceivably be attained by the appropriately placed transposition from the new host's genome of an IS element with an outward-reading promoter. Many IS elements can activate transcription of adjacent genes (e.g. IS*1*, IS*2*, IS*3* and IS*5*: Kleckner, 1981) and indeed IS*50L* of Tn*5* contains the promoter that activates the Tn*5 neo* gene (Berg, this volume). This hypothesis has the merit of combining gene activation with acquisition of part of the transposition apparatus. An alternative route for gene activation could be transcriptional readthrough from sequences adjacent to the position of transposon insertion. It is interesting to note that the properties of some insertion sequences and Tn*3*-like (type II) transposons would readily allow these situations to occur. Thus, the *Streptococcus* transposon Tn*917* and its *S. fradiae* 'relative' Tn*4556* contain internal sites into which promoterless genes have been placed *in vitro* and shown to be transcriptionally activated by readthrough from promoters adjacent to the transposon (Youngman *et al.*, 1985; Chung, 1987). This appears also to be true for the *Bcl*I site of the 2.6 kb minicircle (Fig. 4: Lydiate *et al.*, 1987).

Such a mechanism cannot account for the expression of all resistance genes in transposons and plasmids: for example, Tn*10 tet* is sandwiched between genes reading in the opposite direction to it (Bertrand *et al.*, 1983; Schollmeier *et al.*, 1985) and the β-lactamase gene of Tn*3* reads *towards* the nearest transposon end (Heffron *et al.*, 1979).

If it is assumed that some of these genes originate from streptomycetes, one might suppose that the promoters have evolved mainly by point mutations to function more efficiently in their new hosts, and thus that the spread of such genes from streptomycetes would occur progressively from organisms with relatively [G + C]-rich DNA to those with DNA less rich in [G + C]. Consistent with this, the first clinical isolates of enteric bacteria containing a fosfomycin resistance gene related to one from *Streptomyces* (see above) were strains of *Serratia marcescens* (which contains 59% [G + C] in its DNA), and several years elapsed before the gene appeared in *Klebsiella* isolates (54% [G + C] in their DNA) and (more recently still) in *E. coli* (50% [G + C] in its DNA) (J. E. Suarez, personal communication).

How could genetic contact be made between streptomycetes and other bacteria?

It appears that DNA sequencing has so far provided little evidence to support the recent origin of R-factor and transposon-mediated resistance determinants from *Streptomyces*, with the emerging possible exception of fosfomycin resistance. Nevertheless, this one case could be sufficient to establish that this evolutionary route occurs, and hence to raise the question of how genes could migrate from streptomycetes. While it is possible that random DNA fragments from streptomycetes could become incorporated into a heterologous organism by a process of transformation and illegitimate recombination, it is more attractive to think that resistance genes borne on plasmids or transposons in *Streptomyces* would have a better chance of dissemination. There are very few such situations known. Antibiotic resistance genes associated with production are often embedded in clusters of production genes, which themselves appear generally to be chromosomally determined (with the notable exception of the plasmid-linked methylenomycin production and resistance gene cluster) (Hopwood, 1986). However, in *Streptomyces* it has been seen that the distinction between chromosome and plasmid and transposable element is not always unambiguous since a fluctuating linkage between various DNA elements is often seen, and since biosynthetic and resistance genes for several antibiotics are located in unstable regions of DNA, the exact nature of which is still unclear (see earlier sections). There is no compelling reason to believe that any of these gene sets are transposable (except for the general observation that they occur sporadically in various relatively unrelated streptomycetes and actinomycetes). Among resistance genes that are not associated with production, one – the amplifiable determinant for spectinomycin-resistance in an *S. achromogenes* strain (Hornemann *et al.*, 1987; see above) – is placed between repeated DNA sequences, and might conceivably prove to be transposable.

Thus, existing information, while not disproving it, does not provide strong encouragement for horizontal mobility of *Streptomyces* resistance genes, and if the evolutionary hypothesis for resistance genes has any validity, either the relevant mobile *Streptomyces* genes have yet to be discovered or the spread of resistance genes is associated with significant changes in the organisation of their flanking regions. With respect to the latter possibility, we note that, in the evolution of Tn*3*-related transposons, different resistance genes have become incorporated into the basic transposon structure without leaving any trace of the means by which this occurred; and that a Tn*3*-like transposon has now been described in *Streptomyces* (see above).

The ecological/physiological problem of how *Streptomyces* DNA could gain entry into the cytoplasms of other bacteria is even more open to unsubstantiated conjecture than the molecular aspects. It may be relevant that, during development of a *Streptomyces* colony, many of its own original vegetative hyphae (and quite possibly nearby cells of other organisms) become lysed and provide nutrients for aerial mycelium and spore development (Chater, 1984). *Streptomyces* protoplasts and free DNA from *Streptomyces* and other organisms may therefore be produced wherever streptomycetes grow in nature, and may not be solely a laboratory artefact. Since it is known from laboratory experiments that, within species, artificially produced protoplasts of many Gram-positive bacteria can fuse with each other, or take up externally added DNA to give recombinant organisms, it seems entirely plausible that intergeneric gene transfer could happen occasionally in the natural environment, and provide a route by which resistance genes could escape from streptomycetes. (Of course, some bacteria are able to take up external DNA by specific 'competence' mechanisms which do not involve protoplast formation at all.) It is worth noting that direct acquisition of an expressible resistance gene from a streptomycete could immediately provide the recipient with a marked selective advantage: if it resisted killing by the antibiotic being produced by the organism from which the resistance gene came, not only would it be able to divide while its siblings would not, but it would be situated in the nutritionally favourable environment provided by the lysing substrate mycelium.

ACKNOWLEDGEMENTS

We thank colleagues in many laboratories who responded to our request for information, often by the provision of unpublished results (which are acknowledged in the text). We are grateful to H. A. Baylis and T. Kieser for comments on the manuscript and to Anne Williams for long hours spent at the word processor in the preparation of successive versions of the manuscript.

REFERENCES

ALTENBUCHNER, J. & CULLUM, J. (1985). Structure of an amplifiable DNA sequence in *Streptomyces lividans* 66. *Molecular and General Genetics*, **201**, 192–7.

ALTON, N. K. & VAPNEK, D. (1979). Nucleotide sequence analysis of the chloramphenicol resistance transposon Tn9. *Nature*, **282**, 864–9.

BECK, E., LUDWIG, G., AUERSWALD, E. A., REISS, B. & SCHALLER, H. (1982). Nucleotide sequence and exact localization of the neomycin phosphotransferase gene from transposon Tn5. *Gene*, **19**, 327–36.

BENVENISTE, R. & DAVIES, J. (1973). Aminoglycoside antibiotic-inactivating enzymes in Actinomycetes similar to those present in clinical isolates of antibiotic-resistant bacteria. *Proceedings of the National Academy of Science, USA*, **70**, 2276–80.

BERTRAND, K. P., POSTLE, K., WRAY, L. V., JR. & REZNIKOFF, W. S. (1983). Overlapping divergent promoters control expression of Tn10 tetracycline resistance. *Gene*, **23**, 149–56.

BIBB, M. J., BIBB, M. J., WARD, J. M. & COHEN, S. N. (1985). Nucleotide sequences encoding and promoting expression of three antibiotic resistance genes indigenous to *Streptomyces*. *Molecular and General Genetics*, **199**, 26–36.

BIBB, M. J., CHATER, K. F. & HOPWOOD, D. A. (1983). Developments in *Streptomyces* cloning. In *Experimental Manipulation of Gene Expression*, ed. M. Inouye, pp. 53–82. Academic Press, New York.

BIBB, M. J. & COHEN, S. N. (1982). Gene expression in *Streptomyces*: construction and application of promoter-probe plasmid vectors in *Streptomyces lividans*. *Molecular and General Genetics*, **187**, 265–77.

BIBB, M. J., FREEMAN, R. F. & HOPWOOD, D. A. (1977). Physical and genetical characterisation of a second sex factor, SCP2, for *Streptomyces coelicolor* A3(2). *Molecular and General Genetics*, **154**, 155–66.

BIBB, M. J., WARD, J. M., KIESER, T., COHEN, S. N. & HOPWOOD, D. A. (1981). Excision of chromosomal DNA sequences from *Streptomyces coelicolor* forms a new family of plasmids detectable in *Streptomyces lividans*. *Molecular and General Genetics*, **184**, 230–40.

BIRCH, A. & CULLUM, J. (1985). Temperature-sensitive mutants of the *Streptomyces* plasmid pIJ702. *Journal of General Microbiology*, **131**, 1299–303.

BOER, P. H. & HICKEY, D. A. (1986). The α-amylase gene in *Drosophila melanogaster*: nucleotide sequence, gene structure and expression motifs. *Nucleic Acids Research*, **14**, 8399–411.

BRAWNER, M. E., AUERBACH, J. I., FORNWALD, J. A., ROSENBERG, M. & TAYLOR, D. P. (1985). Characterization of *Streptomyces* promoter sequences using the *Escherichia coli* galactokinase gene. *Gene*, **40**, 191–201.

BRÜCKNER, R. & MATZURA, H. (1985). Regulation of the inducible chloramphenicol acetyltransferase gene of the *Staphylococcus aureus* plasmid pUB112. *The EMBO Journal*, **4**, 2295–300.

BRUTON, C. J. & CHATER, K. F. (1987). Nucleotide sequence of IS*110*, an insertion sequence of *Streptomyces coelicolor* A3(2). *Nucleic Acids Research*, **15**, 7053–65.

BUTTNER, M. J. & BROWN, N. L. (1987). Two promoters from the *Streptomyces* plasmid pIJ101 and their expression in *Escherichia coli*. *Gene*, **51**, 179–86.

CALOS, M. P. & MILLER, J. M. (1980). Transposable elements. *Cell*, **20**, 579–95.

CAMPBELL, A. M. & BOTSTEIN, D. (1983). Evolution of the lambdoid phages. In *Lambda II*, eds. R. W. Hendrix, J. W. Roberts, F. W. Stahl & R. A. Weisberg, pp. 365–80. Cold Spring Harbor: Cold Spring Harbor Laboratory.

CHAN, P. T. (1986). Nucleotide sequence of the *Staphylococcus aureus* PC1 β-lactamase gene. *Nucleic Acids Research*, **14**, 5940.

CHATER, K. F. (1984). Morphological and physiological differentiation in *Streptomyces*. In *Microbial Development*, eds. R. Losick & L. Shapiro, pp. 89–115. Cold Spring Harbor: Cold Spring Harbor Laboratory.

CHATER, K. F. (1986). *Streptomyces* phages and their application to *Streptomyces* genetics. In *The Bacteria*, vol. IX, Antibiotic-producing *Streptomyces*, eds. S. E. Queener & L. E. Day, pp. 119–58. Orlando, Florida: Academic Press.

CHATER, K. F., BRUTON, C. J., FOSTER, S. G. & TOBEK, I. (1985). Physical and genetic analysis of IS*110*, a transposable element of *Streptomyces coelicolor* A3(2). *Molecular and General Genetics*, **200**, 235–9.

CHATER, K. F. & HOPWOOD, D. A. (1984). *Streptomyces* genetics. In *The Biology of the Actinomycetes*, eds. M. Goodfellow, M. Mordarski & S. T. Williams, pp. 229–86. London: Academic Press.

CHATER, K. F. & L. C. WILDE (1980). *Streptomyces albus* G mutants defective in the *Sal*GI restriction-modification system. *Journal of General Microbiology*, **116**, 323–34.

CHINENOVA, T. A., MKRTUMIAN, N. M. & LOMOVSKAYA, N. D. (1982). Genetic study of a novel phage resistance character in *Streptomyces coelicolor* A3(2). *Genetika*, **18**, 1945–52.

CHUNG, S. T. (1987). Tn*4556*, a 6.8 kb transposable element of *Streptomyces fradiae*. *Journal of Bacteriology*, **169**, 4436–41.

COURVALIN, P., WEISBLUM, B. & DAVIES, J. (1977). Aminoglycoside-modifying enzyme of an antibiotic-producing bacterium acts as a determinant of antibiotic resistance in *Escherichia coli*. *Proceedings of the National Academy of Sciences, U.S.A.*, **74**, 999–1003.

CULLUM, J., ALTENBUCHNER, J., FLETT, F., PIENDL, W. & PLATT, J. (1987). Genetic instability and DNA amplification in *Streptomyces*. In *Genetics of Industrial Microorganisms*, part A, eds. M. Alacevic, D. Hranueli & Z. Toman, pp. 237–46. Zagreb: Pliva.

DANILENKO, V. N., PUZYNINA, G. G. & LOMOVSKAYA, N. D. (1977). Multiple drug resistance in Actinomycetes. *Genetika*, **10**, 1831–42.

DANILENKO, V. N., STARODUBTSEVA, L. I. & NAVASHIN, S. M. (1986). Regulation of expression of kanamycin and chloramphenicol resistance determinants in *S. lividans* 66. In *Biological, Biochemical and Biomedical Aspects of Actinomycetes*, eds. G. Szabo, S. Biro & M. Goodfellow, pp. 79–81. Budapest: Akademiai Kiado.

DEBOUCK, C., RICCIO, A., SCHUMPERLI, D., MCKENNEY, K., JEFFERS, J., HUGHES, C., ROSENBERG, M., HEUSTERSPREUTE, M., BRUNEL, F. & DAVISON, J. (1985). Structure of the galactokinase gene of *Escherichia coli*, the last (?) gene of the *gal* operon. *Nucleic Acids Research*, **13**, 1841–53.

DEHOTTAY, P., DUSART, J., DE MEESTER, F., JORIS, B., VAN BEEUMEN, J., ERPICUM, T., FRERE, J-M. & GHUYSEN, J-M. (1987). Nucleotide sequence of the gene encoding the *Streptomyces albus* G β-lactamase precursor. *European Journal of Biochemistry*, **166**, 345–50.

DENG, Z., KIESER, T. & HOPWOOD, D. A. (1986). Expression of a *Streptomyces* plasmid promoter in *E. coli*. *Gene*, **43**, 295–300.

DISTLER, J., BRAUN, C., EBERT, A. & PIEPERSBERG, W. (1987). Gene cluster for streptomycin biosynthesis in *Streptomyces griseus*: analysis of a central region including the major resistance gene. *Molecular and General Genetics*, **208**, 204–10.

DOOLITTLE, R. F. (1981). Similar amino acid sequences: chance or common ancestry? *Science*, **214**, 149–59.

DYSON, P., BETZLER, M., KUMAR, T. & SCHREMPF, H. (1987). Biochemical and genetic analysis of spontaneous genetic instability and DNA amplification in *Streptomyces lividans*. In *Genetics of Industrial Microorganisms, part B*, eds. M. Alacevic, D. Hranueli & Z. Toman, pp. 57–65. Zagreb: Pliva.

FEDORENKO, V. A., DANILENKO, V. N. & LOMOVSKAYA, N. D. (1988). Genetic mapping of unstable chloramphenicol resistance determinant in *Streptomyces coelicolor* A3(2). *Genetika*, in press.

FISHER, S. H., BRUTON, C. J. & CHATER, K. F. (1987). The glucose kinase gene of *Streptomyces coelicolor* and its use in selecting deletions from defined end-points. *Molecular and General Genetics*, **206**, 35–44.

FISHMAN, S. E., ROSTECK, P. R., JR. & HERSHBERGER, C. L. (1985). A 2.2-kilobase repeated DNA segment is associated with DNA amplification in *Streptomyces fradiae*. *Journal of Bacteriology*, **161**, 199–206.

FLETT, F. & CULLUM, J. (1987). DNA deletions in spontaneous chloramphenicol-sensitive mutants of *Streptomyces coelicolor* A3(2) and *Streptomyces lividans* 66. *Molecular and General Genetics*, **207**, 499–502.

FLETT, F., PLATT, J. & CULLUM, J. (1987). DNA rearrangements associated with instability of an arginine gene in *Streptomyces coelicolor* A3(2). *Journal of Basic Microbiology*, **27**, 3–10.

FOSTER, S. G. (1983). A search for transposons active on *Streptomyces* DNA. Ph.D. Thesis, University of East Anglia, Norwich.

FREEMAN, R. F., BIBB, M. J. & HOPWOOD, D. A. (1977). Chloramphenicol acetyl-transferase-independent chloramphenicol resistance in *Streptomyces coelicolor* A3(2). *Journal of General Microbiology*, **98**, 453–65.

GENTHNER, F. J., HOOK, L. A. & STROHL, W. R. (1985). Determination of the molecular mass of bacterial genomic DNA and plasmid copy number by high pressure liquid chromatography. *Applied and Environmental Microbiology*, **50**, 1007–13.

GLADEK, A. & ZAKRZEWSKA, J. (1984). Genome size of *Streptomyces*. *FEMS Microbiology Letters*, **24**, 73–6.

GRYCZAN, T., ISRAELI-REICHES, M., DEL BUE, M. & DUBNAU, D. (1984). DNA sequence and regulation of *ermD*, a macrolide-lincosamide-streptogramin B resistance element from *Bacillus licheniformis*. *Molecular and General Genetics*, **194**, 349–56.

HAGENBUECHLE, O., BOVEY, R. & YOUNG, R. A. (1980). Tissue specific expression of mouse α-amylase genes: nucleotide sequence of isozyme mRNAs from pancreas and salivary gland. *Cell*, **21**, 179–87.

HARWOOD, C. R., WILLIAMS, D. M. & LOVETT, P. S. (1983). Nucleotide sequence of a *Bacillus pumilus* gene specifying chloramphenicol acetyltransferase. *Gene*, 24, 163–9.

HEFFRON, F., MCCARTHY, B. J., OHTSUBO, H. & OHTSUBO, E. (1979). DNA sequence analysis of the transposon Tn*3*: three genes and three sites involved in transposition of Tn*3*. *Cell*, 18, 1153–63.

HENDERSON, G., KRYGSMAN, P., LIU, C. J., DAVEY, C. C. & MALEK, L. T. (1987). Characterization and structure of genes for proteases A and B from *Streptomyces griseus*. *Journal of Bacteriology*, 169, 3778–84.

HERBERT, C. J., SARWAR, M., NER, S. S., GILES, I. G. & AKHTAR, M. (1986). Sequence and interspecies transfer of an aminoglycoside phosphotransferase gene (APH) of *Bacillus circulans*. *Biochemical Journal*, 233, 383–93.

HODGSON, D. A. (1982). Glucose repression of carbon source uptake and metabolism in *Streptomyces coelicolor* and its perturbation in mutants resistant to 2-deoxyglucose. *Journal of General Microbiology*, 128, 2417–30.

HODGSON, D. A. & CHATER, K. F. (1981). A chromosomal locus controlling extracellular agarase production by *Streptomyces coelicolor* A3(2), and its inactivation by chromosomal integration of plasmid SCP1. *Journal of General Microbiology*, 124, 339–48.

HOLLIDAY, R. & PUGH, J. E. (1975). DNA modification mechanisms and gene activity during development. *Science*, 187, 226–32.

HOPWOOD, D. A. (1967). Genetic analysis and genome structure in *Streptomyces coelicolor*. *Bacteriological Reviews*, 31, 373–403.

HOPWOOD, D. A. (1986). Cloning and analysis of antibiotic biosynthetic genes in *Streptomyces*. In *Biological, Biochemical and Biomedical Aspects of Actinomycetes*, eds. G. Szabo, S. Biro & M. Goodfellow, pp. 3–14. Budapest: Akademiai Kiado.

HOPWOOD, D. A., BIBB, M. J., CHATER, K. F., KIESER, T., BRUTON, C. J., KIESER, H. M., LYDIATE, D. J., SMITH, C. P., WARD, J. M. & SCHREMPF, H. (1985). *Genetic Manipulation of Streptomyces: a Laboratory Manual*. Norwich: The John Innes Foundation.

HOPWOOD, D. A., CHATER, K. F., DOWDING, J. E. & VIVIAN, A. (1973). Recent advances in *Streptomyces coelicolor* genetics. *Bacteriological Reviews*, 37, 371–405.

HOPWOOD, D. A., KIESER, T., LYDIATE, D. & BIBB, M. J. (1986). *Streptomyces* plasmids: their biology and use as cloning vectors. In *The Bacteria. A Treatrise on Structure and Function*, Vol. IX, Antibiotic-producing *Streptomyces*, eds. S. W. Queener & L. E. Day, pp. 159–229. Orlando, Florida: Academic Press.

HOPWOOD, D. A., KIESER, T., WRIGHT, H. M. & BIBB, M. J. (1983). Plasmids, recombination and chromosome mapping in *Streptomyces lividans* 66. *Journal of General Microbiology*, 129, 2257–69.

HOPWOOD, D. A. & WRIGHT, H. M. (1976). Integration of the plasmid SCP1 with the chromosome of *Streptomyces coelicolor* A3(2). In *Second International Symposium on the Genetics of Industrial Microorganisms*, ed. K. D. Macdonald, pp. 607–19. London: Academic Press.

HORINOUCHI, S., BYEON, W. & WEISBLUM, B. (1983). A complex attenuator regulates inducible resistance in macrolide, lincosamide, and streptogramin B antibiotics in *Streptococcus sanguis*. *Journal of Bacteriology*, 154, 1252–62.

HORINOUCHI, S. & WEISBLUM, B. (1982). Nucleotide sequence and functional map of pE194, a plasmid that specifies inducible resistance to macrolide, lincosamide and streptogramin type B antibiotics. *Journal of Bacteriology*, 150, 804–14.

HORNEMANN, U., OTTO, C. J., HOFFMAN, G. G. & BERTINUSON, A. C. (1987). Spectinomycin resistance and associated DNA amplification in *Streptomyces achromogenes* subsp. *rubradiris*. *Journal of Bacteriology*, **169**, 2360–9.

HUNTER, I. S. (1985). Gene cloning in *Streptomyces*. In *DNA cloning*, vol. II, ed. D. M. Glover, pp. 19–44. Oxford: IRL Press.

HÜTTER, R. & ECKHARDT, T. (1988). Genetic Manipulation. In *Actinomycetes in Biotechnology*, eds. M. Goodfellow, S. T. Williams & M. Mordarski. London: Academic Press, in press.

IIDA, S., MEYER, J. & ARBER, W. (1983). Prokaryotic IS elements. In *Mobile Genetic Elements*, ed. J. A. Shapiro, pp. 159–221. New York: Academic Press.

ISHIHARA, H., NAKANO, N. M. & OGAWARA, H. (1985). Cloning of a gene from *Streptomyces* species complementing *argG* mutations. *Journal of Antibiotics*, **38**, 787–94.

JAURIN, B. & COHEN, S. N. (1985). *Streptomyces* contain *Escherichia coli*-type A + T-rich promoters having novel structural features. *Gene*, **39**, 191–201.

KANEHISA, M. I. (1982). Los Alamos sequence analysis package for nucleic acids and proteins. *Nucleic Acids Research*, **10**, 183–96.

KATZ, E., THOMPSON, C. J. & HOPWOOD, D. A. (1983). Cloning and expression of the tyrosinase gene from *Streptomyces antibioticus* in *Streptomyces lividans*. *Journal of General Microbiology*, **129**, 2703–14.

KATZ, L., BROWN, D. P., TUAN, J. S. & CHIANG, S. D. (1987). Autonomous replication and site-specific integration in *Streptomyces lividans* of plasmid pSE101 that exists in the chromosome of *Saccharopolyspora erythraea*. *Journal of Bacteriology*, submitted.

KENDALL, K., ALI-DUNKRAH, U. & CULLUM, J. (1987). Cloning of the galactokinase gene (*galK*) from *Streptomyces coelicolor*. *Journal of General Microbiology*, **133**, 721–5.

KENDALL, K. & CULLUM, J. (1984). Cloning and expression of an extracellular agarase gene from *Streptomyces coelicolor* A3(2) in *Streptomyces lividans* 66. *Gene*, **29**, 315–21.

KENDALL, K. & CULLUM, J. (1986). Identification of a DNA sequence associated with plasmid integration in *Streptomyces coelicolor* A3(2). *Molecular and General Genetics*, **202**, 240–5.

KIESER, T., HOPWOOD, D. A., WRIGHT, H. M. & THOMPSON, C. J. (1982). pIJ101, a multi-copy broad host-range *Streptomyces* plasmid: functional analysis and development of DNA cloning vectors. *Molecular and General Genetics*, **185**, 223–38.

KINASHI, H., SHIMAJI, M. & SAKAI, A. (1987). Giant linear plasmids in *Streptomyces* which code for antibiotic biosynthesis genes. *Nature*, **328**, 454–6.

KLECKNER, N. (1981). Transposable elements in prokaryotes. *Annual Review of Genetics*, **15**, 341–404.

KOLLER, K.-P. (1986). Over-production of a polypeptide α-amylase inhibitor (tendamistat) in *Streptomyces tendae*: genetic and regulatory aspects. In *Biological, Biochemical and Biomedical Aspects of Actinomycetes*, eds. G. Szabo, S. Biro & M. Goodfellow, pp. 177–183. Budapest: Akademiai Kiado.

LEMARIE, H-G. & MULLER-HILL, B. (1986). Nucleotide sequences of the *galE* gene and the *galT* gene of *E. coli*. *Nucleic Acids Research*, **14**, 7705–11.

LESKIW, B. K., MEVARECH, M., MYERS, L. A. & JENSEN, S. E. (1986). Spontaneous insertion of *Streptomyces clavuligens* genomic DNA into the *Streptomyces* plasmid pIJ702. *Abstract Book, Fifth International Symposium on the Genetics of Industrial Microorganisms*, p. 25.

LIPMAN, D. J. & PEARSON, W. R. (1985). Rapid and sensitive protein similarity searches. *Science*, **227**, 1435–41.

LOMOVSKAYA, N. D., CHATER, K. F. & MKRTUMIAN, N. M. (1980). Genetics and molecular biology of *Streptomyces* bacteriophages. *Microbiological Reviews*, **44**, 206–29.

LONG, C. M., VIROLLE, M. J., CHANG, S. Y., CHANG, S. & BIBB, M. J. (1987). The α-amylase gene of *Streptomyces limosus*: nucleotide sequence, expression motifs and amino acid sequence homology to mammalian and invertebrate α-amylases. *Journal of Bacteriology*, **169**, in press.

LYDIATE, D. J., HENDERSON, D. J., ASHBY, A. M. & HOPWOOD, D. A. (1987). Transposable elements of *Streptomyces coelicolor* A3(2). In *Genetics of Industrial Microorganisms*, Part B, eds. M. Alacevic, D. Hranueli & Z. Toman, pp. 49–56. Zagreb: Pliva.

LYDIATE, D. J., IKEDA, H. & HOPWOOD, D. A. (1986). A 2.6 kb DNA sequence of *Streptomyces coelicolor* A3(2) which functions as a transposable element. *Molecular and General Genetics*, **203**, 79–88.

LYDIATE, D. J., MENDEZ, C., KIESER, H. M. & HOPWOOD, D. A. (1987). Mutation and cloning of clustered *Streptomyces* genes essential for sulphate metabolism. *Molecular and General Genetics*, submitted.

MCLAUGHLIN, J. R., MURRAY, C. L. & RABINOWITZ, J. C. (1981). Unique features of the ribosome binding site sequence of the Gram-positive *Staphylococcus aureus* B-lactamase gene. *Journal of Biological Chemistry*, **256**, 11283–91.

MADON, J., MORETTI, P. & HÜTTER, R. (1987). Site-specific integration and excision of pMEA100 in *Nocardia mediterranei*. *Molecular and General Genetics*, **209**, 257–64.

MAZODIER, P., COSSART, P., GIRAUD, E. & GASSER, F. (1985). Completion of the nucleotide sequence of the central region of Tn*5* confirms the presence of three resistance genes. *Nucleic Acids Research*, **13**, 195–205.

NEUGEBAUER, K., SPRENGEL, R. & SCHALLER, H. (1981). Penicillinase from *Bacillus licheniformis*; nucleotide sequence of the gene and implications for the biosynthesis of a secretory protein in a Gram-positive bacterium. *Nucleic Acids Research*, **9**, 2577–88.

ODELSON, D. A., RASMUSSEN, J. L., SMITH, C. J. & MACRINA, F. L. (1987). Extrachromosomal systems and gene transmission in anaerobic bacteria. *Plasmid*, **17**, 87–109.

OHNUKI, T., KATOH, T., IMANAKA, T. & AIBA, S. (1985). Molecular cloning of tetracycline resistance genes from *Streptomyces rimosus* in *Streptromyces griseus* and characterisation of the cloned genes. *Journal of Bacteriology*, **161**, 1010–16.

OKA, A., SUGISAKI, H. & TAKANAMI, M. (1981). Nucleotide sequence of the kanamycin resistance transposon Tn*903*. *Journal of Molecular Biology*, **147**, 217–26.

OLSON, M. O. J., NAGABHUSHAN, N., DZWINIEL, M. & SMILLIE, L. B. (1970). Primary structure of α-lytic protease: a bacterial homologue of the pancreatic serine proteases. *Nature*, **228**, 438–42.

OMER, C. A. & COHEN, S. N. (1984). Plasmid formation in *Streptomyces*: excision and integration of the SLP1 replicon at a specific chromosomal site. *Molecular and General Genetics*, **196**, 429–38.

OMER, C. A. & COHEN, S. N. (1986). Structural analysis of plasmid and chromosomal loci involved in site-specific excision and integration of the SLP1 element of *Streptomyces coelicolor*. *Journal of Bacteriology*, **166**, 999–1006.

OMER, C. A. & COHEN, S. N. (1987). Site-specific excision and integration of the *Streptomyces* transmissible element SLP1. In *Genetics of Industrial Microorganisms*, Part A, eds. M. Alacevic, D. Hranueli & Z. Toman, pp. 95–101. Zagreb: Pliva.

ORGEL, L. E. & CRICK, F. H. C. (1980). Selfish DNA: the ultimate parasite. *Nature*, **284**, 604–6.

ORLOVA, V. A. & DANILENKO, V. N. (1983). DNA multiplication fragment in *Streptomyces antibioticus* – a producer of oleandomycin. *Antibiotiki*, **28**, 163–76.

PERNODET, J. L., SIMONET, J. M. & GUERINEAU, M. (1984). Plasmids in different strains of *Streptomyces ambofaciens*: free and integrated forms of the plasmid pSAM2. *Molecular and General Genetics*, **198**, 35–41.

POTEKHIN, Y. A. & DANILENKO, V. N. (1985). The determinant of kanamycin resistance of *Streptomyces rimosus*: amplification in the chromosome and reversed gentic instability. *Molekularnaya Biologiya*, **19**, 805–17, English translation, vol. 19, 672–83.

RASMUSSEN, J., ODELSON, D. & MACRINA, F. (1986). Complete nucleotide sequence and transcription of *ermF*: a macrolide-lincosamide-streptogramin B resistance determinant from *Bacteroides fragilis*. *Journal of Bacteriology*, **168**, 523–33.

ROBERTS, A., HUDSON, G. & BRENNER, S. (1985). An erythromycin-resistance gene from an erythromycin-producing strain of *Arthrobacter* spp. *Gene*, **35**, 259–70.

RODGERS, W. H., SPRINGER, W. & YOUNG, F. H. C. (1982). Cloning and expression of a *Streptomyces fradiae* neomycin resistance gene in *Escherichia coli*. *Gene*, **18**, 133–41.

RODICIO, M. R. & CHATER, K. F. (1987). Clonación y caracterización de los genes que codifican el sistema de rescricción-modificación *Sal*I (*Sal*GI). *Abstract book, 11th Meeting of the Spanish Society for Microbiology*, Gijon, Spain, June 1987: pp. 235–9.

SCHOLLMEIER, K., GÄRTNER, D. & HILLEN, W. (1985). A bidirectionally active signal for termination of transcription is located between *tetA* and *orfL* on transposon Tn*10*. *Nucleic Acids Research*, **13**, 4227–37.

SCHREMPF, H., BUJARD, H., HOPWOOD, D. A. & GOEBEL, W. (1975). Isolation of covalently closed circular deoxyribonucleic acid from *Streptomyces coelicolor* A3(2). *Journal of Bacteriology*, **121**, 416–21.

SCHREMPF, H., DYSON, P., BETZLER, M., KUMAR, T. & GROITL, P. (1987). Amplification and deletion of DNA sequences in *Streptomyces*. In *Genetics of Industrial Microorganisms*, part A, eds. M. Alacevic, D. Hranueli & Z. Toman, pp. 177–84. Zagreb: Pliva.

SERMONTI, G., PETRIS, A., MICHELI, M. & LANFALONI, L. (1978). Chloramphenicol resistance in *Streptomyces coelicolor* A3(2): possible involvement of a transposable element. *Molecular and General Genetics*, **164**, 99–103.

SKEGGS, P. A., HOLMES, D. J. & CUNDLIFFE, E. (1987). Cloning of aminoglycoside-resistance determinants from *Streptomyces tenebrarius* and comparison with related genes from other actinomycetes. *Journal of General Microbiology*, **133**, 915–23.

SLADKOVA, I. A. (1986). Physical mapping of IS*281* of *Streptomyces*. *Molekularnaya Biologiya*, **20**, 1079–83.

SLADKOVA, I. A., KLOCHKOVA, O. A., CHINENOVA, T. A. & LOMOVSKAYA, N. D. (1984). Physical mapping of *Streptomyces coelicolor* A3(2) actinophages. VII. Generation of the deletions in the region of φC43 insertion sequence. *Molekularnaya Biologiya*, **18**, 497–503.

SLOMA, A. & GROSS, M. (1983). Molecular cloning and nucleotide sequence of the type I B-lactamase gene from *Bacillus cereus*. *Nucleic Acids Research*, **11**, 4997–5004.

SUAREZ, J. E., ARCA, P., LLANEZA, J., VILLAR, C., HARDISSON, C. & MENDOZA, M. C. (1987). Naturaleza y epidemiología de la resistencia plasmídica a fosfomycin. In *Abstract Book, 11th Meeting of the Spanish Society for Microbiology*, Gijon, Spain, June 1987: pp. 91–8.

SUTCLIFFE, J. G. (1978). Nucleotide sequence of the ampicillin resistance gene of *Escherichia coli* plasmid pBR322. *Proceedings of the National Academy of Science, USA*, **75**, 3737–41.

THOMPSON, C. J. & GRAY, G. S. (1983). Nucleotide sequence of a streptomycete aminoglycoside phosphotransferase gene and its relationship to phosphotransferase encoded by resistance plasmids. *Proceedings of the National Academy of Science, USA*, **80**, 5190–4.

TRIEU-CUOT, P., ARTHUR, M. & COURVALIN, P. (1987). Origin, evolution and dissemination of antibiotic resistance genes. *Microbiological Sciences*, **4**, 263–6.

TRIEU-CUOT, P. & COURVALIN, P. (1983). Nucleotide sequence of the *Streptococcus faecalis* plasmid gene encoding the 3′5″-aminoglycoside phosphotransferase type III. *Gene*, **23**, 331–41.

UCHIYAMA, H. & WEISBLUM, B. (1985). N-methyltransferase of *Streptomyces erythraeus* that confers resistance to the macrolide–lincosamide–streptogramin B antibiotics: amino acid sequence and its homology to cognate R-factor enzymes from pathogenic bacilli and cocci. *Gene*, **38**, 103–10.

VIVIAN, A. & HOPWOOD, D. A. (1973). Genetic control of fertility in *Streptomyces coelicolor* A3(2): new kinds of donor strains. *Journal of General Microbiology*, **76**, 147–62.

WALKER, M. S. & WALKER, J. B. (1970). Streptomycin biosynthesis and metabolism. Enzymatic phosphorylation of dihydrostreptobiosamine moieties of dihydro-streptomycin-(streptidino)-phosphate and dihydrostreptomycin by *Streptomyces* extracts. *Journal of Biological Chemistry*, **245**, 6683–9.

WESTPHELING, J., RANES, M. & LOSICK, R. (1985). RNA polymerase heterogeneity in *Streptomyces coelicolor*. *Nature*, **313**, 22–7.

YOUNG, M. & CULLUM, J. (1987). A plausible mechanism for large-scale chromosomal DNA amplification in streptomycetes. *FEBS Leters*, **212**, 10–14.

YOUNGMAN, P., ZUBER, P., PERKINS, J. B., SANDMAN, K., IGO, M. & LOSICK, R. (1985). New ways to study developmental genes in spore-forming bacteria. *Science*, **228**, 285–91.

TRANSPOSITION IN *STREPTOCOCCUS:* STRUCTURAL AND GENETIC PROPERTIES OF THE CONJUGATIVE TRANSPOSON Tn*916*

DON B. CLEWELL, ELISABETH SENGHAS, JOANNE M. JONES, SUSAN E. FLANNAGAN, MITSUYO YAMAMOTO and CYNTHIA GAWRON-BURKE

Departments of Oral Biology and Microbiology/Immunology, Schools of Dentistry and Medicine, and The Dental Research Institute, The University of Michigan, Ann Arbor, MI, 48109, USA

INTRODUCTION

Streptococci are Gram-positive bacteria closely related phylogenetically to the major genera *Bacillus, Clostridium, Staphylococcus* and *Lactobacillus* (Ludwig *et al.*, 1985), all of which possess markedly [A + T]-rich genomes. Many streptococci are clinically important, and treatment of streptococcal infections is often impeded by the occurrence of antibiotic resistance. Two general groups of transposable genetic elements have been implicated in this resistance. One class is exemplified by Tn*917* (Tomich, An & Clewell, 1979; 1980), which confers erythromycin resistance. Tn*917* closely resembles class II transposons of the Tn*3* family (Perkins & Youngman, 1984; Shaw & Clewell, 1985), and its adaptation as a tool for transposon mutagenesis, especially in *Bacillus subtilis*, has been reviewed by Youngman (1987). Transposons of the other group, one of which (Tn*916*) is the subject of this article, are non-plasmid elements that exhibit intercellular conjugal transfer. (For a recent review of these systems see Clewell & Gawron-Burke (1986).) These elements are generally found on the bacterial chromosome and are able to transpose intracellularly to resident plasmids. Their conjugative transfer to a new bacterial host in the absence of plasmid DNA corresponds to an intercellular transposition event, and insertions into the recipient chromosome occur somewhat randomly. Such elements are therefore referred to as conjugative transposons. In some cases multiple resistance determinants are involved, and these can exist together on segments of DNA up to 60 kilobases (kb) in size. Transfer experiments are generally carried out on solid surfaces (e.g. filter membranes) and frequencies of 10^{-8} to 10^{-5} per donor have been observed. Some elements can transfer to a number of different streptococcal species and even to other gram-positive genera.

Tn916 is a 16.4 kb conjugative transposon originally identified in *Streptococcus (Enterococcus) faecalis* DS16 (Franke & Clewell, 1981). It encodes resistance to tetracycline (Tc) and is closely related to Tn918 of *S. faecalis* RC73 (Clewell *et al.*, 1985) and Tn919 of *S. sanguis* FC1 (Fitzgerald & Clewell, 1985). A Tn916-like element has also been observed on the *S. faecalis* plasmid pCF10 (Christie & Dunny, 1986) and has recently been designated Tn925 (G. Dunny, personal communication). In addition, the 25.3 kb multiple resistance element Tn1545 from *Streptococcus pneumoniae* has properties very similar to Tn916 (Courvalin & Carlier, 1987).

Strain DS16 harbours a conjugative haemolysin/bacteriocin plasmid pAD1 (58 kb) as well as a non-conjugative R-plasmid pAD2 (25 kb); a plasmid-free derivative obtained by curing techniques revealed the chromosome-borne Tc-resistance element (Tn916). The transposon could be transferred in filter matings to plasmid-free recipients, which in turn could pass the resistance on in subsequent matings. The frequency of transfer varied (e.g. 10^{-8} to 10^{-5} per donor) and was characteristic for the particular donor strain (Gawron-Burke & Clewell, 1982). The location of Tn916 in the chromosome and influence from adjacent sequences appear to affect the frequency of transfer. There is a quantitative correlation between the conjugative donor potential and the frequency of transposition to a subsequently introduced plasmid such as pAD1 (Gawron-Burke & Clewell, 1982); thus, the two phenomena appear to share a common step. In addition, transposition to a resident plasmid and conjugative transfer in the absence of plasmid DNA are both Rec-independent events (Franke & Clewell, 1981).

It has been proposed that movement of Tn916 occurs by an excision/insertion mechanism (Fig. 1). Excision is viewed as the rate-limiting step which in turn triggers an efficient expression of functions necessary for conjugation and insertion. Support for an excision/insertion mechanism is based on the behaviour of Tn916 when residing on a conjugative plasmid such as pAD1 or the broad host range erythromycin (Em)-resistance plasmid pAM81 (26 kb) (Gawron-Burke & Clewell, 1982). Transfer of such a plasmid results in a 'zygotic induction' in the recipient, which leads to an excision of the element from the plasmid and its subsequent loss (segregation) or insertion into the chromosome. The excision can be precise, since

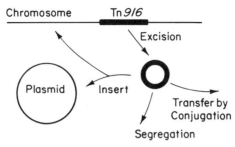

Fig. 1. Working hypothesis for behaviour of Tn*916*.

in the case of a pAD1::Tn*916* derivative it restored expression of an insertionally inactivated haemolysin gene. A similar phenomenon occurs when plasmid DNA bearing the transposon is introduced into *S. sanguis* Challis by natural transformation. A zygotically induced transposition results in insertions at different sites in the recipient chromosome.

DNA fragments containing Tn*916* have been cloned on a plasmid vector (pGL101, a pBR322 derivative encoding resistance to ampicillin; Lauer *et al.*, 1981) in *Escherichia coli* DH1 by selection for Tc-resistance (Gawron-Burke & Clewell, 1984). The transposon, however, is unstable and it readily excises and segregates during growth in the absence of Tc. (Overnight growth in the absence of Tc results in greater than 90% of the cells becoming sensitive to Tc while maintaining the vector.) This behaviour is RecA-independent. The DNA that flanked the transposon is spliced together in the course of the excision event. Although excision occurs at a high frequency, insertion into the chromosome occurs less frequently (Gawron-Burke & Clewell, 1984). It is conceivable that the tendency for Tn*916* to excise represents an aberrant effort by the element to transpose, while on the plasmid vector in the *E. coli* host, a fully balanced expression of all the required genes for transposition and its control may not occur. It is noteworthy that plasmid chimeras generated in *E. coli* can be used to readily transform streptococci (naturally transformable streptococci or *S. faecalis* protoplasts), a process that readily leads to zygotically induced insertions of the transposon into recipient chromosomal DNA (Gawron-Burke & Clewell, 1984; Yamamoto *et al.*, 1987). When such chimeras are used to transform *E. coli* DH1 selecting for Tc-resistance, a small percentage of the transformants do not contain the resistance marker of the plasmid vector and are plasmid-free. Such derivatives have Tn*916* inserted on the chromosome (J. Jones, unpublished data).

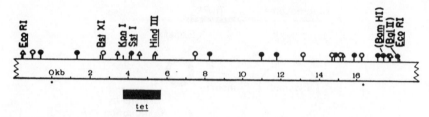

Fig. 2. Restriction map of Tn916. The flanking *Eco*RI sites define the *Eco*RI fragment F::Tn916 from pAD1::Tn916 (pAM211) (Clewell *et al.*, 1982). The analyses were performed on the chimeric plasmid pAM120, which corresponds to the above F::Tn916 fragment cloned in pGL101 (Gawron-Burke & Clewell, 1984). The asterisks mark the ends of the transposon. Restriction enzymes having single sites are named on the diagram. Multiple-site enzymes are shown as follows: solid circle, *Sau*3A; open circle, *Hinc*II; and triangle, *Hpa*II.

STRUCTURE OF Tn916 AND NUCLEOTIDE SEQUENCE OF TERMINI

The ability to clone Tn916 in *E. coli* has greatly facilitated its characterisation. Fig. 2 shows a restriction map of Tn916 based on analyses of the chimeric plasmid, pAM120. A comparison of restriction patterns derived from pAM120 with those obtained from pAM120LT (a derivative from which Tn916 had excised) allowed the identification of Tn916-specific fragments. The Tc-resistance determinant (*tet*) is believed to contain the single *Hind*III site, since insertion of a fragment into this site results in loss of resistance (Gawron-Burke & Clewell, 1984). Subcloning experiments localising *tet* in the 4.8 kb *Hinc*II fragment B (which contains the *Hind*III site) of pAM120 are consistent with this view.

The sequence of the ends of Tn916 was determined using a chimera designated pAM160. We have sequenced about 230 nucleotides of the left end of the transposon and about 260 nucleotides of the right end (Fig. 3). The left end had an overall G + C content of 32% (18% in the first 50 base pairs); whereas the right end was 28%. The two boxed-in decanucleotide segments mark the junction regions of the transposon. The left segment TTAAACTAAA corresponds to the target site (Fig. 4, 'before') into which the transposon originally inserted. The sequence on the right end TTAACTAAAA is similar but not identical to the left segment. This is particularly interesting in the light of additional sequencing data on DNA corresponding to the target region after spontaneous excision of the transposon in *E. coli*. The latter DNA was subcloned in M13 from pAM160LT, a plasmid from a Tc-sensitive segregant of pAM160 containing the regenerated *Eco*RI H fragment of pAD1. Two

A

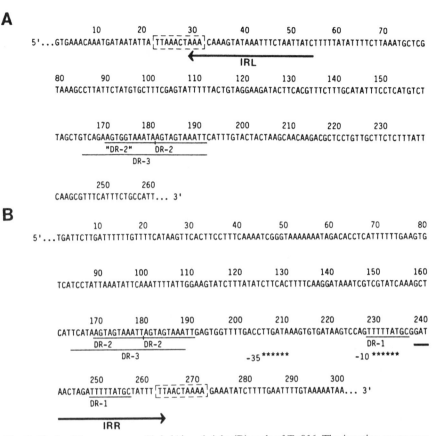

Fig. 3. Nucleotide sequences of left (A) and right (B) ends of Tn*916*. The junction sequences are boxed. The inverted repeat sequences (IRL and IRR) and the direct repeat sequences DR-1, DR-2 ('DR-2') and DR-3 are underlined, and potential promoter hexamers (-10 and -35 sequences) within the right end are indicated by asterisks. The sequencing approach was as follows. pAM160 (pGL101 with the pAD1 *Eco*RI H fragment containing a Tn*916* insert) was used as a source of DNA for subcloning into M13 vectors (M13mp18 and M13mp19). Plaques containing inserts were screened for the presence of Tn*916* sequences (using an appropriate ³²P-labelled Tn*916* probe) as well as pAD1 sequences (probing with pAM717 (Ike & Clewell, 1984), a pAD1 derivative containing a large deletion not affecting the *Eco*RI H fragment). A chimera representing pACYC184 containing a pAD1 *Bam*HI-*Sal*I fragment, which included the *Eco*RI fragment H was used as a source of DNA representative of the original target for Tn*916* insertion. The pAD1 *Eco*RI H fragment was sub-cloned from this source into M13, and sequence comparisons with the clones harbouring fragments of Tn*916* facilitated identification of the transposon termini. The DNA sequencing protocol was as described elsewhere (Shaw & Clewell, 1985), except that both an M13 universal primer (New England Biolabs, Inc.) as well as specific synthetic oligomeric primers (Systec, Inc.; Minneapolis, MN) were used.

independently derived M13 clones yielded two different sequences (Fig. 4, 'after'). Interestingly, one of them contains TTAAAC-TAAA, and the other contains TTAACTAAAA (i.e. corresponding to the sequences located at the two ends of the Tn*916* insertion).

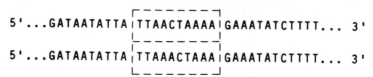

Before

5'...GATAATATTA ⌐TTAAACTAAA¬ GAAATATCTTTT... 3'

After

5'...GATAATATTA ⌐TTAACTAAAA¬ GAAATATCTTTT... 3'

5'...GATAATATTA |TTAAACTAAA| GAAATATCTTTT... 3'

Fig. 4. Sequence of target DNA prior to insertion (before) and after excision (after) of Tn*916*. Two different sequences were observed for two independently obtained excision products.

The data are suggestive of a non-replicative insertion mechanism where Tn*916* contains a sequence at least partially homologous with the target sequence, which facilitates insertion of a circular intermediate of the transposon via a reciprocal recombination. To generate the configuration shown in Fig. 3, however, the recombination event must occur on one side or the other of the dissimilar bases in the decanucleotide. During excision in *E. coli*, which may resemble the first step of a transposition event, either of the two decamers is lost with comparable efficiency. In this regard, when a sequencing analysis was performed directly on a mixture of Tc-sensitive segregants using a method reported by Chen & Seeburg (1985) and a synthetic primer corresponding to a sequence adjacent to the target site, we observed an approximately equal mixture of the two decamer sequences.

With regard to the transposon itself, the two termini contain inverted repeat sequences (designated IRL and IRR) with identity at 20 out of 26 base pairs (Fig. 3). Within the right terminus are two sets of short direct repeats designated DR-1 and DR-2. The two DR-1 sequences are 9 base pairs long and are separated by 11 base pairs; these repeats are close to the end of the transposon. The DR-2 sequences are 11 nucleotides long and are contiguous. Another set of contiguous DR-2 repeats appear near the left end; however, the first segment (indicated as 'DR-2') differs by two base pairs. Within each end of the transposon is a direct repeat designated DR-3 which is 27 nucleotides long with differences at only three positions. These two repeats contain the DR-2 sequences within them.

Potential outwardly reading promoter sites occur in the right end of the transposon. One of these is noted in Fig. 3 with a -10 hexamer sequence within DR-1. In this regard it is noteworthy that Tn916 and a number of other conjugative transposons have been shown to enhance the expression of pAD1 haemolysin when inserted adjacent to the Hly determinant (Franke & Clewell, 1981; Clewell & Gawron-Burke, 1986).

GENETIC ANALYSIS OF Tn916

The gram negative transposon Tn5 has been useful in genetic analyses of Tn916 (Yamamoto et al., 1987; and Senghas et al., 1988) using the following approach. First, the EcoRI fragment containing Tn916 in pAM120 (Fig. 2) was subcloned to the single EcoRI site within the chloramphenicol-resistance determinant of the vector pVA891 (Macrina et al., 1983). The plasmid pVA891 is a deletion derivative of the E. coli – streptococcus shuttle vector pVA838 (Macrina et al., 1982) which is no longer able to replicate in streptococci. It contains an Em-resistance determinant able to express in streptococci. The new chimera, designated pAM620, exhibited a lower frequency of Tn916 excision in the absence of selective pressure than did pAM120. (Overnight growth in the absence of Tc resulted in only 5%, or less, of the E. coli cells segregating Tn916 in contrast to >90% for pAM120.) When pAM620 was used to transform plasmid-free S. faecalis OG1X using a recently developed protoplast transformation system (Wirth, An & Clewell, 1986), Tc-resistant transformants arose at a frequency of about 30 per μg of DNA. The transformants were Em-sensitive indicating that zygotically induced Tn916 insertions had occurred. The transformants could donate Tn916 in filter matings, and when pAD1 was introduced it could be observed to acquire insertions. pAM620 is therefore a useful chimera for investigating the genetics of Tn916 transposition, since mutants can be examined for their ability to: (i) excise in E. coli; (ii) transform S. faecalis and exhibit zygotically induced transposition; (iii) transfer by conjugation from transformants; and (iv) transpose to subsequently introduced pAD1.

Tn5 (encodes resistance to kanamycin (Km)), residing on a bacteriophage lambda delivery vehicle, was used to generate insertions in pAM620 by a modification (Yamamoto et al., 1987) of the procedure of de Bruijn & Lupski (1984). About 58% of Tn5 insertions occurring

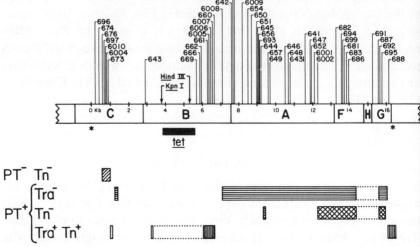

Fig. 5. Tn916 restriction map showing insertions of Tn5 and resulting effects on behaviour. The asterisks mark the ends of the transposon. The segments marked A (5.5 kb), B (4.8 kb), C (3.6 kb), F (1.6 kb), G (1.1 kb), and H (0.4 kb) correspond to HincII fragments originally identified in pAM120 (Gawron-Burke & Clewell, 1984; Yamamoto et al., 1987). Mapping of Tn5 inserts was as reported elsewhere (Yamamoto et al.,1987). The numbers reflect the corresponding pAM derivatives (Table 1) generated. The different behaviours are indicated in Table 1 as: PT$^-$, unable to transform S. faecalis OG1X protoplasts; Tra$^-$, unable to transfer conjugatively; Tn$^-$, unable to detect zygotically induced or intracellular transposition; and PT$^+$ Tra$^+$ Tn$^+$, behaviour similar to wild type Tn916.

in pAM620 were in Tn916 and covered most of the transposon, except for a segment between 1.0 and 5.5 kb (Fig. 5) for which only one insert was observed. As noted earlier, the tet determinant is located in this region. (Tc was present during selection of the Tn5-derivatives and thus precluded Tn5 insertion in tet.)

Among 49 Tn916::Tn5 derivatives, plasmid DNA from 33 was isolated and used to transform S. faecalis OG1X protoplasts selecting for Tc-resistant transformants. As shown in Table 1, Tc-resistant transformants arose in all but three cases, at frequencies ranging from 0.1 to 37 transformants/μg of DNA (Yamamoto et al., 1987; Senghas et al., 1988). All the transformants were sensitive to Em indicating excision from the plasmid vector. Southern blot hybridisation experiments on DNA from a number of representative derivatives using pVA891 as a probe showed that the cells were devoid of vector sequences (Yamamoto et al., 1987). The ability to transform protoplasts was designated PT$^+$ (for protoplast transformation positive). The three derivatives that failed to transform (pAM696, pAM674, and pAM676) are thus PT$^-$ (Table 1). These Tn5 insertions mapped close to the left end of Tn*916* (Fig. 5). Two others with insertions close to the left end (pAM697 and pAM673) transformed poorly (at 0.1 and 0.2 transformants per μg, respectively), and three with insertions near the right end of Tn*916* (pAM687, pAM688, and pAM695) transformed at 2 to 5 per μg.

The results of efforts to conjugatively transfer the Tn*916*::Tn*5* elements that had been introduced into *S. faecalis* are summarised in Table 1. Since the frequency of transfer of Tn*916* is greatly influenced by the position of the insert in the chromosome (Gawron-Burke & Clewell, 1982), at least two transformants generated from each derivative were examined. In each case, the recipient was the plasmid-free *S. faecalis* FA2-2 [mutational resistances to rifampin (Rif) and fusidic acid (Fus)], and the matings were overnight incubations on filter membranes (Clewell *et al.*, 1985). Only two derivatives (originating from pAM643 and pAM697) gave rise to transconjugants at a frequency $>5 \times 10^{-9}$ per donor (wild type is generally $>10^{-8}$). These Tn*5* insertions mapped about 2 kb apart on the left side of *tet* (Fig. 5). The remaining derivatives were unable to transfer (Tra⁻) or donated at $<10^{-9}$ per donor.

To examine intracellular transposability, pAD1 was introduced from *S. faecalis* JH2-2(pAD1) into each of the Tn*916*::Tn*5*-containing OG1X derivatives in overnight filter matings screening for haemolytic, Tc-resistant transconjugants. The pAD1-containing OG1X derivatives (designated chr::Tn*916* (pAD1) donors) were then used as donors in filter matings with FA2-2 recipients selecting for Rif- , Fus- , and Tc-resistant transconjugants on blood agar (Gawron-Burke & Clewell, 1982). Transposition of Tn*916* from the chromosome to pAD1 was based on the following rationale. pAD1 generally enhances conjugative transfer of chromosome-borne Tn*916* by approximately two orders of magnitude (Gawron-Burke & Clewell, 1982), a phenomenon proposed to involve a transient association of the transposon with the plasmid (Clewell & Gawron-Burke, 1986). In our view, the process may involve an insertion in a site or region on pAD1 from which a subsequent zygotically induced transposition occurs at high frequency upon transfer of the pAD1::Tn*916* to a recipient cell. The majority (>90%) of transconjugants arising from matings between *S. faecalis* chr::Tn*916* (pAD1) donors and plasmid-free recipients have Tn*916* on the recipient chromosome and an extrachromosomal pAD1. About 1–5% of the transconjugants are non-haemolytic, and 1–5% are hyperhaemolytic. The latter phenotype (as noted above in the section dealing with Tn*916* sequences)

Table 1. *Effects of Tn5-insertions on Tn916-behaviour*

HincII fragment	Mutants[a]	Map position[a] (kb)	PT[b]	Tra[c]	Tn[d]	Excision[e] from pAM620::Tn5 (%)	pGL101-subclone (%)
C	pAM696	0.2	−	n.d.	n.d.	n.d.	5
	pAM674	0.4	−	n.d.	n.d.	0	0
	pAM676	0.6	−	n.d.	n.d.	0	0
	pAM697	0.7	+	++	++	n.d.	7
	pAM673	0.9	+	−	+	2	37
B	pAM643	2.9	+	+++	++	29	50
	pAM669[g]	5.7	+	+	+	n.d.	68
	pAM666[g]	5.9	+	+	+	n.d.	75
	pAM662[g]	6.0	+	+	+	28	77
	pAM661	6.3	+	+	++	2	72
	pAM660	6.7	+	−	++	28	66
A	pAM642	7.6	n.d.[f]	n.d.	n.d.	n.d.	76
	pAM698	7.8	+	−	+	n.d.	n.d.
	pAM654	8.5	n.d.	n.d.	n.d.	n.d.	71
	pAM651	8.9	n.d.	n.d.	n.d.	n.d.	68
	pAM645	9.0	+	−	−	7	n.d.
	pAM693	9.1	+	−	++	n.d	n.d.
	pAM646	10.5	+	−	+	7	64
	pAM6431	10.8	+	−	+	3	68
	pAM641	11.6	+	−	++	3	74
	pAM647	11.9	+	−	−	3	n.d.
	pAM6001	12.1	n.d.	n.d.	n.d.	n.d.	65
F	pAM682	13.3	n.d.	n.d.	n.d.	8	48
	pAM694	13.6	+	−	−	n.d.	48
	pAM699	13.7	+	−	−	n.d.	37
	pAM681	13.8	n.d.	n.d.	n.d.	1	48
	pAM683	13.9	+	−	+	4	44
	pAM686	14.0	+	−	−	n.d.	n.d.
G	pAM691	15.3	+	−	−	n.d.	n.d.
	pAM687	15.6	+	−	−	n.d.	19
	pAM692	15.7	+	−	+	10	24
	pAM695	15.8	+	+	+	n.d.	45
	pAM688	16.2	+	+	+	n.d.	40
control	pAM620		+	+++	+++	5	
	pAM120		n.d.	n.d.	n.d.		90

[a] The pAM numbers relate to those corresponding positions on the map (Fig. 5) without the pAM designation. (The map shows some additional inserts.)

[b] PT, protoplast transformation. + indicates transformants were obtained. The frequency ranged from 0.1 to 37 transformants per μg of plasmid DNA (11). Those derivatives failing to transform (indicated as -) were negative with up to 150 μg plasmid DNA.

[c] Tra, conjugative transfer (overnight filter matings) to *S. faecalis* FA2-2 with selection on plates containing Tc, Rif and Fus. (OG1X is sensitive to Rif and Fus but has a mutational

is the result of a Tn*916* inserted near the haemolysin determinant. The appearance of hyperhaemolytic pAD1::Tn*916* derivatives among transconjugants confirms that Tn*916* was able to transpose. The results shown in Table 1 indicate that among the 16 derivatives that did not transfer Tc-resistance in the absence of pAD1, nine were able to transfer it in the presence of the plasmid. For seven of these (641, 646, 660, 673, 683, 693, and 698), the numbers of transconjugants were high enough to detect hyperhaemolytic pAD1::Tn*916* derivatives; these occurred in proportions typical of the wild type. For the remainder (692 and 6431), the numbers were too low to detect hyperhaemolytics. Although transfer in the latter cases may have involved a transient association of Tn*916* with pAD1, we cannot rule out other mechanisms; a plasmid-mobilisation of chromosomal markers in these low frequency ranges is known to occur by a mechanism not yet elucidated (Franke *et al.*, 1978; Franke & Clewell, 1981). However, the fact that pAD1 was not able to mobilise *tet* in the case of a number of derivatives is supportive, but not conclusive, of a transposition (hitch-hiking) mechanism in those cases where transfer was observed. For the purpose of Table 1 and Fig. 5, we have assumed that transfer of *tet* in the presence of pAD1 implies an intracellular transposition.

Eight Tn*916*::Tn*5* derivatives exhibited a $PT^+ Tra^+ Tn^+$ phenotype (Table 1), but only two (643 and 697) showed transfer- and transposition-frequencies comparable to the wild type transposon. These derivatives mapped to the left of *tet* (Fig. 5), implicating a region

Notes to Table 1 (*cont.*)
streptomycin-resistance determinant.) Transfer potential (per donor cell) is indicated as +, $>10^{-10}$; ++, $>10^{-9}$; +++, $>10^{-8}$. Because transfer potential is influenced by the particular location on the chromosome, at least two donors (independent OG1X-transformants of each derivative) were examined; the 'frequencies' reflect the average values.
[d] Tn, transposition to pAD1. Transposition potential (per cell per generation) is indicated as : +, $>10^{-8}$; ++, $>10^{-7}$; +++, $>10^{-6}$. Estimates were based on the frequency of Tc-resistant transconjugants in matings between OG1X::Tn*916*::Tn*5* (pAD1) derivatives and FA2-2; the frequency of Tc-resistant transconjugants (overnight filter matings) were divided by the frequency of transfer of pAD1 (taken as 10^{-2}). Because transposition is influenced by the location on the chromosome, at least two isolates (independent OG1X transformants of each derivative) were examined; the 'frequencies' reflect the average values.
[e] Excision in *E. coli* DH1. Cells were cultured overnight in the absence of Tc and then inoculated into fresh media and allowed to grow 3–4 generations in exponential phase before plating on media (devoid of Tc). Colonies were then tested for resistance to Tc by streaking on plates containing 5 μg/ml Tc. At least 200 colonies were checked in each case.
[f] n.d., not determined.
[g] These mutants were generated at 37 °C in contrast to the other insertions which were generated at 30 °C.

presumably along with *tet* that is not essential for normal movement of Tn*916*.

As noted above, *E. coli* plasmid vectors containing Tn*916* lose the transposon by excision when the cells are grown in the absence of Tc. To examine the effects of the Tn*5* insertions on Tn*916* excision, the *Eco*RI fragments of 27 of the pAM620 derivatives were subcloned to pGL101 where the excision rate is much greater and therefore easier to analyse. The results of segregation experiments involving both the original plasmid (pAM620 derivatives) and the subclones (pGL101 vector) are shown in Table 1. The majority of the mutants exhibited high excision frequencies; however, insertions near the left end were associated with significantly reduced or undetectable excision. It is particularly noteworthy that the insertion derivatives that were unable to excise were those that were PT⁻. For three representative mutants (subclones from pAM674, pAM676, and pAM697) the excision function could be restored (>95% segregation in overnight cultures) by complementation in *trans* with pAM6015. The latter is a pACYC184-chimera containing the *Hind*III-fragment representing the left end of Tn*916* (subcloned from pAM182; (Gawron-Burke & Clewell, 1984)).

CONCLUDING REMARKS

The above-described information on the structure and behaviour of Tn*916* begins to provide some insight into the nature of this interesting transposon. The generation of an exact duplication of the target sequence distinguishes the element from a number of other transposons (Kleckner, 1981; and Grindley & Reed, 1985). In this regard, however, there is resemblance to the staphylococcal transposon Tn*554* (Murphy & Lofdahl, 1984; Murphy *et al.*, 1986; Murphy, this volume) and the *Streptomyces* SLP1 element (Omer & Cohen, 1986; see review by Chater *et al.*, this volume). Transposition would appear to involve a reciprocal recombination between a circular Tn*916* intermediate containing a 'core' sequence (separating the two ends) and a target sequence having at least partial homology. Insertion would result in an appearance of these two sequences (i.e. 'core' and target) at opposite ends of the element, and a subsequent transposition would begin with an excision leading to an intermediate that could contain either of the two sequences. The specific flanking ('core') sequence to accompany the 'looped-out' intermediate would depend on which side of the 'paired sequences' the reciprocal recombination event occurred. It is conceivable that the inverted repeat sequences or even the 27 base-pair direct repeats (DR-3) are involved in bringing the two ends close to each other. (One could envision protein subunits that recognise this sequence and then dimerise.)

Although the proposed core region is observed here as a decamer sequence, it is conceivable that the essential component of this region is actually smaller and that some of the bases observed here represent non-essential (but possibly helpful) homology between the ends of the transposon and the target. Analyses of additional inserts (currently ongoing) may show segments of homology both larger and smaller. There are bacterial strains (non-*S. faecalis*) where a large degree of site or regional specificity for chromosomal insertion of Tn*916* and the similar element Tn*919* have been observed (Gawron-Burke *et al.*, 1986; Hill, Daly & Fitzgerald, 1985). Possibly these represent sites where the chromosome has a large degree of homology with the ends of the transposon.

The genetic studies identify specific regions which contribute to different aspects of the behaviour of Tn*916*. A region near the left end appears essential for zygotically induced transpositions. It was suggested previously (Gawron-Burke & Clewell, 1984) that the high-frequency excision that occurs in *E. coli* may be a manifestation of the first step in transposition and that an unbalanced expression of Tn*916* genes results in this behaviour. The fact that those variants unable to transform *S. faecalis* were also unable to excise in *E. coli* is supportive of this view. It is conceivable that a specific recombinase (excisase) determinant is located near the left end of the transposon.

It is evident that a major segment of Tn*916* is devoted specifically to conjugative functions, as Tn*5* insertions in over half of the element (to the right of *tet*) exhibit a Tra⁻ phenotype. Within this region is a segment necessary for intracellular transposition. Requirement for the latter, however, would appear to be bypassed in situations involving a zygotically induced transposition. Otherwise it would not have been possible to introduce the derivatives into *S. faecalis*. It would be expected that such mutants would not be able to conjugatively transfer, since transposition functions are an integral part of conjugation. In the case of an established element in a streptococcal host, these Tn-related gene product(s) may be involved in triggering the initial steps in transposition, perhaps by sensing stimulatory factors originating outside the transposon.

Conjugative transposons are ubiquitous in *Streptococcus* and appear to have played a major role in the spread of antibiotic resistance in this genus (Clewell & Gawron-Burke, 1986). In some species of streptococci (e.g. *S. pneumoniae, S. pyogenes*, and the viridans streptococci), they may be more prevalent than conjugative R-plasmids. A partial characterisation of multiple-resistance, conjugative transposons from *S. pneumoniae* and *S. agalactiae*, has been reported (Vijayakumar *et al*.,1986 *a,b*; Inamine & Burdett, 1985; Courvalin & Carlier, 1987), and one such element (Tn*1545*) has been found to have terminal sequences almost identical to those of Tn*916* (Caillaud & Courvalin, 1987). It is likely that many of the streptococcal conjugative non-plasmid elements have a common origin. Consistent with this view is the observation that all such elements reported thus far contain a Tc-resistance determinant of the *tetM* class (Burdett, Inamine, & Rajagopalan, 1982).

ACKNOWLEDGEMENTS

The authors wish to acknowledge support from the National Institutes of Health (Grant AI10318). They thank D. Friedman and D. Berg for helpful discussions and strains. They also thank P. Schultz for technical assistance. C. G.-B. was the recipient of a Junior Faculty Research Award from the American Cancer Society while at the University of Michigan.

REFERENCES

BURDETT, V., INAMINE, J. & RAJAGOPALAN, S. (1982). Heterogeneity of tetracycline resistance determinants in *Streptococcus. Journal of Bacteriology*, **149**, 995–1004.

CAILLAUD, F. & COURVALIN, P. (1987). Nucleotide sequence of the ends of the conjugative shuttle transposon Tn*1545*. *Molecular and General Genetics*, **209**, 110–15.

CHEN, E. Y. & SEEBURG, P. H. (1985). Supercoil sequencing: a fast and simple method for sequencing plasmid DNA. *DNA*, **4**, 165–70.

CHRISTIE, P. J. & DUNNY, G. M. (1986). Identification of regions of the *Streptococcus faecalis* plasmid pCF10 that encode antibiotic resistance and pheromone response functions. *Plasmid*, **15**, 230–41.

CLEWELL, D. B., AN, F., WHITE, B. A. & GAWRON-BURKE, C. (1985). *Streptococcus faecalis* sex pheromone (cAM373) also produced by *Staphylococcus aureus* and identification of a conjugative transposon (Tn*918*). *Journal of Bacteriology*, **162**, 1212–20.

CLEWELL, D. B. & GAWRON-BURKE, C. (1986). Conjugative transposons and the dissemination of antibiotic resistance in streptococci. *Annual Review of Microbiology*, **40**, 635–59.

CLEWELL, D. B., TOMICH, P. E., GAWRON-BURKE, M. C., FRANKE, A. E., YAGI, Y. & AN, F. (1982). Mapping of *Streptococcus faecalis* plasmids pAD1 and pAD2 and studies relating to transposition of Tn*917*. *Journal of Bacteriology*, **152**, 1220–30.

COURVALIN, P. & CARLIER, C. (1987). TN*1545:* A conjugative shuttle transposon. *Molecular and General Genetics*, **206**, 259–64.

DE BRUIJN, F. J. & LUPSKI, J. R. (1984). The use of transposon Tn*5* mutagenesis in the rapid generation of correlated physical and genetic maps of DNA segments cloned into multicopy plasmids – a review. *Gene*, **27**, 131–49.

FITZGERALD, G. F. & CLEWELL, D. B. (1985). A conjugative transposon (Tn*919*) in *Streptococcus sanguis*. *Infection and Immunity*, **47**, 415–20.

FRANKE, A. E. & CLEWELL, D. B. (1981). Evidence for a chromosomeborne resistance transposon (Tn*916*) in *Streptococcus faecalis* that is capable of 'conjugal' transfer in the absence of a conjugative plasmid. *Journal of Bacteriology*, **145**, 494–502.

FRANKE, A. E., DUNNY, G. M., BROWN, B., AN, F., OLIVER, D., DAMLE, S. & CLEWELL, D. B. (1978). Gene transfer in *Streptococcus faecalis*. Evidence for the mobilization of chromosomal determinants by transmissible plasmids. In, *Microbiology-1978*, (ed. Schlessinger, D.), pp. 45–7 Washington, DC, Am. Soc. Microbiol.

GAWRON-BURKE, C. & CLEWELL, D. B. (1982). A transposon in *Streptococcus faecalis* with fertility properties. *Nature*, **300**, 281–4.

GAWRON-BURKE, C. & CLEWELL, D. B. (1984). Regeneration of insertionally inactivated streptococcal DNA fragments after excision of Tn*916* in *Escherichia coli:* strategy for targeting and cloning genes from gram-positive bacteria. *Journal of Bacteriology*, **159**, 214–21.

GAWRON-BURKE, C., WIRTH, R., YAMAMOTO, M., FLANNAGAN, S., FITZGERALD, G., AN, F. & CLEWELL, D. B. (1986). Properties of the streptococcal transposon Tn*916*. In, *Molecular Microbiology and Immunobiology of Streptococcus mutans*, (ed. Hamada, S., *et al.*), pp. 191–7 New York: Elsevier.

GRINDLEY, N. D. F. & REED, R. R. (1985). Transpositional recombination in prokaryotes. *Annual Review of Biochemistry*, **54**, 863–96

HILL, C., DALY, C. & FITZGERALD, G. F. (1985). Conjugative transfer of the transposon Tn*919* to lactic acid bacteria. *FEMS Microbiology Letters*, **30**, 115–19.

IKE, Y. & CLEWELL, D. B. (1984). Genetic analysis of the pAD1 pheromone response in *Streptococcus faecalis*, using transposon Tn*917* as an insertional mutagen. *Journal of Bacteriology*, **158**, 777–83.

INAMINE, J. & BURDETT, V. (1985). Structural organization of a 67-kilobase streptococcal conjugative element mediating multiple antibiotic resistance. *Journal of Bacteriology*, **161**, 620–6.

KLECKNER, N. (1981). Transposable elements in prokaryotes. *Annual Review of Genetics*, **15**, 341–404.

LAUER, G., PASTRANA, R., SHERLEY, J. & PTASHNE, M. (1981). Construction of overproducers of the bacteriophage 434 repressor and cro proteins. *Journal of Molecular and Applied Genetics*, **1**, 139–47.

LUDWIG, W., SEEWALDT, E., KILPPER-BALZ, R., SCHLEIFER, K. H., MAGRUM, L., WOESE, C. R., FOX, G. E. & STACKEBRANDT, E. (1985). The phylogenetic position of *Streptococcus* and *Enterococcus*. *Journal of General Microbiology* , **131**, 543–51.

MACRINA, F. L., EVANS, R. P., TOBIAN, J. A., HARTLEY, D. L., CLEWELL, D. B. & JONES, K. R. (1983). Novel shuttle plasmid vehicles for *Escherichia-Streptococcus* transgeneric cloning. *Gene*, **25**, 145–50.

MACRINA, F. L., TOBIAN, J. A., JONES, K. R., EVANS, R. P. & CLEWELL, D. B. (1982). A cloning vector able to replicate in *Escherichia coli* and *Streptococcus sanguis*. *Gene*, **19**, 345–53.

MURPHY, E. & LOFDAHL, S. (1984). Transposition of Tn*554* does not generate a target duplication. *Nature*, **307**, 292–4.

MURPHY, E., HUWYLER, L. & BASTOS, M. (1986). Transposon Tn*554*: Complete nucleotide sequence and isolation of transposition-defective and antibiotic-sensitive mutants. *EMBO Journal*, **4**, 3357–65.

OMER, C. A. & COHEN, S. N. (1986). Structural analysis of plasmid and chromosomal loci involved in site-specific excision and integration of the SLP1 element of *Streptomyces coelicolor*. *Journal of Bacteriology*, **166**, 999–1006.

PERKINS, J. & YOUNGMAN, P. J. (1984). A physical and functional analysis of Tn*917*, a *Streptococcus* transposon in the Tn*3* family that functions in *Bacillus*. *Plasmid*, **12**, 119–38.

SENGHAS, E., JONES, J. M., YAMAMOTO, M., GAWRON-BURKE, C. & CLEWELL, D. B. (1988). Genetic organization of the bacterial transposon Tn*916*. *Journal of Bacteriology*, in press.

SHAW, J. H. & CLEWELL, D. B. (1985). Complete nucleotide sequence of macrolide-lincosamide-streptogramin B-resistance transposon Tn*917* in *Streptococcus faecalis*. *Journal of Bacteriology*, **164**, 782–96.

TOMICH, P., AN, F. & CLEWELL, D. B. (1979). A transposon (Tn*917*) in *Streptococcus faecalis* that exhibits enhanced transposition during induction of drug resistance. *Cold Spring Harbor Symposium on Quantitative Biology*, **43**, 1217–21.

TOMICH, P., AN, F. & CLEWELL, D. B. (1980). Properties of erythromycin-inducible Tn*917* in *Streptococcus faecalis*. *Journal of Bacteriology*, **141**, 1366–74.

VIJAYAKUMAR, M. N., PRIEBE, S. D. & GUILD, W. R. (1986*a*). Structure of a conjugative element in *Streptococcus pneumoniae*. *Journal of Bacteriology*, **166**, 978–84.

VIJAYAKUMAR, M. N., PRIEBE, S. D., POZZI, G., HAGEMAN, J. M. & GUILD, W. R. (1986*b*). Cloning and physical characterization of chromosomal conjugative elements in streptococci. *Journal of Bacteriology*, **166**, 972–7.

WIRTH, R., AN, F. Y. & CLEWELL, D. B. (1986). A highly efficient protoplast transformation system for *Streptococcus faecalis* and a new *Escherichia coli* – *S. faecalis* shuttle vector. *Journal of Bacteriology*, **165**, 831–6.

YAMAMOTO, M., JONES, J. M., SENGHAS, E., GAWRON-BURKE, C. & CLEWELL, D. B. (1987). Generation of Tn*5* insertions in the streptococcal conjugative transposon Tn*916*. *Applied and Environmental Microbiology*, **53**, 1069–72.

YOUNGMAN, P. (1987) Plasmid vectors for recovering and exploiting Tn*917* transpositions in *Bacillus* and other Gram-positive bacteria. In, *Plasmids, a practical approach*. (ed. Hardy, K. G.), pp. 79–104, Oxford, IRL Press.

TRANSPOSABLE ELEMENTS IN *STAPHYLOCOCCUS*

ELLEN MURPHY

The Public Health Research Institute, 455 First Avenue, New York, NY 10016, USA

INTRODUCTION

Despite significant advances in the prevention and treatment of infectious diseases, *Staphylococcus aureus* remains a major cause of nosocomial infection and death in hospitals worldwide. The emergence of multiply antibiotic-resistant strains is one of the chief reasons. The indiscriminate use of antibiotics has contributed to the dissemination of resistance genes, primarily those that are carried by transposons and/or plasmids. An important factor appears to be the exchange of genetic information between *S. aureus* and the ubiquitous, but relatively innocuous, coagulase-negative species such as *S. epidermidis*, *S. haemolyticus* and *S. hominis* (Archer & Johnston, 1983; McDonnell, Sweeney & Cohen, 1983). An example is the recent emergence of both plasmid- and transposon-encoded constitutive erythromycin resistance in *S. aureus* and *S. epidermidis* following the introduction of clindamycin into clinical medicine (Lampson & Parisi, 1986; Thakker-Varia *et al.*, 1987). Prior to this, most erythromycin-resistant strains expressed inducible resistance and were therefore sensitive to clindamycin (Mitsuhashi *et al.*, 1973; Lacey, Keyworth & Lincoln, 1984). Another serious problem is that of methicillin-resistant *S. aureus*, which commonly carry multiple resistance genes and are responsive only to vancomycin, a costly and toxic drug. Methicillin resistance is also believed to be carried by a mobile element. Thus, transposition in *Staphylococcus* is assuming considerable medical importance. This review will focus on the molecular biology of transposition in *Staphylococcus*. For a recent review of antibiotic resistance in *S. aureus*, see Lyon & Skurray (1987).

An understanding of transposition in *Staphylococcus* is just emerging. Table 1 lists those elements known or thought to be transposable. Transposable elements are clearly implicated in resistance to antibiotics such as erythromycin, spectinomycin, gentamycin and penicillin.

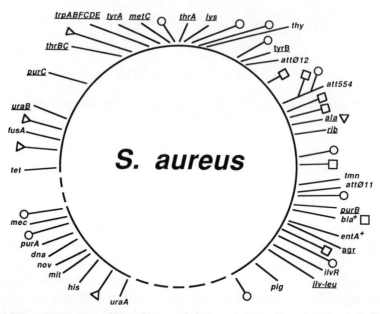

Fig. 1. Chromosomal map of *S. aureus*, modified from Pattee *et al.* (1984) and Lyon & Skurray (1987). Underlined markers are those for which auxotrophic mutants were obtained by Tn551 mutagenesis. ○, representative Tn551 insertions; △, Tn916 insertions; □, Tn4001 insertions. Dotted lines represent gaps in the circular map.

Other elements, carrying resistance to trimethoprim and methicillin, possess some properties of transposons; however, key experiments, such as the laboratory demonstration of serial transposition of a discrete DNA segment and transposition in a recombination-deficient host, remain to be done. Elements similar to insertion sequences (IS elements) described in Gram-negative organisms have also been detected in both *S. aureus* and coagulase-negative species. For most of these, definitive evidence for transposition is lacking. However, the appearance of these sequences in multiple chromosomal and plasmid sites, especially flanking resistance markers, suggests that they play a role in promoting the recombination that leads to plasmid and chromosome rearrangements and deletions. In addition to drug resistance, transposable elements may also be involved in the virulence of staphylococci through the secretion of a wide variety of extracellular proteins. These exoproteins, which are not required for viability, are highly variable among isolates of *S. aureus* and the properties of their genetic determinants suggest the involvement of mobile elements. The genes for toxic shock syndrome toxin

Table 1. *Probable transposable elements from* Staphylococcus

Designation	Resistances conferred	Size	Terminal repeats	Target dupl.	Source	References
Transposons						
Tn551	Em	5.3 kb	40 bp IR	5 bp	pI258	(Khan & Novick, 1980)
Tn554	Em Sp	6691 bp	none	none	chromosome	(Murphy & Lofdahl, 1984, Murphy et al., 1985)
Tn3853	Em Sp	nd	nd	nd	pWG4, pWG25	(Townsend et al., 1986)
Tn4001	GmTbKm	4.7 kb	IS256 (IR)	nd	pSK1	(Lyon et al., 1984a)
Tn3851	GmTbKm	5.2 kb	nd	nd	chromosome	(Townsend et al., 1984b)
Tn552	Pc	6.7 kb	nd	nd	pI524, chromosome	(Asheshov 1969; Shalita et al., 1980)
Tn4201	Pc	6.7 kb	nd	nd	pCRG1600	(R. V. Goering, personal communication)
Tn4002	Pc	6.7 kb	80 bp IR	nd	pSK4	(Lyon & Skurray, 1987)
Tn3852	Pc	7.3 kb	nd	nd	chromosome	(Kigbo et al., 1985)
Tn4003	Tp	3.6 kb	IS257 (DR)	nd	pSK1	(Gillespie et al., 1987a)
IS-like elements						
IS256		1.35 kb	nd	nd	Tn4001	(Lyon et al., 1987)
IS431-L		800 bp	22 bp		pI524 (*mer*)	(Barberis-Maino et al., 1987)
IS431-R		786 bp	14 bp		pI524 (*mer*)	(Barberis-Maino et al., 1987)
IS431 mec		0.8 kb	nd	nd	chromosome	(Barberis-Maino et al., 1987)
IS257		0.9 kb	nd	nd	Tn4003	(Gillespie et al., 1987a)

IR, inverted repeats; DR, direct repeats; nd, not determined; Em, erythromycin; Sp, spectinomycin; Gm, gentamicin; Tb, tobramycin; Km, kanamycin; Pc, penicillin; Tp, trimethoprin.

and enterotoxin B are two examples that have been suggested to be transposable.

Tn551

Introduction

Tn551 is a 5.3 kb Tn3-like transposon that specifies constitutive resistance to the macrolide-lincosamide-streptogramin B (MLS) antibiotics (Novick et al., 1979). It is found on the plasmid pI258, an incI penicillinase plasmid isolated in Japan (Mitsuhashi et al., 1963). pI258 is atypical in that many closely related plasmids have been found in strains from all over the world, but only pI258 carries macrolide resistance. Tn551 has been extensively used as a mutagenic tool for chromosome and plasmid mapping in S. aureus (Pattee et al., 1977; Novick et al., 1979; Pattee, Jones & Yost, 1984), much in the way that Tn5 has been used in Gram-negative bacteria. Although the sequence specificity for Tn551 insertion is low, it has clear regional specificities for the S. aureus chromosome. Thus, some regions have been saturated while others contain large gaps. Other transposons of Gram-positive bacteria, such as Tn917, Tn916 (see chapter by Clewell et al., this volume) and Tn4001 (see below) have the potential to fill these gaps (Jones, Yost & Pattee, 1987), but their usefulness may be limited: Tn916 tends to undergo deletion and re-insertion at other locations (Gawron-Burke & Clewell, 1984), and Tn4001 exhibits a very low transposition frequency and a tendency to insert into hotspots (P. A. Pattee, personal communication).

Tn551 is nearly identical in restriction pattern to Tn917, a transposon isolated from Streptococcus faecalis and used extensively as a mutagenic tool in genetic analysis of Bacillus subtilis (Weisblum, Holder & Halling, 1979; Tomich, An & Clewell, 1980; Youngman, Perkins & Losick, 1984). The resistance genes ermB (Tn551) and ermAm (Tn917) are the only two of the family of MLS resistance genes that cross-hybridise. Since Tn917 has been more extensively characterised than Tn551, its characteristics are included in the following discussion.

Mechanism of erythromycin resistance specified by Tn551

The mechanism and regulation of macrolide resistance has been studied in detail with the ermC (pE194) and ermA (Tn554) genes, both of which are related to *ermB* and *ermAm*. Resistance is mediated by the action of an rRNA methylase that demethylates a specific residue of 23s rRNA, reducing the affinity of erythromycin for the ribosome (Lai *et al.*, 1973). Expression of methylase is regulated at the translational level by alterations in the conformation of the leader region of the methylase mRNA (Gryczan *et al.*, 1980; Horinouchi & Weisblum, 1980; Dubnau, 1985). For *ermAm* and *ermB* the leader sequences are longer and the postulated secondary structures are more complex than those of *ermC* (Horinouchi, Byeon & Weisblum, 1983; Dubnau & Monod, 1986). Tn*551* is considered to express constitutive erythromycin resistance, which is defined by growth in the presence of a non-inducing MLS antibiotic such as tylosin (Garrod, 1957; Weaver & Pattee, 1964). Nevertheless, *ermB* can be induced about 4-fold with erythromycin (Dubnau & Monod, 1986). The phenotypic differences between *ermB* and *ermAm* can be explained by alterations in the nucleotide sequences of their regulatory regions (Horinouchi *et al.*, 1983; Dubnau & Monod, 1986).

Relationship of Tn551 and Tn917 to Tn3

Tn*551* and Tn*917* are related to the Tn*3* family of transposons (see chapter by Hatfull *et al.*, this volume). Like the other members of this family, Tn*551* and Tn*917* contain homologous short terminal inverted repeats and generate a 5 bp duplication of the target DNA during transposition (Khan & Novick, 1980; Perkins & Youngman, 1984). The genetic map of Tn*917* places it in the Tn*501* subclass of the Tn*3* family, where the gene order is resistance marker – resolvase (*res*) – transposase (*tnpA*) and all three genes are transcribed from the same strand (Fig. 2); in contrast, in Tn*3* *res* and *tnpA* are divergently expressed from overlapping promoters (Chou *et al.*, 1979; Heffron *et al.*, 1979).

Both Tn*501* and Tn*917* can be induced for transposition by the substrate of the resistance gene (Tomich *et al.*, 1980; Sherratt, Arthur & Burke, 1980). Induction of the upstream resistance gene could cause expression of transposase via increased transcriptional readthrough transcription of the *tnpA* transcript. Such a mechanism is supported by the finding of increased levels of a transcript corresponding to the full length of Tn*917* following induction with erythromycin (Shaw & Clewell, 1985). The high basal level of expression of *ermB* in Tn*551* (Dubnau & Monod, 1986) might mask such an effect and explain why it has not been demonstrated with this element.

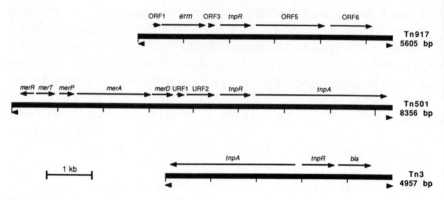

Fig. 2. Genetic organisation of Tn917 (Shaw & Clewell, 1985), Tn501 (Brown et al., 1985; Brown et al., 1986; Misra et al., 1984; Schmitt et al., 1981) and Tn3 (Heffron et al., 1979). ORF, open reading frame; res, site of cointegrate resolution; tnpA and tnpR, genes encoding transposase and resolvase, respectively; arrowheads below the elements, terminal inverted repeats; bla, β-lactamase gene; erm, erythromycin resistance gene; merR, merT, merP, merA, merD, genes of the mercury resistance operon. In Tn917, ORF-1 is the erm leader peptide, ORF3 is a short ORF of unknown function, and ORF5 and ORF6 together probably encode the transposase (see text). In Tn501, URF1 and URF2 are ORFs of unknown function.

Tn917, and by implication, Tn551, are probably more closely related to the Tn3 family than has been previously apparent. The published sequence of Tn917 reveals an open reading frame whose predicted amino acid sequence is homologous to the resolvase family of proteins (Shaw & Clewell, 1985). Mutations in two additional reading frames, ORFs-5 and 6, downstream from res, result in loss of transposition function (Perkins & Youngman, 1984; Shaw & Clewell, 1985). Although Shaw and Clewell reported no homology between ORFs-5 and 6 and the tnpA proteins of Tn3/Tn501 (Heffron et al., 1979; Brown et al., 1985), if one assumes that a sequence determination error was made that changes the reading frame, the two ORFs merge and exhibit 32.5% identity with the deduced amino acid sequence of the Tn501 transposase.

Despite their similarities, members of the Tn3 family display significant differences in insertion specificity. Tn3 rarely transposes to the bacterial chromosome (Kretschmer & Cohen, 1977), Tn917 transposes both to chromosomal and plasmid sites, with a preference for at least some plasmids (Weaver & Clewell, 1987), and Tn551 preferentially transposes to the chromosome (Pattee et al., 1977). These differences in specificity may reflect a subtle mechanistic difference in transposition, or they may be due to other factors such as chromosome organisation that make one replicon more accessible than another as a target for insertion.

Tn554

Introduction

Tn554 is a site-specific transposon that encodes resistance to the MLS antibiotics and to spectinomycin (Phillips & Novick, 1979). Historically, erythromycin resistance in *S. aureus* has most commonly been attributable to plasmid markers, typically the *ermC* gene carried by pE194 or pNE131-like plasmids (Iordanescu, 1976; Lacey, 1984; Lampson & Parisi, 1986) (Tn551 [*ermB*], found on pI258, has been isolated only once among MLS-resistant strains of *S. aureus*). In contrast, Tn554 is chromosomally located and carries the *ermA* gene. *ermA* specifies a methylase that is 58% identical in amino acid sequence to *ermC* (Murphy, 1985b). Tn554 is not a unique isolate; first isolated in 1969 in Wisconsin (Weisblum & Demohn, 1969), similar elements have recently been increasingly detected in North American and Australian clinical isolates of both *S. aureus* and *S. epidermidis*. Some of these carry constitutive rather than inducible MLS resistance and some do not express spectinomycin resistance, but all are similar in overall structure, size and properties to the prototype Tn554 (Townsend *et al.*, 1986; Thakker-Varia *et al.*, 1987).

Tn554 has several features that distinguish it from most other prokaryotic and eukaryotic transposable elements. Its ends are asymmetric, lacking either inverted or direct terminal repeats; it does not generate a duplication of a target sequence upon transposition, and it is extremely site-specific, usually inserting at the same location in the *S. aureus* chromosome, called *att554* (Fig. 1; Krolewski *et al.*, 1981; Murphy & Löfdahl, 1984). Insertions at this site occur preferentially in one orientation, designated (+), and rarely in the opposite (−) orientation. Tn554 also exhibits a very high frequency of transposition: the transposition frequency is essentially 100% into DNA carrying *att554* (Murphy *et al.*, 1981). If *att554* is absent either by deletion (Murphy & Löfdahl, 1984) or naturally as in *S. epidermidis* (Townsend *et al.*, 1986), or if it is blocked by a pre-existing insertion of Tn554 (Phillips & Novick, 1979; Murphy *et al.*, 1981), transposition of Tn554 is markedly diminished.

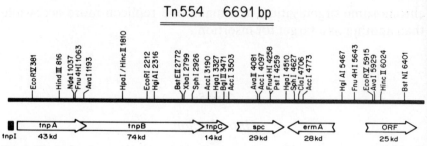

Fig. 3. Physical and genetic map of Tn*554*. reprinted from Murphy *et al.* (1985). For details see the text.

Genetic analysis of Tn554 transposition

Tn*554* is 6691 bp in length and contains six open reading frames, five of them reading from left to right (Fig. 3; Murphy, Huwyler & Bastos, 1985). Three, designated *tnpA, tnpB* and *tnpC*, are required for transposition; they encode proteins of 43, 74 and 14 kD respectively. Two genes correspond to the antibiotic resistance markers *ermA* (Murphy 1985*b*), mentioned above, and *spc*. The latter encodes a spectinomycin adenyltransferase that, unlike the streptomycin–spectinomycin adenyltransferase from Gram-negative bacteria, to which it is related, does not specify resistance to strepto-mycin (Davis & Smith, 1978; Murphy, 1985*a*). The function of the sixth reading frame, designated ORF, is unknown; mutations in this ORF affect neither the frequency nor specificity of transposition (Murphy *et al.*, 1985). An additional locus, *tnpI*, located at the left terminus of Tn*554*, is involved in *trans*-inhibition of transposition (Murphy, 1983).

Mutational and complementation analyses implicate three genes in Tn*554* transposition. All mutations created at ten restriction sites within *tnpA, tnpB* or *tnpC* reduced or abolished the ability of Tn*554* to transpose, whereas mutations at several other locations (e.g. at the *Acc*I site just 3' to *tnpC* and at three sites in the 3' ORF) had no effect upon transposition (Murphy *et al.*, 1985). The product of each of these genes appears to function efficiently in *trans*, since all *tnp⁻* mutants can be complemented for transposition, with fre-quencies ranging from 7–85%, by plasmids carrying cloned inserts containing the appropriate wild-type gene. For every mutant, com-plementation results in insertion of the mutant copy of Tn*554* into *att554* in the (+) orientation (M. Bastos & E. Murphy, unpublished data).

```
              *   *                                                               *
Flp         H I G R H L M T S F L S M K G L T E L T N V V G N W S D K R A S A V A R T T Y T H
P1          H S A R V G A A R D M A R A G - V S I P E I M Q A G G W T N - V N I V - M N Y I R
P4          H G F R T M A R G A L G E S G L W S D D A I E R Q L S H S E - R N N V R A A Y I H
Phi80       H D M R R T I A T N L S E L G - C P P H V I E K L L G H Q M - V G V M - A H Y N L
Lambda      H E L R S L S A - R L Y E K Q - I S D K F A Q H L L G H K S - D T M A - S Q Y R -
P22         H D L R H T W A S W L V Q A G - V P I S V L Q E M G G W E S - I E M V - R R Y A H
P2          H A L R H S F A T H F M I N G - G S I I T L Q R I L G H T R - I E Q T - M V Y A H
186         H V L R H T F A S H F M M N G - G N I L V L Q R V L G H T D - I K M T - M R Y A H
Tn554-TnpA  H M L R H T H A T Q L I R E G - W D V A F V Q K R L G H A H V Q T T L - N T Y V H
Tn554-TnpB  H A F R H T V G T R M I N N G - M P Q H I V Q K F L G H E S - P E M T - S R Y A H
```

Fig. 4. Amino acid homology among the Int family of site-specific recombinases aligned according to Argos *et al.* (1986). Identical residues are shaded; residues identical in all ten proteins are marked with an asterisk. Sequences of the eight bacterial integrases are from Argos *et al.* (1986); Tn*554* *tnp*A and *tnp*B are from Murphy *et al.* (1985).

*tnp*A and *tnp*B appear to specify proteins (TnpA and TnpB respectively) that are members of the Int family of site-specific recombinases, based on their similarity to a short consensus sequence located near the C-termini of these proteins (Fig. 4). The conserved region includes three residues (His-396, Arg-399 and Tyr-433 in the relative alignment of Argos *et al.*, 1986) that are identical in all ten proteins. The conserved tyrosine (corresponding to Tyr-338 and Tyr-500 of TnpA and TnpB, respectively) is likely to be the residue that forms the phosphodiester linkage to the DNA during recombination. This suggests that both TnpA and TnpB behave as site-specific topoisomerases during Tn*554* transposition, and that one or both of them will be found to contain site-specific DNA-binding domains which recognise *att554*. The product of *tnpC*, on the other hand, is required for *orientation* specificity. Mutants in *tnpC* almost always insert in *att554* in the (−) orientation (M. Bastos & E. Murphy, unpublished data). Mutations in *tnpC* also reduce the transposition frequency to <1% of wild-type, indicating that *tnpC* is also required for efficient transposition *per se* in addition to its role in specificity.

Secondary site insertions of Tn554

Insertion of Tn*554* into locations other than *att554* is extremely rare; such insertions can be isolated only in the absence of *att554*. Those into pI524 and related plasmids occur repeatedly at the same site (Murphy & Löfdahl 1984; unpublished observations). None of the secondary site insertions contains a mutation that alters its specificity; all re-insert at *att554* upon transfer to a new host (Murphy *et al.*, 1981), as do most secondary site chromosomal insertions (unpublished observations). A naturally occurring variant of Tn*554*, designated Tn*3853*, is found on closely related plasmids in *S. aureus* and

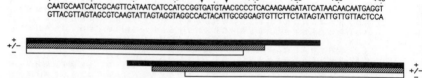

Fig. 5. Sequence and mutational analysis of *att554* (Murphy & Lofdahl, 1984; unpublished data). Arrows indicate the site of insertion of Tn*554* in the usual (+) and opposite (−) orientation. Solid bars, extent of DNA present in *att554* deletions retaining wild-type activity; hatched bars, DNA present in deletions with reduced activity. Open bars, DNA present in deletions with no *att554* activity.

S. epidermidis. Upon transfer it re-inserts into the chromosome of *S. aureus* but not *S. epidermidis* (which apparently lacks *att554*) with concomitant loss of the donor plasmid (Townsend *et al.*, 1986), similarly to the behaviour of the pI524 insertions (Murphy *et al.*, 1981).

About half of all chromosomal secondary site insertions, isolated in a host carrying a deletion of *att554*, are found at a single site in both orientations, while the remaining secondary site insertions appear to be unique isolates. The major secondary site, which is separated by at least 2 kb from *att554* (E. Murphy, unpublished observations), is identical to that found in some clinical *S. aureus* isolates that contain two copies of Tn*554* (Thakker-Varia *et al.*, 1987; D. Dubin, personal communication). Neither the plasmid nor chromosomal secondary sites are similar in DNA sequence to *att554* (Murphy & Löfdahl, 1984; D. Dubin, personal communciation).

Mechanism of Tn554 transposition

Analysis of att554

Analysis of the junction sequences of insertions of Tn*554* in both orientations in *att554* and of two secondary site insertions into plasmid pI524 reveals that Tn*554* does not generate a duplication of the target DNA and does not contain inverted or direct repeats at its termini (Murphy & Löfdahl, 1984). A 6 base 'core' sequence, GATGTA, is present at both the right end of Tn*554* and *att554* but lacking in the secondary *att* sites. Insertion in the (+) orientation occurs 3′ to this sequence (Fig. 5). Insertion in the (−) orientation, which occurs rarely in the *S. aureus* chromosome but frequently (≈10%) in cloned *att554* sites, is displaced 6 bp upstream, just 5′ to the core sequence. Secondary site insertions in the chromosome as well as in pI524 are found in both orientations. In the latter case

they are displaced 7 bp; the 'core' GATGTA in the wild-type *att554*
site is replaced by the sequence TAAAAT in the chromosomal
secondary site (D. Dubin, personal communication) or by the 7 bp
sequence TTTAATT in the pI524 insertion site (Murphy & Lofdahl,
1984).

A study of mutant and deleted *att554* sites has provided some
insight into the mechanism of Tn*554* site-specific recombination. We
previously reported the cloning of a 3.8 kb chromosomal fragment
containing *att554* into which Tn*554* transposed with high efficiency
and in both orientations (Murphy & Löfdahl, 1984). We have subcloned *att554* on a 171 bp *Taq*I–*Mbo*I fragment and constructed a
set of derivatives containing deletions from either end toward the
core insertion sequence (Fig. 5). We analysed the transposition of
Tn*554* into such plasmids by scoring transduction of erythromycin
resistance into recipient strains each containing an *att554* deletion
and also carrying a chromosomal deletion of *att554*. The results indicated that a fully functional *att554* site is between 31–53 nucleotides
in length. Deletions extending closer than 10–25 bp on either side
of the 'core' resulted in partial or complete loss of *att554* function.
Those deletions retaining only 7 bp on either side of the GATGTA
'core' sequence were totally inactive even though they still contain
the actual insertion site, while those that are deleted to within 8–12 bp
on the right, or 15 bp on the left, have reduced activity compared
to wild-type. Insertion of Tn*554* into these deleted sites often resulted
in aberrant junction sequences, which always involved the replacement of the GATGTA at the right end of Tn*554* with a different
6 base sequence (E. Murphy, unpublished data).

We have also begun to analyse point mutants in *att554* in order
to define the nucleotides important in site-specific recognition. Mutations that inactivate *att554* are located in the regions −5 to −15
and +5 to +19. The only mutant isolated that contains a point mutation in the core (+1 G→A) has wild-type *att554* function. Transposition into this mutant results in conversion to the mutant sequence
GATATA at both junctions, independently of insertion orientation
(E. Murphy, unpublished data).

Comparison with bacteriophage integration systems
Our analysis of *att554*, although preliminary, together with the
implied relationship of TnpA and TnpB with the Int family of site-
specific recombination proteins, suggest that comparison with other
site-specific recombination systems such as λ integration may be

informative in thinking about the mechanism of Tn*554* transposition. One similarity is that sequences outside the actual recombination site are required; presumably they contain important recognition sites that may stabilise the contacts with the transposition proteins, increasing both the frequency and specificity of the recombination reaction. It is not known whether these sites are an integral part of the recognition sequence, analogous to the arm binding sites of bacteriophage λ (Ross & Landy, 1982), or whether they are functionally similar to the recombination enhancer sequences of the invertases, which are active in either orientation and at great distances from the recombination site (Johnson & Simon, 1985; Kahmann *et al.*, 1985; see also the chapters in this volume by van de Putte and Meyer & Haas).

A significant difference between Tn*554* and λ recombination emerges from analysis of an *att554* mutant containing an altered core sequence, GATATA (+1G→A) (E. Murphy, unpublished data). This mutation has no effect on transposition frequency or specificity, implying that homology between the recombining partners may not be as important for Tn*554* transposition as for recombination mediated by λ Int or P1 Cre, in which sequence homology in the overlap region (i.e. the sequence separating the pair of single-stranded nicks made in each duplex partner) is more critical than the specific sequence (Weisberg *et al.*, 1983; De Massy *et al.*, 1984; Hoess *et al.*, 1987). Examination of the junction sequences of five insertions indicated that the *att554* mutant exhibits 100% conversion to the mutant (target) sequence, regardless of orientation. In bacteriophage integrative recombination the direction and frequency of mismatch conversion following the infrequent recombination between mismatched sites have provided genetic evidence in support of the formation of a Holliday structure as an intermediate in recombination (De Massy *et al.*, 1984; Leong, Nunes-Duby & Landy, 1985; Iida & Hiestand-Nauer, 1986). The available data for Tn*554*, although limited to a single mutation, rule out a flush, double-stranded break in *att554*, the simplest model consistent with the lack of target duplication. The data can be explained by a mechanism in which a heteroduplex joint is created by a pair of single-stranded exchanges on either side of position +1, with directed conversion of the mismatched joints either during or after resolution of the Holliday intermediate. The requirement for more than one recombination protein could be a factor in the observed directionality of conversion.

Regulation of Tn554 transposition

Any transposable element must be capable of regulating its own transposition to achieve the dual function of ensuring its own survival while at the same time minimising the probability of insertions lethal to the host. Transposition of Tn554 is highly efficient, but the choice of insertion sites is very limited, thus guaranteeing transposition under permissive conditions (site available) but precluding it otherwise. Thus, transposition is inhibited 100–1000 fold if the cell already contains Tn554 inserted at att554, while there is a burst of transposition, similar to zygotic induction of a prophage, upon transfer to a cell lacking a resident copy (Phillips & Novick, 1979; Murphy *et al.*, 1981).

Tn554 transposition is also regulated by a locus called *tnpI*, which, when present on a high copy plasmid vector, acts in *trans* to inhibit transposition of Tn554 to a vacant chromosomal att554 site. *tnpI* was mapped to the leftmost 89 bp of Tn554 (Murphy, 1983). There are no potential open reading frames in this region, nor were any transcripts such as those that regulate Tn10 transposition (Simons & Kleckner, 1983) detected by hybridisation analysis (Murphy *et al.*, 1985). Thus it is likely that inhibition due to *tnpI* is due to titration of one or more required proteins that bind at the left end of Tn554. The *tnpA* coding region begins immediately downstream (nt 134) from *tnpI*, suggesting that *tnpI* may overlap or be contiguous with the *tnpA* promoter. Thus an attractive model for inhibition mediated by *tnpI* would be one in which binding of the *tnpA* product to its own promoter would serve both to titrate the free transposase as well as to autoregulate its production.

Tn4001

Genetic organisation of Tn4001

Tn4001 is a recently discovered transposon that encodes resistance to the aminoglycoside antibiotics gentamycin, tobramycin and kanamycin (GmTbKm). Serial transposition from plasmid pSK1 to the chromosome and from there to another plasmid, pII147, in a rec⁻ host establishes Tn4001 as a *bona fide* transposable element (Lyon, May & Skurray, 1984; Lyon *et al.*, 1987). Tn4001 is probably identical to Tn3851, also isolated in Australia, although the reported sizes differ slightly (Townsend *et al.*, 1984b).

Tn*4001* is a composite transposon 4.7 kb in length; a 2 kb unique region specifying the antibiotic resistance is flanked by 1.35 kb inverted repeats designated IS*256* (Lyon *et al.*, 1984*a*). The resistance gene specifies a 56 kD bifunctional modifying enzyme with both aminoglycoside 2″-phosphotransferase (*aphD*, Gm) and aminoglycoside 6′-acetyltransferase (*aacA*, Tb-Km) activities [APH(2″)–AAC(6′)] (Martel *et al.*, 1983; Ubukata *et al.*, 1984). The coding region for the modifying enzyme would be large enough to account for most of the unique portion of Tn*4001*. It is likely that, as for other composite transposons, the recombination functions are encoded by one or both of the flanking inverted repeat sequences. Sequences homologous to IS*256* have been detected at multiple sites in some highly resistant clinical isolates (Lyon *et al.*, 1987); whether this represents independent transposition of IS*256* is not known.

Evolution and geographical distribution of gentamicin resistance

Distinct geographical differences still exist among gentamicin-resistant strains. Plasmid-linked GmTbKm resistance is now common in both Australian and North American isolates of *S. aureus* but is carried by unrelated plasmids. The plasmids isolated in Australia all closely resemble pSK1, an *incI*, nonconjugative plasmid also carrying resistances to trimethoprim, quaternary ammonium compounds, and ethidium bromide (Lyon *et al.*, 1984*b*; Townsend *et al.*, 1984*a*). The North American Gm^r plasmids, typified by pGO1, are conjugative and compatible with the *incI* plasmids (Archer, Dietrick & Johnston, 1985). They are widely distributed among *S. aureus* and *S. epidermidis* strains and can be transferred between these species by conjugation (Archer & Johnson, 1983; McDonnell *et al.*, 1983). Although the North American and Australian plasmids are unrelated to one another, all contain a common 2.5 kb *Hind*III fragment which includes the entire unique region of Tn*4001* including the resistance gene and part of each IS*256* sequence (Jaffe *et al.*, 1982; Gray, Huang & Davies, 1983; Mandel *et al.*, 1984; Lyon *et al.*, 1987). However, the Gm^{r}Tb^{r}Km^r marker of the American plasmids is not transposable (Archer & Johnston, 1983; Lyon *et al.*, 1987). This is almost certainly due to the fact that the IR sequences flanking the Gm^{r}Tb^{r}Km^r gene on these plasmids are truncated, about 0.7 kb in length compared to the 1.35 kb of IS*256*, and are therefore like to be defective (Lyon *et al.*, 1987; G. Archer, personal communication).

In contrast, most European strains carry chromosomal resistance. Although there are no data on their relatedness at the DNA level nor on the physical structure of the Gm^r element, the APH(2″)–AAC(6′) activities present in the European Gm^{r}Tb^{r}Km^r strains appear similar to those specified by Tn*4001* (Kayser, Homberger & Devaud, 1981; El Solh, Moreau & Ehrlich, 1986). Tn*4001* may have evolved from a chromosomally encoded Gm^{r}Tb^{r}Km^r gene. In the late 1970s, most of the resistant Australian isolates carried chromosomal resistance (Lyon, May & Skurray, 1983). These isolates have been re-examined and found to contain sequences homologous to, but 1.3 kb larger than, Tn*4001*. By the mid-1980s, Gm resistance in Australia was usually found associated with Tn*4001* on pSK1 or related plasmids. Recent chromosomal isolates, now accounting for only 25% of all resistant strains, also appear to contain elements identical to Tn*4001* (Gillespie *et al.*, 1987*b*). The simultaneous appearance of the 4.7 kb Tn*4001* element and plasmids carrying Gm resistance suggests that the earlier, larger, chromosomal element was incapable of transposition; this might explain the absence of plasmid-linked resistance in Europe (Gillespie *et al.*, 1987*b*).

TENTATIVE TRANSPOSABLE ELEMENTS

Transposition of penicillin resistance

In *S. aureus* resistance to penicillin is mediated by β-lactamase, the product of *blaZ*. This gene is generally carried by the large, nonconjugative penicillinase plasmids belonging to incompatibility groups I and II (Shalita, Murphy & Novick, 1980). The prototype penicillinase plasmid, pI524, carries a *rec*-independent, 2.2 kb invertible region, flanked by 0.65 kb inverted repeat sequences, adjacent to the *bla* operon (Murphy & Novick, 1979). Related plasmids such as pI258 lack the invertible segment, carry the *bla* operon in inverted orientation relative to pI524, and possess a single copy of the inverted repeat sequence upstream from *bla* (Murphy & Novick, 1980). Such plasmids express penicillinase normally, implying that neither the 2.2 kb invertible region nor inversion *per se* is required for β-lactamase expression.

Transposition of penicillin resistance was first detected from the chromosome of *S. aureus* strain 9789 to the plasmid resident in that strain, pI9789 (Asheshov, 1969). pI9789 is closely related to pI524 and pI258, but is naturally deleted for *bla* and the invertible segment (Shalita *et al.*, 1980). The *bla–inv* region of the plasmid obtained after 'transposition' of the chromosomal *bla* gene to pI9789 is indistinguishable by restriction analysis from that of pI524 (Shalita *et al.*, 1980). pI9789 contains a deletion of 6.1 kb relative to pI524 and appears to contain one copy of the pI524 inverted repeat, making the total size of the 9789 *bla* element (designated 'Tn*552*') 6.7 kb. Transposition of this element has not been rigorously demonstrated, and recombination involving these IR sequences is known to be both highly efficient (Murphy & Novick, 1980) and site-specific (Shalita *et al.*, 1980). Thus the 'transposition' of Tn*552* may actually involve events mediated by homologous recombination between inverted repeat sequences present on pI9789 and on the chromosome adjacent to *bla*. Whether the host recombination system is required for these events is not known.

Recently, several other elements carrying penicillin resistance, all of them with overall similarity in size and structure to Tn*552*/pI524, have been isolated. The structural similarities include the presence of the invertible DNA segment, although not all of them undergo inversion. For one of these, Tn*4201*, found on the 52 kb conjugative plasmid pCRG1600 (Asch, Goering & Ruff, 1984), there is convincing evidence for transposition: the element transposes from pCRG1600 to the chromsome in a *rec⁻* host (R. V. Goering, personal communication). Transposition is site-specific: insertion occurs at the same location between *purB* and *ilv* at which *bla* was previously mapped in a naturally occurring chromosomally resistant strain (Pattee *et al.*, 1977; R. V. Goering, personal communication). Tn*4201* also undergoes deletion and reinsertion into pCRG1600 in a *rec*-dependent fashion. Reinsertion occurs in both orientations, at both the original site or displaced by about 800 bp, as if insertion were occurring at either end of an IS-like sequence (R. V. Goering, personal communication). Thus, this recombination is probably similar to the events involving Tn*552*. It is likely that Tn*4002* and Tn*3852*, elements homologous to Tn*552* and Tn*4201* that have been detected in plasmids isolated in Australia (Kigbo *et al.*, 1985; Lyon & Skurray, 1987), will be found to exhibit similar properties.

Transposition of methicillin resistance

Methicillin(Mc) is a semi-synthetic penicillin not susceptible to inactivation by β-lactamase. Resistant strains of *S. aureus* have become a serious problem in hospitals worldwide, and, since these strains often carry multiple antibiotic resistances, they can be treated only with vancomycin, a costly and toxic drug (Parker & Hewitt, 1970; Haley *et al.*, 1982). Although resistant strains of *S. aureus* were first detected in 1961 (Barber, 1961), our understanding of the molecular basis of resistance is still incomplete. Recent studies indicate that all Mc-resistant strains contain an additional penicillin binding protein, PBP2a or PBP2′, that has very low affinity for Mc and other β-lactams (Hartman & Tomasz, 1984; Hartman & Tomasz, 1986; Reynolds & Fuller, 1986). PBP2a is believed to be a transpeptidase enzyme (Reynolds & Fuller, 1986); presumably substitution of PBP2a for one of the normal cellular transpeptidases plays a role in resistance.

Genetics of methicillin resistance
Both environmental and genetic factors influence the expression of Mc resistance. Temperature and ionic strength, for example, affect resistance in strains expressing heterogeneous resistance, in which only 10^{-6} cells in the population are resistant (Annear, 1968; Dyke, 1969). Homogeneous strains, in which all cells are resistant, are unaffected by changes in these variables. The genetic factors include several loci. The resistance gene, *mec*, has been mapped to a specific location on the *S. aureus* chromosome, linked by transformation to *purA* and *nov* (Kuhl, Pattee & Baldwin, 1978; Fig. 1). Mutations in at least two other loci, unlinked to *mec* but not mapped relative to one another, produce Mc sensitivity or convert a homogeneously resistant strain to heterogeneous expression (Berger–Bachi, 1983; Kornblum *et al.*, 1986). A gene for PBP2a has been cloned (Matsuhashi *et al.*, 1986), but its relationship to *mec* or to either of these regulatory loci is unknown. PBP2a is not, however, a sufficient condition for expression of resistance, since it is present in equivalent amounts in homogeneous and heterogeneous Mc^r cultures (Kornblum *et al.*, 1986).

In transductional crosses, in which the DNA introduced into the recipient is limited in length, Mc^r and Mc^s did not behave as allelic markers, that is, Mc sensitivity, but not resistance, was co-transducible with *purA* (Stewart & Rosenblum, 1980). This suggests that the gene for Mc resistance is part of an inserted DNA fragment for which there is no homologue in isogenic Mc^s strains. This idea is supported by the cloning of 24 kb of DNA unique to resistant strains (Matthews, Reed & Stewart, 1987); so far it is not known

if this clone carries sufficient information to produce a Mc^r phenotype. A smaller insert contained within the 24 kb (Beck, Berger–
Bachi & Kayser, 1986) does not express methicillin resistance when
re-introduced into an *S. aureus* host (J. Kornblum & R. Novick,
personal communciation), nor is it known whether either of these
clones direct the synthesis of PBP2a or are homologous with the
gene for PBP2a cloned previously (Matsuhashi *et al.*, 1986).

In addition to the presence of additional DNA in resistant strains,
the ability of *mec* to be transferred to a *rec⁻* recipient (Cohen,
Sweeney & Basu, 1977; Shafer & Iandolo, 1980) suggests that *mec*
might be carried by a transposable element. Although Mc resistance
has often been reported to be associated with markers such as enterotoxin B or with resistance to tetracycline, erythromycin or penicillin
(Dornbusch, Hallander & Lofquist, 1969; Cohen & Sweeney, 1970;
Dornbusch & Hallander, 1973; Shalita, Hertman & Sarid, 1977;
D. L. Trees & J. J. Iandolo, Abstracts of the Annual Meeting of
the American Society for Microbiology, 1987), in no case could a
physical linkage be demonstrated. The two Mc^r-specific clones that
have been isolated both contain copies of an IS-like element,
IS*431*/IS*257* (discussed in more detail below). This element is also
found flanking the mercury resistance operon of pI524 and related
penicillinase plasmids (Barberis-Maino *et al.*, 1987). This fact may
explain the results of a study in which transduction of Mc^r to strain
8325, which lacks IS*431*-like sequences (G. Archer, personal communication), was found to be dependent upon the presence of pI524
in the recipient strain (Cohen & Sweeney, 1970). Thus, the apparent
transposability of methicillin, like that of penicillin resistance, could
be explained on the basis of homology-dependent, *recA*-independent
recombination between IS*431*-like sequences, rather than by transposition.

OTHER IS-LIKE ELEMENTS

In addition to IS*256* and homologous sequences (see Tn*4001*, above),
a second family of closely related IS-like elements, about 800 bp
in length, has been detected in multiple plasmid and chromosomal
locations in *S. aureus* and in several coagulase-negative species.
The sequences known as IS*431*-L and IS*431*-R are found as direct
repeats flanking the mercury resistance operon of pI524 and related
plasmids (Barberis-Maino *et al.*, 1987). The methicillin-specific

DNA cloned by Matthews *et al.* (1987) contains four copies of a sequence homologous to IS*431*, one of which (IS*431mec*) is located on the 3.5 kb *Bgl*II fragment contained within this clone (Barberis-Maino *et al.*, 1987); the remaining three copies flank the *mer* and *tet* markers also present on this clone (Matthews *et al.*, 1987). A sequence homologous to IS*431*, designated IS*257* and estimated by electron microscopy to be 920 bp in length, is found as direct repeats flanking the trimethoprim (Tp) resistance gene of plasmid pSK1; this 'composite transposon' has been designated Tn*4003* (Gillespie *et al.*, 1987*a*).

Probes specific for IS*431mec* indicate that homologous sequences are widely distributed among clinical isolates of *S. aureus* and coagulase-negative staphylococci (G. Archer, personal communication; B. Berger-Bachi, personal communication). The conjugative plasmid pGO1, found in both *S. aureus* and *S. epidermidis*, contains eight regions that hybridise to IS*431*, flanking the Tp and Gm resistance genes and the *tra* operon. (Plasmid pGO1 also contains sequences hybridising to IS*256*/Tn*4001* (Gillespie *et al.*, 1987*a*) but it is not known if the same regions on pGO1 hybridise to both IS*256* and IS*431*/IS*257*, or if these are distinct elements.) Four of the eight copies on pGO1 contain deletions relative to IS*431*, as they fail to hybridise to a 180 bp probe specific for the end of the element (G. Archer, personal communication). Chromosomal copies of sequences homologous to IS*431* are also widely distributed in multiply antibiotic-resistant strains, and are both more frequently present and more abundant in coagulase-negative isolates than in *S. aureus*. These, too, are also often degenerate in relation to IS*431* (G. Archer, personal communication).

Relationship of IS431/IS257 to IS26, an insertion sequence from Gram-negative bacteria
Nucleotide sequence analysis of IS*431* suggests an evolutionary relationship to other IS sequences, in both Gram-negative and Gram-positive organisms. IS*431*-L, IS*431*-R and IS*431mec*, which are > 99% homologous in DNA sequence (Barberis-Maino *et al.*, 1987) contain a long open reading frame whose deduced amino acid sequence is 40% identical to the putative transposase encoded by IS*26*, an IS element originally isolated in *Proteus vulgaris* as direct repeats flanking the Km[r] transposon Tn*2680* (Iida *et al.*, 1982). IS*431*-L is 800 bp in length and contains a perfect 22 bp terminal inverted repeat; IS*431*-R is deleted at each end for 8 bp relative to IS*431*-L.

```
                      **  *************                    **********  **
IS431-L    gttatattAATTCTGGTTCTGTTGCAAAGT--aaa--800bp--ttcaACTTTGCAACAGAACCAGAATTttgatata
IS431-R    aaatgaatttttaggggTTCTGTTGCAAAGTaaaaa--786bp--ttcaACTTTGCAACAGAAtcttttatagtatcaa
IS431mec   ----------------------------------------------------ttcaactttgcaacagaaccagaacctattatgg
ISS1                  GGTTCTGTTGCAAAGTTT-----808bp----AAACTTTGCAACAGAACC
IS26         tggtgcacGGCACTGTTGCAAAgt-------820bp------taTTTGCAACAGTGCCtggtgcac
```

Fig. 6. Terminal inverted repeats of IS431-L, IS431-R, IS*mec* (Barberis-Maino *et al.*, 1987), IS26 (Mollet *et al.*, 1983) and a related element from *Streptococcus* (Polzin & Shimizu-Kadota, 1987). Inverted repeats are shown in upper case; bases common to all of the known sequences are indicated by asterisks.

IS26 is 820 bp in length, and produces 8 bp target repeats upon transposition (Mollet *et al.*, 1983). ISS1, an element found in *Streptococcus lactis*, shares 44% amino acid identity with IS26 and 57% with IS431; like IS26, ISS1 produces 8 bp direct repeats (Polzin & Shimizu-Kadota, 1987). Although the terminal inverted repeats for all these elements are very highly conserved (Fig. 6) no directly repeated sequences suggestive of a target duplication were detected adjacent to IS431, suggesting that these elements may be vestiges of once-active insertion sequences (Barberis-Maino *et al.*, 1987).

Recombination promoted by defective IS-like elements

Although IS431-like sequences appear to be widespread in *S. aureus* and *S. epidermidis* and are similar in both structure and sequence to insertion sequences from other organisms, rigorous evidence for transposition (transfer to a recombination-deficient host or demonstration of serial transfer in the laboratory) is still lacking. Thus, the designation of Tn registry numbers for resistance markers flanked by IS431-like repeats is probably premature. An example is the plasmid-linked *mer* operon specifying resistance to mercury that is found on most *incI* and *incII* penicillinase plasmids. Spontaneous, site-specific deletion of 6.5 kb associated with loss of mercury resistance (Shalita *et al.*, 1980) plus the finding of apparently identical elements in at least two locations in the *S. aureus* chromosome (Witte *et al.*, 1986) suggest that *mer* may be part of a mobile element. However, attempts in several laboratories to detect transposition of *mer* have been unsuccessful. The observed deletions are now understood to be a consequence of recombination between the flanking directly repeated copies of IS431, an event that is likely to be independent of both transposon-specific and host recombination functions (Glickman & Ripley, 1984), and the nucleotide sequences themselves suggest that IS431-R, at least, is defective (Barberis-Maino *et al.*, 1987; Laddaga *et al.*, 1987). The constant location of *mer* in the penicillinase plasmids (Shalita *et al.*, 1980) suggests that the loss of transposition function pre-dated the proliferation of these plasmids.

Even if, as seems likely, these sequences represent remnants of once active IS elements and are incapable of independent transposition, they may play an important role in the evolution of multiply resistant strains via homologous recombination between plasmids or chromosomal markers containing these sequences. Indeed, high-frequency recombination involving repeated sequences present on these plasmids, resulting in deletions, duplications and other rearrangements, has been observed in several laboratories (Murphy & Novick, 1980; G. Archer, personal communication; R. V. Goering, personal communication). At least some of these recombination events are known to be *rec*-dependent and therefore do not fall into the category of transposon-mediated recombination. On the other hand, these sequences are likely to be more stable than are plasmids, and as such may prove useful as genomic markers in epidemiological studies.

EXOPROTEINS AND PATHOGENICITY

Infections due to staphylococci

The hallmark of infection due to S. *aureus* is suppuration and abscess formation. In a healthy individual the skin and mucous membranes provide excellent physical barriers to infection; when these barriers are compromised, the bacterium is able to gain access to the underlying tissue and initiate abscess formation. If the organism penetrates the local barrier presented by the abscess, sepsis and metastatic foci, such as osteomyelitis, may also develop. S. *aureus* also causes toxinoses, including food poisoning, toxic shock syndrome, and staphylococcal scalded skin syndrome. In these diseases the symptoms are due entirely to the effects of a circulating toxin which may have been ingested or produced at the site of a minor focal infection.

Coagulase-negative staphylococci such as S. *epidermidis*, S. *hominis* and S. *haemolyticus*, previously dismissed as non-pathogenic contaminants, are now recognised as the causative agents of serious nosocomial infections involving indwelling foreign plastic bodies including catheters, shunts, pacemakers and prosthetic devices such as artificial joints and cardiac valves. Infection with one of these organisms, most commonly S. *epidermidis*, now accounts for more than one-third of all cases of prosthetic valve endocarditis (Dismukes, 1981; Karchmer, Archer & Dismukes, 1983) and half of all intravenous catheter infections (Pulverer, 1985). Although the overall incidence of these infections is low, they always lead to failure of the implanted device. In addition to their importance as occasional pathogens, coagulase-negative staphylococci, which are carried on the skin of all healthy persons, are known to exchange plasmids with *S. aureus* and are therefore considered to be an important reservoir of genetic information for the development of antibiotic resistance in *S. aureus* (Archer & Johnston, 1983; McDonnell, Sweeney & Cohen, 1983).

Role of exoproteins in virulence of staphylococci

The ability of *S. aureus* to invade and destroy tissue is in large measure due to the plethora of extracellular products elaborated by the bacterium. None of these exoproteins is required for viability. These accessory proteins include leucocidin, phospholipase, staphylokinase (fibrinolysin), hyaluronidase, protein A, staphylococcal nuclease and at least four distinct haemolysins. In addition, some strains of *S. aureus* secrete circulating toxins, as noted above. These exoproteins are rarely produced by coagulase-negative species, which are generally much less invasive than *S. aureus*.

The role of these various exoproteins in the disease process is not clearly understood, except in the case of the circulating toxins. Specific exoproteins are commonly associated with certain types of infections. For example, strains producing lipase but lacking hyaluronidase tend to produce localised abscesses, while those producing hyaluronidase but not lipase are involved in spreading infections (Jessen & Bulow, 1967), and strains producing γ-toxin are associated with experimental osteomyelitis (Kurek *et al.*, 1977). Other exoproteins, such as leucocidin, contribute to the ability of *S. aureus* to evade the host's immune system, in this case via destruction of polymorphonuclear leucocytes and macrophages.

Variability of exoprotein expression: are exoprotein genes transposable?

Early genetic studies had suggested that a number of the genes specifying these exoproteins did not behave as if they were simple chromosomal markers. Expression varied from strain to strain or was lost upon subculture or following treatments intended to cure strains of plasmids. For example, while α-haemolysin and protein A are produced by most strains (Langone, 1982), enterotoxins are produced by about half of *S. aureus* isolates (Šimkovičová, Schulz & Bergdoll, 1985), and β-haemolysin and TSST-1 by only 20% (Freer & Arbuthnott, 1983; Melconiam, Brun & Fleurette, 1983). The production of the capsular polysaccharide, which inhibits phagocytosis by macrophages, is frequently lost on subculture (Smith *et al.*, 1977).

There are several hypotheses to explain the variability in exoprotein expression among different strains. These include bacteriophage conversion (lipase, enterotoxin A, staphylokinase, β-haemolysin: Winkler *et al.*, 1965; Sako *et al.*, 1983; Betley & Mekalanos, 1985; Lee & Iandolo, 1986), plasmid linkage (exfoliatin B: Wiley & Rogolsky, 1977), and possibly transposition. The genes for enterotoxin B and TSST-1 are the best candidates for transposable elements among the exoproteins. The TSST-1 determinant maps at more than one chromosomal location (Kreiswirth, Kornblum & Novick, 1985), *entB* can be transferred to a recombination-deficient strain (Shafer & Iandolo, 1980), and the DNA for both is generally absent from non-producing strains, as would be expected for a transposable element (Kreiswirth *et al.*, 1985; S. Khan, personal communciation). The availability of cloned DNA for these and a number of other exoprotein genes is now making possible experiments to determine whether the absence of a particular exoprotein is correlated with the absence of the structural gene and unique flanking sequences. At the same time, manipulation of the cloned genes to insert selectable markers will allow direct tests of their transposability.

ACKNOWLEDGEMENTS

I would like to thank my colleagues who provided information in advance of publication, Dr R. Weisberg for pointing out to me the Int-Tn*554* homology, and Dr K. Drlica for his critical reading of the manuscript. Work in my laboratory was supported by a grant from the National Institutes of Health (GM27253).

REFERENCES

ANNEAR, D. I. (1968). The effect of temperature on resistance of *Staphylococcus aureus* to methicillin and some other antibiotics. *Medical Journal of Australia*, i, 444–6.

ARCHER, G. L. & JOHNSTON, J. L. (1983). Self-transmissible plasmids in staphylococci that encode resistance to aminoglycosides. *Antimicrobial Agents and Chemotherapy*, 24, 70–7.

ARCHER, G. L., DIETRICK, D. R. & JOHNSTON, J. L. (1985). Molecular epidemiology of transmissible gentamicin resistance among coagulase-negative staphylococci in a cardiac surgery unit. *Journal of Infectious Diseases*, 151, 243–51.

ARGOS, P., LANDY, A., ABREMSKI, K., EGAN, J. B., HAGGARD-LJUNGQUIST, E., HOESS, R. H., KAHN, M. L., KALIONIS, B., NARAYANA, S. B. L., PIERSON III, L. S., STERNBERG, N. & LEONG, J. M. (1986). The integrase family of site-specific recombinases: regional similarities and global diversity. *The EMBO Journal*, 5, 433–40.

ASCH, D. K., GOERING, R. V. & RUFF, E. A. (1984). Isolation and preliminary characterization of a plasmid mutant derepressed for conjugal transfer in *Staphylococcus aureus* . *Plasmid*, 12, 197–202.

ASHESHOV, E. H. (1969). The genetics of penicillinase production in *Staphylococcus aureus* strain PS80. *Journal of General Microbiology*, 59, 289–301.

BARBER, M. (1961). Methicillin-resistant staphylococci. *Journal of Clinical Pathology*, 14, 385–93.

BARBERIS-MAINO, L., BERGER-BACHI, B., WEBER, H., BECK, W. D. & KAYSER, F. H. (1987). IS431, a staphylococcal insertion sequence-like element related to IS26 from *Proteus vulgaris*. *Gene*, 59, 107–13.

BECK, W. D., BERGER-BACHI, B. & KAYSER, F. H. (1986). Additional DNA in methicillin-resistant *Staphylococcus aureus* and molecular cloning of *mec*-specific DNA. *Journal of Bacteriology*, 165, 373–8.

BERGER-BACHI, B. (1983). Insertional inactivation of staphyloccal methicillin resistance by Tn551. *Journal of Bacteriology*, 154, 479–87.

BETLEY, M. J. & MEKALANOS, J. J. (1985). Staphylococcal enterotoxin A is encoded by phage. *Science*, 229, 185–7.

BROWN, N. L., WINNIE, J. N.., FRITZINGER, D. & PRIDMORE, R. D. (1985). The nucleotide sequence of the *tnp*A gene completes the sequence of the *Pseudomonas* transposon Tn501. *Nucleic Acids Research*, 13, 5657–69.

BROWN, N. L., MISRA, T. K., WINNIE, J. N., SCHMIDT, A., SEIFF, M. & SILVER, S. (1986). The nucleotide sequence of the mercuric resistance operons of plasmid R100 and transposon Tn501: further evidence for *mer* genes which enhance the activity of the mercuric ion detoxification system. *Molecular and General Genetics*, 202, 143–51.

CHOU, J., CASADABAN, M. J., LEMAUX, P. G. & COHEN, S. N. (1979). Identification and characterization of a self-regulated repressor of translocation of the Tn3 element. *Proceedings of the National Academy of Sciences, USA*, 76, 4020–4.

COHEN, S. & SWEENEY, H. M. (1970). Transduction of methicillin resistance in *Staphylococcus aureus* dependent on an unusual specificity of the recipient strain. *Journal of Bacteriology*, 104, 1158–67.

COHEN, S. & SWEENEY, H. M. (1973). Effect of the prophage and penicillinase plasmid of the recipient strain upon the transduction and the stability of methicillin resistance in *Staphylococcus aureus*. *Journal of Bacteriology*, 116, 803–11.

COHEN, S., SWEENEY, H. & BASU, S. (1977). Mutations in prophage ⌀11 that impair the transducibility of their *Staphylococcus aureus* lysogens for methicillin resistance. *Journal of Bacteriology*, 129, 237–45.

Davies, J. & Smith, D. I. (1978). Plasmid-determined resistance to antimicrobial agents. *Annual Review of Microbiology*, 32, 469–518.

DE MASSY, B., STUDIER, F. W., DORGAI, L., APPELBAUM, E. & WEISBERG, R. A. (1984). Enzymes and sites of genetic recombination: studies with gene-3 endonuclease of phage T7 and with site-affinity mutants of phage λ. *Cold Spring Harbor Symposia on Quantitative Biology*, **49**, 715–26.

DISMUKES, W. E. (1981). Prosthetic valve endocarditis: factors influencing outcome and recommendations for therapy. In *Treatment of Infective Endocarditis.*, ed. A. L. Bisno, pp. 167. New York: Grune and Stratton.

DORNBUSCH, K., HALLANDER, H. O. & LOFQUIST, F. (1969). Extrachromosomal control of methicillin resistance and toxin production in *Staphylococcus aureus*. *Journal of Bacteriology*, **98**, 351–8.

DORNBUSCH, K. & HALLANDER, H. O. (1973). Transduction of penicillinase production and methicillin resistance–enterotoxin B production in strains of *Staphylococcus aureus*. *Journal of General Microbiology*, **76**, 1–11.

DUBNAU, D. (1985). Induction of *ermC* requires translation of the leader peptide. *The EMBO Journal*, **4**, 533–7.

DUBNAU, D. & MONOD, M. (1986). The regulation and evolution of MLS resistance. In *Antibiotic resistance genes: Ecology, transfer and expression.* (Banbury Report vol. 24), ed. S. B. Levy & R. P. Novick, pp. 369–85. Cold Spring Harbor, New York: Cold Spring Harbor Laboratory.

DYKE, K. G. H. (1969). Penicillinase production and intrinsic resistance to penicillins in methicillin-resistant cultures of *Staphylococcus aureus*. *Journal of Medical Microbiology*, **2**, 261–78.

EL SOLH, N., MOREAU, N. & EHRLICH, S. D. (1986). Molecular cloning and analysis of *Staphylococcus aureus* chromosomal aminoglycoside resistance genes. *Plasmid*, **15**, 104–18.

FREER, J. H. & ARBUTHNOTT, J. P. (1983). Toxins of *Staphylococcus aureus*. *Pharmacology and Therapeutics*, **19**, 55–106.

GARROD, L. P. (1957). The erythromycin group of antibiotics. *British Medical Journal*, **2**, 57–63.

GAWRON-BURKE, C. & CLEWELL, D. B. (1984). Regeneration of insertionally inactivated streptococcal DNA fragments after excision of transposon Tn*916* in *Escherichia coli*: strategy for targeting and cloning of genes from Gram-positive bacteria. *Journal of Bacteriology*, **159**, 214–21.

GILLESPIE, M. T., LYON, B. R., LOO, L. S. L., MATTHEWS, P. R., STEWART, P. R. & SKURRAY, R. A. (1987*a*). Homologous direct repeat sequences associated with mercury, methicillin, tetracycline and trimethoprim resistance determinants in *Staphylococcus aureus*. *FEMS Microbiology Letters*, **43**, 165–71.

GILLESPIE, M. T., LYON, B. R., MESSEROTTI, L. J. & SKURRAY, R. A. (1987*b*). Chromosome- and plasmid-mediated gentamicin resistance in *Staphylococcus aureus* encoded by Tn*4001*. *Journal of Medical Microbiology*, **24**, 139–44.

GLICKMAN, B. W. & RIPLEY, L. S. (1984). Structural intermediates of deletion mutagenesis: a role for palindromic DNA. *Proceedings of the National Academy of Sciences, USA*, **81**, 512–16.

GRAY, G. S., HUANG, R. T. & DAVIES, J. (1983). Aminocyclitol resistance in *Staphylococcus aureus*: presence of plasmids and aminocyclitol-modifying enzymes. *Plasmid*, **9**, 147–58.

GRYCZAN, T. J., GRANDI, G., HAHN, J., GRANDI, R. & DUBNAU, D. (1980). Conformational alteration of mRNA structure and the posttranscriptional regulation of erythromycin-induced drug resistance. *Nucleic Acids Research*, **8**, 6081–97.

HALEY, R. W., HIGHTOWER, A. W., KHABBAZ, R. F., THORNSBERRY, C., MARTONE, W. J., ALLEN, J. R. & HUGHES, J. M. (1982). The emergence of methicillin-resistant *Staphylococcus aureus* infections in United States hospitals. *Annals of Internal Medicine*, **97**, 297–303.

HARTMAN, B. J. & TOMASZ, A. (1984). Low-affinity penicillin binding protein associated with β-lactam resistance in *Staphylococcus aureus*. *Journal of Bacteriology*, **158**, 513–16.

HARTMAN, B. J. & TOMASZ, A. (1986). Expression of methicillin resistance in heterogeneous strains of *Staphylococcus aureus*. *Antimicrobial Agents and Chemotherapy*, **29**, 85–92.

HEFFRON, F., McCARTHY, B. J., OHTSUBO, H. & OHTSUBO, E. (1979). DNA sequence analysis of the transposon Tn*3*: three genes and three sites involved in transposition of Tn*3*. *Cell*, **18**, 1153–63.

HOESS, R., ABREMSKI, K., FROMMER, B., WIERZBICKI, A. & KENDALL, M. (1987). The *lox*-Cre site-specific recombination system of bacteriophage P1. In *Mechanisms of DNA Replication and Recombination. UCLA Symposia on Molecular and Cellular Biology* (vol. 47), ed. T. Kelly & R. McMacken, pp. 745–56. New York: Alan R. Liss, Inc.

HORINOUCHI, S. & WEISBLUM, B. (1980). Posttranscriptional modification of mRNA conformation: mechanism that regulates erythromycin-induced resistance. *Proceedings of the National Academy of Sciences, USA*, **77**, 7079–83.

HORINOUCHI, S., BYEON, W. & WEISBLUM, B. (1983). A complex attenuator regulates inducible resistance to macrolides, lincosamides, and streptogramin type B antibiotics in *Streptococcus sanguis*. *Journal of Bacteriology*, **154**, 1252–62.

IIDA, S. MEYER, J., LINDER, P., GOTO, N., NAKAYA, R., REIF, H. & ARBER, W. (1982). The kanamycin resistance transposon Tn*2680* derived from the R plasmid Rts1 and carried by phage P1Km has flanking 0.8-kb-long direct repeats. *Plasmid*, **8**, 187–98.

IIDA, S. & HIESTAND-NAUER, R. (1986). Localized conversion at the crossover sequences in the site-specific DNA inversion system of bacteriophage P1. *Cell*, **45**, 71–9.

IORDANESCU, S. (1976). Three distinct plasmids originating in the same *Staphylococcus aureus* strain. *Archives Roumaines de Pathologie Experimentale et de Microbiologie*, **35**, 111–18.

JAFFE, H. W., SWEENEY, H. M., WEINSTEIN, R. A., KABINS, S. A., NATHAN, C. & COHEN, S. (1982). Structural and phenotypic varieties of gentamicin resistance plasmids in hospital strains of *Staphylococcus aureus* and coagulase-negative staphylococci. *Antimicrobial Agents and Chemotherapy*, **21**, 773–9.

JESSEN, O. & BULOW, P. (1967). Changes of pathogenicity of *Staphylococcus aureus* by lysogenic conversion influencing lipase production, as evidenced by experimental skin infection in pigs. *Acta Pathologica, Microbiologica et Immunologica, Scandinavica*, **187** Suppl., 48–9.

JOHNSON, J. C. & SIMON, M. I. (1985). Hin-mediated site-specific recombination requires two 29 bp recombination sites and a 60 bp recombinational enhancer. *Cell*, **41**, 781–91.

JONES, J., YOST, S. & PATTEE, P. (1987). Transfer of the conjugal tetracycline resistance transposon Tn*916* from *Streptococcus faecalis* to *Staphylococcus aureus* and identification of some insertion sites in the staphylococcal chromosome. *Journal of Bacteriology*, **169**, 2121–31.

KAHMANN, R., RUDT, F., KOCH, C. & MERTENS, G. (1985). G inversion in bacteriophage Mu DNA is stimulated by a site within the invertase gene and a host factor. *Cell*, **41**, 771–80.

KARCHMER, A. W., ARCHER, G. L. & DISMUKES, W. E. (1983). *Staphylococcus epidermidis* prosthetic valve endocarditis: microbiological and clinical observations as guides to therapy. *Annals of Internal Medicine*, **98**, 447–55.

KAYSER, F. H., HOMBERGER, F. & DEVAUD, M. (1981). Aminocyclitol-modifying enzymes specified by chromosomal genes in *Staphylococcus aureus*. *Antimicrobial Agents and Chemotherapy*, **19**, 766–72.

KHAN, S. & NOVICK, R. P. (1980). Terminal nucleotide sequences of Tn*551*, a transposon specifying erythromycin resistance in *Staphylococcus aureus*: homology with Tn*3*. *Plasmid*, **4**, 148–54.

KIGBO, E. P., TOWNSEND, D. E., ASHDOWN, N. & GRUBB, W. B. (1985). Transposition of penicillinase determinants in methicillin-resistant *Staphylococcus aureus*. *FEMS Microbiology Letters*, **28**, 39–43.

KORNBLUM, J., HARTMAN, B. J., NOVICK, R. P. & TOMASZ, A. (1986). Conversion of a homogeneously methicillin-resistant strain of *Staphylococcus aureus* to heterogeneous resistance by Tn*551*-mediated insertional inactivation. *European Journal of Clinical Microbiology*, **5**, 714–18.

KREISWIRTH, B. N., KORNBLUM, J. S. & NOVICK, R. P. (1985). Genotypic variability of the toxic shock syndrome exoprotein determinant. In *The Staphylococci: Proceedings of the Vth International Symposium on Staphylococci and staphylococcal infections*, Warsaw, June 26–30, 1984. *Zentralblatt fuer Bakteriologie, Parasitenkunde, Infektionskrankheiten und Hygiene*, supplement 14, ed. J. Jeljaszewicz, pp. 105–9. Stuttgart: Gustav Fischer Verlag.

KRETSCHMER, P. J. & COHEN, S. N. (1977). Selected translocation of plasmid genes: frequency and regional specificity of translocation of the Tn*3* element. *Journal of Bacteriology*, **130**, 888–99.

KROLEWSKI, J. J., MURPHY, E., NOVICK, R. P. & RUSH, M. G. (1981). Site-specificity of the chromosomal insertion of *Staphylococcus aureus* transposon Tn*554*. *Journal of Molecular Biology*, **152**, 19–33.

KUHL, S. A., PATTEE, P. A. & BALDWIN, J. N. (1978). Chromosomal map location of the methicillin resistance determinant in *Staphylococcus aureus*. *Journal of Bacteriology*, **135**, 460–5.

KUREK, M., PRYJMA, K., BARTOWSKI, S. & HECZKO, P. B. (1977). Anti-staphylococcal gamma haemolysin antibodies in rabbits with staphylococcal osteomyelitis. *Medical Microbiology and Immunity*, **163**, 61–5.

LACEY, R. W. (1984). Antibiotic resistance in *Staphylococcus aureus* and streptococci. *British Medical Bulletin*, **40**, 77–83.

LACEY, R. W., KEYWORTH, N. & LINCOLN, C. (1984). Staphylococci in the UK: a review. *Journal of Antimicrobial Chemotherapy*, **14** (Suppl. D), 19–25.

LADDAGA, R. A., CHU, L., MISRA, T. K. & SILVER, S. (1987). Nucleotide sequence and expression of the mercurial resistance operon from *Staphylococcus aureus* plasmid pI258. *Proceedings of the National Academy of Sciences, USA*, **84**, 5106–10.

LAI, C., WEISBLUM, B., FAHNESTOCK, S. R. & NOMURA, M. (1973). Alteration of 23S ribosomal RNA and erythromycin-induced resistance to lincomycin and spiramycin in *Staphylococcus aureus*. *Journal of Molecular Biology*, **74**, 67–72.

LAMPSON, B. C. & PARISI, J. T. (1986). Naturally occurring *Staphylococcus epidermidis* plasmid expressing constitutive macrolide–lincosamide–streptogramin B resistance contains a deleted attenuator. *Journal of Bacteriology*, **166**, 479–83.

LANGONE, J. J. (1982). Protein A of *Staphylococcus aureus* and related immunoglobulin receptors produced by streptococci and pneumococci. *Advances in Immunology*, **32**, 157–252.

LEE, C. Y. & IANDOLO, J. J. (1986). Lysogenic conversion of staphylococcal lipase caused by insertion of the bacteriophage phage L54a genome into the lipase structural gene. *Journal of Bacteriology*, **166**, 385–91.

LEONG, M. M., NUNES-DUBY, S. E. & LANDY, A. (1985). Generation of single base-pair deletions, insertions, and substitutions by a site-specific recombination system. *Proceedings of the National Academy of Sciences, USA*, **82**, 6990–4.

LYON, B. R., MAY, J. W. & SKURRAY, R. A. (1983). Analysis of plasmids in nosocomial strains of multiple-antibiotic-resistant *Staphylococcus aureus*. *Antimicrobial Agents and Chemotherapy*, **23**, 817–26.

Lyon, B. R., May, J. W. & Skurray, R. A. (1984a). Tn*4001*: A gentamicin and kanamycin resistance transposon in *Staphylococcus aureus*. *Molecular and General Genetics*, **193**, 554–6.

LYON, B. R., IUORIO, J. L., MAY, J. W. & SKURRAY, R. A. (1984b). Molecular epidemiology of multiresistant *Staphylococcus aureus* in Australian hospitals. *Journal of Medical Microbiology*, **17**, 79–89.

LYON, B. R., GILLESPIE, M. T., BYRNE, M. E., MAY, J. W. & SKURRAY, R. A. (1987). Plasmid-mediated resistance to gentamicin in *Staphylococcus aureus*: the involvement of a transposon. *Journal of Medical Microbiology*, **23**, 105–14.

LYON, B. R. & SKURRAY, R. A. (1987). Antimicrobial resistance of *Staphylococcus aureus*: genetic basis. *Microbiological Reviews*, **51**, 88–134.

MANDEL, L. J., MURPHY, E., STEILBIGEL, N. H. & MILLER, M. H. (1984). Gentamicin uptake in *Staphylococcus aureus* possessing plasmid-encoded, aminoglycoside-modifying enzymes. *Antimicrobial Agents and Chemotherapy*, **26**, 563–9.

MARTEL, A., MASSON, M., MOREAU, N. & LE GOFFIC, F. (1983). Kinetic studies of aminoglycoside acetyltransferase and phosphotransferase from *Staphylococcus aureus* RPAL. Relationship between the two activities. *European Journal of Biochemistry*, **133**, 515–21.

MATSUHASHI, M., SONG, M. D., ISHIMO, F., WACHI, M., DOI, M., INOUE, M., UBUKATA, K., YAMASHITA, N. & KONNO, M. (1986). Molecular cloning of the gene of a penicillin-binding protein supposed to cause high resistance to β-lactam antibiotics in *Staphylococcus aureus*. *Journal of Bacteriology*, **167**, 975–80.

MATTHEWS, P. R., REED, K. C. & STEWART, P. R. (1987). The cloning of chromosomal DNA associated with methicillin and other resistances in *Staphylococcus aureus*. *Journal of General Microbiology*, 133 (in press).

McDONNELL, R. W., SWEENEY, H. M. & COHEN, S. (1983). Conjugational transfer of gentamicin-resistance plasmids intra- and interspecifically in *Staphylococcus aureus* and *Staphylococcus epidermidis*. *Antimicrobial Agents and Chemotherapy*, **23**, 151–60.

MELCONIAM, A. K., BRUN, Y. & FLEURETTE, J. (1983). Enterotoxin production, phage typing and serotyping of *Staphylococcus aureus* strains isolated from clinical materials and food. *Journal of Hygiene*. **91**, 235–42.

MISRA, T. K., BROWN, N. L., FRITZINGER, D. C., PRIDMORE, R. D., BARNES, W. M., HABERSTROH, L. & SILVER, S. (1984). Mercuric ion-resistance operons of plasmid R100 and transposon Tn*501*: The beginning of the operon including the regulatory region and the first two structural genes. *Proceedings of the National Academy of Sciences, USA*, **81**, 5975–9.

MITSUHASHI, S., MORIMURA, M., KONO, K. & OSHIMA, H. (1963). Elimination of drug resistance of *Staphylococcus aureus* by treatment with acriflavine. *Journal of Bacteriology*, **86**, 162–4.

MITSUHASHI, S., INOUYE, M., KAWABE, K., OSHIMA, H. & OKUBO, T. (1973). Genetic and biochemical studies of drug resistance in staphylococci. In *Staphylococci and staphylococcal infections*, ed. J. Jeljaszewicz, pp. 144–65. Warsaw: Polish Medical Publishers.

MOLLET, B., IIDA, S., SHEPHERD, J. & ARBER, W. (1983). Nucleotide sequence of IS26, a new prokaryotic mobile genetic element. *Nucleic Acids Research*, **11**, 6319–30.

MURPHY, E. & NOVICK, R. (1979). Physical mapping of *Staphylococcus aureus* penicillinase plasmid pI524: Characterization of an invertible region. *Molecular and General Genetics*, 175, 19–30.

MURPHY, E. & NOVICK, R. (1980). Site-specific recombination between plasmids of *Staphylococcus aureus*. *Journal of Bacteriology*, 141, 316–26.

MURPHY, E., PHILLIPS, S., EDELMAN, I. & NOVICK, R. P. (1981). Tn554: Isolation and characterization of plasmid insertions. *Plasmid*, 5, 292–305.

MURPHY, E. (1983). Inhibition of Tn554 transposition: deletion analysis. *Plasmid*, 10, 260–9.

MURPHY, E. & LÖFDAHL, S. (1984). Transposition of Tn554 does not generate a target duplication. *Nature*, 307, 292–4.

MURPHY, E., HUWYLER, L. & BASTOS, M. (1985). Transposon Tn554: complete nucleotide sequence and isolation of transposition-defective and antibiotic-sensitive mutants. *The EMBO Journal*, 4, 3357–65.

MURPHY, E. (1985a). Nucleotide sequence of a spectinomycin adenyltransferase AAD(9) determinant from *Staphylococcus aureus* and its relationship to AAD(3″)(9). *Molecular and General Genetics*, 200, 33–9.

MURPHY, E. (1985b). Nucleotide sequence of *ermA*, a macrolide-lincosamide-streptogramin B (MLS) determinant in *Staphylococcus aureus*. *Journal of Bacteriology*, 162, 633–40.

NOVICK, R., EDELMAN, I., SCHWESINGER, M., GRUSS, A., SWANSON, E. & PATTEE, P. A. (1979). Genetic translocation in *Staphylococcus aureus*. *Proceedings of the National Academy of Sciences, USA*, 76, 400–4.

PARKER, M. T. & HEWITT, J. H. (1970). Methicillin resistance in *Staphylococcus aureus*. *Lancet*, i, 800–4.

PATTEE, P. A., THOMPSON, N. E., HAUBRICH, D. & NOVICK, R. P. (1977). Chromosomal map locations of integrated plasmids and related elements in *Staphylococcus aureus*. *Plasmid*, 1, 38–51.

PATTEE, P. A., JONES, J. M. & YOST, S. C. (1984). Chromosome map of *Staphylococcus aureus*. In *Genetic maps 1984: a compilation of linkage and restriction maps of genetically studied organisms*, vol. 3, ed. S. J. O'Brien, pp. 126–130. Cold Spring Harbor, New York: Cold Spring Harbor Laboratory.

PERKINS, J. B. & YOUNGMAN, P. J. (1984). A physical and functional analysis of Tn917, a streptococcus transposon in the Tn3 family that functions in *Bacillus*. *Plasmid*, 12, 119–38.

PHILLIPS, S. & NOVICK, R. (1979). Tn554: a repressible site-specific transposon in *Staphylococcus aureus*. *Nature*, 278, 476–8.

POLZIN, K. M. & SHIMIZU-KADOTA, M. (1987). Identification of a new insertion element, similar to Gram-negative IS26, on the lactose plasmid of *Streptococcus lactis* ML3. *Journal of Bacteriology*, 169, 5481–8.

PULVERER, G. (1985). On the pathogenicity of coagulase-negative staphylococci. In *The Staphylococci: Proceedings of the Vth International Symposium on Staphylococci and staphylococcal infections*, Warszawa, June 26–30, 1984. Supplement 14., ed. J. Jeljaszewicz, pp. 1–9. Stuttgart: Gustav Fischer Verlag.

REYNOLDS, P. E. & FULLER, C. (1986). Methicillin-resistant strains of *Staphylococcus aureus*; presence of additional penicillin-binding protein in all strains examined. *FEMS Microbiology Letters*, 33, 251–4.

ROSS, W. & LANDY, A. (1982). Bacteriophage λ Int protein recognizes two classes of sequence in the phage *att* site: characterization of arm-type sites. *Proceedings of the National Academy of Sciences, USA*, 79, 7724–8.

SAKO, T., SAWAKI, S., SAKURAI, T., ITO, S., YOSHIZAWA, Y. & KONDO, I. (1983). Cloning and expression of the staphylokinase gene of *Staphylococcus aureus* in *Escherichia coli. Molecular and General Genetics*, **190**, 271–7.

SCHMITT, R., ALTENBUCHNER, J., WIEBAUER, K., ARNOLD, W., PUHLER, A. & SCHOFFL, F. (1981). Basis of transposition and gene amplification by Tn*1721* and related tetracycline-resistance transposons. *Cold Spring Harbor Symposia on Quantitative Biology*, **45**, 59–65.

SHAFER, W. M. & IANDOLO, J. J. (1980). Transduction of staphylococcal enterotoxin B synthesis: establishment of the toxin gene in a recombination-deficient mutant. *Infection and Immunity*, **27**, 280–2.

SHALITA, Z., HERTMAN, I. & SARID, S. (1977). Isolation and characterization of a plasmid involved with enterotoxin B production in *Staphylococcus aureus. Journal of Bacteriology*, **129**, 317–25.

SHALITA, Z., MURPHY, E. & NOVICK, R. P. (1980). Penicillinase plasmids of *Staphylococcus aureus*: structural and evolutionary relationships. *Plasmid*, **3**, 291–311.

SHAW, J. H. & CLEWELL, D. B. (1985). Complete nucleotide sequence of macrolide–lincosamide–streptogramin B-resistance transposon Tn*917* in *Streptococcus faecalis. Journal of Bacteriology*, **164**, 782–96.

SHERRATT, D., ARTHUR, A. & BURKE, M. (1980). Transposon-specified site-specific recombination systems. *Cold Spring Harbor Symposia on Quantitative Biology*, **45**, 275–81.

SIMKOVICOVA, M., SCHULZ, F. & BERGDOLL, M. (1985). Long-term analysis of staphylococcal food poisoning outbreaks in Slovakia. Distribution of enterotoxigenic strains. In *The Staphylococci: Proceedings of the Vth International Symposium on Staphylococci and staphylococcal infections*, Warsaw, June 26–30, 1984. *Zentralblatt fuer Bakteriologie, Parasitenkunde, Infektionskrankheiten und Hygiene*, Supplement 14., ed. J. Jeljaszewicz, pp. 616–620. Stuttgart: Gustav Fischer Verlag.

SIMONS, R. W. & KLECKNER, N. (1983). Translational control of IS*10* transposition. *Cell*, **34**, 683–91.

SMITH, R. M., PARISI, J. T., VICAL, L. & BALDWIN, J. N. (1977). Nature of the genetic determinant controlling encapsulation in *Staphylococcus aureus* Smith. *Infection and Immunity*, **17**, 231–4.

STEWART, G. C. & ROSENBLUM, E. D. (1980). Genetic behaviour of the methicillin resistance determinant in *Staphylococcus aureus. Journal of Bacteriology*, **144**, 1200–2.

THAKKER-VARIA, S., JENSSEN, W., MOON-MCDERMOTT, L., WEINSTEIN, M. & DUBIN, D. (1987). Molecular epidemiology of macrolide–lincosamide–streptogramin B resistance in *Staphylococcus aureus* and coagulase-negative-staphylococci. *Antimicrobial Agents and Chemotherapy*, **31**, 735–43.

TOMICH, P., AN, F. & CLEWELL, D. B. (1980). Properties of an erythromycin inducible transposon Tn*917* in *Streptococcus faecalis. Journal of Bacteriology*, **141**, 1366–74.

TOWNSEND, D. E., ASHDOWN, N., GREED, L. C. & GRUBB, W. B. (1984*a*). Analysis of plasmids mediating gentamicin resistance in methicillin-resistant *Staphylococcus aureus. Journal of Antimicrobial Chemotherapy*, **13**, 347–52.

TOWNSEND, D. E., ASHDOWN, N., GREED, L. C. & GRUBB, W. B. (1984*b*). Transposition of gentamicin resistance to staphylococcal plasmids encoding resistance to cationic agents. *Journal of Antimicrobial Chemotherapy*, **14**, 115–24.

TOWNSEND, D. E., BOLTON, S., ASHDOWN, N., ANNEAR, D. I. & GRUBB, W. B. (1986). Conjugative, staphylococcal plasmids carrying hitch-hiking transposons similar to Tn*554*: intra- and interspecies dissemination of erythromycin resistance. *Australian Journal of Experimental Biology and Medical Science*, **64**, 367–79.

similar to Tn*554*: intra- and interspecies dissemination of erythromycin resistance. *Australian Journal of Experimental Biology and Medical Science*, **64**, 367–79.

UBUKATA, K., YAMASHITA, N., GOTOH, A. & KONNO, M. (1984). Purification and characterization of aminoglycoside-modifying enzymes from *Staphylococcus aureus* and *Staphylococcus epidermidis*. *Antimicrobial Agents and Chemotherapy*, **27**, 754–9.

WEAVER, K. E. & CLEWELL, D. B. (1987). Transposon Tn*917* delivery vectors for mutagenesis in *Streptococcus faecalis*. In *Streptococcus Genetics*, ed. J. J. Ferretti & R. Curtiss III. Washington, D.C.: American Society for Microbiology.

WEAVER, J. R. & PATTEE, P. A. (1964). Inducible resistance to erythromycin in *Staphylococcus aureus*. *Journal of Bacteriology*, **88**, 574–80.

WEISBERG, R. A., ENQUIST, L. W., FOELLER, C. & LANDY, A. (1983). Role for DNA homology in site-specific recombination. The isolation and characterization of a site affinity mutant of coliphage λ. *Journal of Molecular Biology*, **170**, 319–42.

WEISBLUM, B. & DEMOHN, V. (1969). Erythromycin-inducible resistance in *Staphylococcus aureus*: survey of antibiotic classes involved. *Journal of Bacteriology*, **98**, 447–52.

Weisblum, B., Holder, S. B. & Halling, S. M. (1979). Deoxyribonucleic acid sequence common to staphylococcal and streptococcal plasmids which specify erythromycin resistance. *Journal of Bacteriology*, **138**, 990–8.

WILEY, B. B. & ROGOLSKY, M. (1977). Molecular and serological differentiation of staphylococcal exfoliative toxin synthesized under chromosomal and plasmid control. *Infection and Immunity*, **18**, 487–94.

WINKLER, K. C., DE WART, J. & GROOTSEN, C. (1965). Lysogenic conversion of staphylococci to loss of β-toxin. *Journal of General Microbiology*, **39**, 321–33.

WITTE, W., GREEN, L., MISRA, T. K. & SILVER, S. (1986). Resistance to mercury and to cadmium in chromosomally resistant *Staphylococcus aureus*. *Antimicrobial Agents and Chemotherapy*, **29**, 663–9.

YOUNGMAN, P., PERKINS, J. B. & LOSICK, R. (1984). Construction of a cloning site near one end of Tn*917* into which foreign DNA may be inserted without affecting transposition in *Bacillus subtilis* or expression of the transposon-borne *erm* gene. *Plasmid*, **12**, 1–9.

MULTIPLE CORRELATIONS AMONG INSERTION SEQUENCES IN THE GENOME OF NATURAL ISOLATES OF *ESCHERICHIA COLI*

DANIEL L. HARTL[1] and STANLEY A. SAWYER[1,2]

[1]*Department of Genetics, Washington University School of Medicine, Saint Louis, Missouri 63110, USA;* [2]*Department of Mathematics, Washington University, Saint Louis, Missouri 63110, USA*

INTRODUCTION

Insertion sequences are transposable DNA sequences which range in size from about one to two kilobase pairs. They are flanked by short perfect or nearly perfect inverted repeat sequences, and they invariably contain one long open reading frame and often one or more shorter ones. The product of the long reading frame is probably the transposase protein involved in transposition. At least in some cases, the shorter reading frames appear to code for proteins that regulate transposition (Calos & Miller, 1980; Kleckner, 1981).

Natural isolates of *Escherichia coli* are polymorphic for the presence or absence of insertion sequences (Hu & Deonier, 1981; Nyman *et al.*, 1983; Green *et al.*, 1984; Dykhuizen *et al.*, 1985). Among strains in which insertion sequences are present, there are large differences in the number of copies and their locations in the genome. Insertion sequences are also found in large and small plasmids, which enables the insertion sequences to be disseminated among strains by means of conjugational and co-conjugational transfer.

Insertion sequences play at least three important roles in the evolution of bacteria and their plasmids. First, insertion sequences may act as agents of insertion mutations or chromosome rearrangements. Secondly, acting in pairs, insertion sequences can mobilise the transposition of arbitrary DNA sequences, including some sequences which are beneficial to the host. Thirdly, insertion sequences may be considered as parasites at the DNA level, and important aspects of their distribution and abundance result from their parasitic characteristics.

As mutagenic agents, the transposition of insertion sequences can interrupt genes and render them dysfunctional (Botstein & Shortle, 1985). Insertions can also influence the level of gene expression and have been shown to reactivate cryptic genes (Saedler *et al.*, 1974; Reynolds *et al.*, 1981; Zafarullah *et al.*, 1981; Charlier *et al.*, 1982, Ciampi *et al.*, 1982, Georgopoulos *et al.*, 1982; Jaurin & Normark, 1983; Scordilis *et al.*, 1987). Insertion sequences serve as agents of chromosome rearrangement by recombination between elements and imprecise excision, and they provide sites for plasmid–plasmid or plasmid–chromosome replicon fusion (Saedler *et al.*, 1980; Miller *et al.*, 1984, Gaffney & Lessie, 1987).

Nearby pairs of insertion sequences can transpose together as a single intact unit and mobilise the DNA between them, which may include selectively important genes, such as those for resistance to antibiotics or heavy metals. Composite transposons containing genes sandwiched between pairs of insertion sequences are well documented, and their ability to become incorporated into transmissible plasmids accounts in large part for the rapid spread of antibiotic resistance among bacteria. As the resistance plasmids become widely spread through the action of natural selection, the insertion sequences that helped create the favoured plasmid, and others perhaps fortuituously present, hitchhike along (Calos & Miller, 1980).

Insertion sequences are selfish or parasitic DNA elements that can be maintained in populations by their ability to replicate and transpose. Possible reductions in fitness resulting from the presence of the elements are offset by the ability of the elements to be disseminated among hosts by means of plasmids (Doolittle & Sapienza, 1980; Orgel & Crick, 1980; Campbell, 1981; Hartl *et al.*, 1984). Evidence has been presented that insertion sequences can be beneficial to the host under certain conditions. For example, the insertion sequences IS*10* and IS*50* have been shown to confer a selective advantage during growth in bacterial chemostats (Chao *et al.*, 1983; Hartl *et al.*, 1983). However, the selective advantage of IS*50* occurs independently of transposition (Hartl *et al.*, 1983), while that of IS*10* involves transposition through the production of particular insertion mutations that are beneficial to the cell (Chao *et al.*, 1983).

The ability of insertion sequences to mobilise selectively advantageous DNA sequences and to act as agents of mutation and chromosome rearrangement is undoubtedly important in the long-term evolution and dissemination of insertion sequences. However, during the short run, the distribution and abundance of insertion sequences among strains of a particular species may be governed primarily by the parasitic properties of plasmid-mediated transfer and transposition.

Focusing on the parasitic characteristics of insertion sequences, we have studied the distribution and abundance of six insertion sequences in the chromosome and plasmids among 71 strains constituting the ECOR reference collection of *E. coli* (Ochman & Selander, 1984). The primary purpose was to determine the manner in which the numbers of copies of insertion sequences affect the probability of transposition and the probability of selective elimination of the host. Accordingly, various models incorporating different functional relations between copy number and transposition, or copy number and fitness, were tested for their goodness of fit to the observed distributions (Sawyer *et al.*, 1987).

During the course of this study, we obseved that unrelated insertion sequences did not occur independently among hosts. Among pairs of insertion sequences, there was a small but significant tendency for the insertion sequences to occur either together or not at all, and a correspondingly small but significant deficiency in the number of strains containing just one of the pair. Among multiple insertion sequences the statistical association was very highly significant.

The conclusion therefore seemed inescapable that insertion sequences have a small but real tendency to occur together. The origin of the correlation might result from some undiscovered biological interaction between insertion sequences, or it might result from some subtle feature of their population biology. A branching-process model for the plasmid-mediated transmission of insertion squences has been studied among hosts and it has been demonstrated that a positive correlation in the presence of unrelated insertion sequences is a consequence of plasmid transmission, even in the absence of interactions in regard to transposition or effects on fitness. Moreover, the predictions of the model are quantitatively in agreement with the observed correlations among insertion sequences. By way of background, here is an examination of the data on the distribution and abundance of insertion sequences in *E. coli*.

DISTRIBUTION OF INSERTION SEQUENCES

The number of copies of insertion sequences present in the chromosome and plasmid pools of 71 strains of *E. coli* in the ECOR reference collection (Ochman and Selander 1984) was estimated by Southern

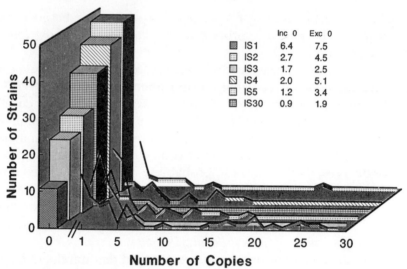

Fig. 1. Distribution of number of copies of insertion sequences in the chromosome of 71 natural isolates of the ECOR reference collection of *E. coli*. The stacked bars at left are the number of strains lacking each insertion sequence, which, from shortest bar to tallest, represent IS*1*, IS*3*, IS*2*, IS*30*, IS*4* and IS*5*. The curves at the right are the distributions of numbers of elements among infected strains. The distributions differ principally in the number of uninfected strains and in the length of the right-hand tail (the longest tail is that of IS*1*). The inset numbers are the mean number of copies of each insertion sequence per strain, either with uninfected strains excluded (Exc 0) or included (Inc 0).

hybridisation as described (Sawyer *et al.*, 1987). The data include 1055 chromosomal occurrences and 118 plasmid occurrences of the six unrelated insertion sequences IS*1*, IS*2*, IS*3*, IS*4*, IS*5* and IS*30*. The distributions of chromosomal copies of the insertion sequences are illustrated in Fig. 1. For each insertion sequence, the distribution of copy numbers was used to obtain maximum likelihood estimates of the parameters in various theoretical models, and goodness of fit tests were performed. Data analysed were for chromosomal copies only, but the conclusions were unaffected by the inclusion of plasmid copies.

The theoretical models are all cases of a general branching-process model in which the rate of transposition of an insertion sequence and the fitness of strains containing it are simple functions of the number of copies. It is assumed that a host with n copies of an insertion sequence has a fitness $R - D(n)$, where $D(0) = 0$ and $D(n) > 0$ for $n > 0$. For a strain with $n > 1$ copies of the element, the probability of transposition creating a strain with $n + 1$ copies is $T(n)$, and uninfected strains become infected at the rate u per

Table 1. *Branching-process models*

Model	$T(n)$	$D(n)$
CC	T	D
HC	T/n	D
CR	T	$D\sqrt{n}$
RC	$T\sqrt{n}$	Dn
RR	$T\sqrt{n}$	$D\sqrt{n}$
DR	$T/\sqrt{n}$	$D\sqrt{n}$
RL	$T\sqrt{n}$	Dn
LL	Tn	Dn
LQ	Tn	Dn^2

generation. Analytical results with this and related models are discussed in Sawyer & Hartl (1986) and Sawyer *et al.* (1987). For the model fitting, the nine models of $T(n)$ and $D(n)$ are considered in Table 1, which are combinations of the functional forms constant (C), harmonic (H), root (R), inverse root (D), linear (L), and quadratic (Q). The constants T/D and u/D were estimated for each model. (Since the models are equilibrium models, only ratios of u, T and D can be estimated.)

Methods of model fitting and assessments of goodness of fit are discussed in Sawyer *et al.* (1987). In summary, all insertion-sequence distributions can be fitted very well by at least one model, and sometimes by several models. The CC model provides an excellent fit to the observed distributions of IS2, IS4 and IS30. For IS1 and IS5, the models LL, LQ, RC and RR fit the best, but CC is still acceptable. Only HC, DR and CR models fit the distribution of IS3 satisfactorily. It is important to note that each insertion sequence is fitted by at least one model in which the fitness of infected strains is independent of copy number. This finding justifies use of the assumption $D(n) = D$ in the analysis of the plasmid-transmission model. When $D(n) = D$ is constant, the ratio u/D determines the relative proportions of infected and uninfected strains, and u/T determines the distribution of copy number among infected hosts. Table 2 provides the maximum likelihood estimates of u/D and the P-value in the χ^2 test for goodness of fit. The χ^2 tests have three degrees of freedom because the tails of the distributions were pooled in the goodness of fit tests; that is, all strains containing ≥ 5 copies of an insertion sequence were lumped into a single category, giving six classes of data, namely $n = 0, 1, 2, 3, 4$, and ≥ 5 copies.

Table 2. *Maximum likelihood estimates and goodness of fit*

IS	Model	Estimated u/D	P value
1	CC	0.85 ± 0.08	0.78
2	CC	0.61 ± 0.11	0.41
3	HC	0.68 ± 0.11	0.57
4	CC	0.39 ± 0.11	0.60
5	CC	0.35 ± 0.11	0.28
30	CC	0.49 ± 0.12	0.32

Table 3. *Distribution of large and small plasmids*

Small	Large plasmids					Total
	0	1	2	3	4	
0	9	7	6	3	1	26
1	2	3	5	1	0	11
2	3	1	1	0	0	5
3	3	5	0	0	0	8
4	2	2	1	2	1	8
5	1	1	1	0		3
6	1	0	2	1		4
7		1	2	1		4
8			1			1
9						0
10						0
11		1				1
Total	21	21	19	8	2	71

DISTRIBUTION OF LARGE AND SMALL PLASMIDS

The number of large and small plasmids among the strains was estimated as described (Hartl *et al.*, 1986). Large plasmids are larger than about 40 kilobase pairs, small ones are smaller than about 7.5 kilobase pairs, and very few are intermediate in size (Silver *et al.*, 1980). Table 3 gives the distribution of numbers of large and small plasmids observed among the 71 strains. These estimates should not be afforded more credibility than warranted. Because the estimates are based on the number of bands appearing in agarose gels

Table 4. *Chromosome and plasmid locations of insertion sequences*

IS	Chromosomal copies	Plasmid copies	Per cent plasmid
1	452	58	11.4
2	193	17	8.1
3	119	13	9.8
4	142	5	3.4
5	84	16	16.0
30	66	8	10.8
Total	1056	117	10.1

after electrophoresis, some large plasmids may migrate nearly together and not be distinguished, and some small plasmids may be overcounted because different physical forms (e.g. supercoiled vs. relaxed) have different mobilities. Therefore, the numbers should more properly be regarded as estimating the number of plasmid bands rather than as the number of plasmids themselves. Nevertheless, taking the numbers at face value, the distribution of large plasmid bands is Poisson with mean of 1.3 per strain. The distribution of small plasmid bands has mean 2.3 and standard deviation 2.6 and is not Poisson. The χ^2 test for presence vs. absence of large and small plasmid bands is non-significant, as is the correlation in number of large and small plasmid bands. We conclude that the typical natural isolate contains 0–2 large plasmid bands and 0–4 small plasmid bands, and that the numbers of large and small plasmid bands are independent. Only 12.7% of the isolates (9/71) contain no detectable plasmids.

PRESENCE OF INSERTION SEQUENCES IN PLASMIDS

Insertion sequences are present in plasmid bands as well as in the chromosome. The data in Table 4 indicate that, among all observed copies of the insertion sequences, the proportion found in plasmids ranged from 3.4–16.0%, averaging 10.1% in plasmids. As might be expected, there is a highly significant correlation between the number of chromosomal copies of elements and the number of plasmid copies, which means that insertion sequences that are abundant in chromosomes are also abundant in the plasmid pools. There is a slight heterogeneity among the insertion sequences, largely because

Table 5. *Number of different IS elements in plasmid pools. (Numbers in parentheses are those expected with a Poisson distribution.)*

Number of IS elements	Number of strains (having one large plasmid)	Number of strains (all strains)
0	8 (8.1)	24 (20.0)
1	7 (7.7)	17 (22.6)
2	5 ($\downarrow$ 5.2)	12 (12.8)
3	1	8 ($\downarrow$ 6.6)
4	0	0
5	0	1

Table 6. *Insertion sequences in chromosomes and plasmid pools*

IS	Neither	Chromosome only	Plasmids only	Both
1	6	34	1	21
2	19	30	5	8
3	20	33	2	7
4	34	23	3	2
5	35	13	2	12
30	29	27	1	5

the incidence of IS4 in plasmids is too low ($\chi^2_5 = 13.1$ with $P = 0.022$ from 2×6 contingency table test; when IS4 is excluded, $P = 0.34$).

Different insertion sequences are approximately independent in their occurrence in the plasmid pools. This point is supported by the data in Table 5, which gives the number of different IS elements in the plasmid pools of strains containing one large plasmid band (second column) or among all 62 strains containing at least one plasmid (third column). The numbers in parentheses are those expected assuming a Poisson distribution (pooling all classes with numbers greater than those indicated by the downward arrow), and the means are essentially the same (0.95 and 1.13, respectively). As suggested by the similarity of the fitted means, the majority (approximately 75–80%) of insertion sequences which occur in plasmid pools are associated with large plasmid bands.

Table 6 shows the manner in which insertion sequences are partitioned between the chromosome and plasmid pools among the 62 strains containing at least one plasmid. Each row of numbers defines a 2×2 contingency table for presence vs. absence of the insertion sequence in the chromosome and plasmid pool. Only the contingency table for IS5 is significant (Fisher exact test). This finding supports the hypothesis that insertion sequences occur approximately independently in the chromosome and plasmid pools. Were it the case, for example, that insertion sequences newly introduced into strains had a strong tendency to transpose into the chromosome, a significant positive association would be expected because insertion sequences would tend to be found either in both the chromosome and plasmid pool or in neither.

CORRELATION IN OCCURRENCE OF INSERTION SEQUENCES

The study of Sawyer *et al.* (1987) revealed significant positive correlations in the presence of unrelated insertion sequences among strains. Part of these associations could be attributed to stratification of samples in the ECOR strains. For example, the ECOR collection contains about equal numbers of three types of strains, designed Type I, Type II, and Type III, which were determined by principal components analysis of measures of genetic distance based on electrophoretic mobilities of 12 enzymes (Whittam *et al.*, 1983). The Types or clusters of related strains can be discriminated in spite of considerable genetic variation within Types.

With respect to the distribution of insertion sequences among the three Types of strains, IS1, IS3, IS5 and IS30 occur more often in Type I strains (P ≤ 0.52), and IS4 more often in Type III strains, than would be expected by chance. All else being equal, these correlations with Type alone would result in positive correlations between the occurrence of IS1, IS3, IS5 and IS30. Other, generally weaker, correlations were found between some insertion sequences and the origin of the isolates (e.g. human vs. animal, or North American vs. Swedish). Attempts to minimise the effects of correlations resulting from stratification in the ECOR sample have been made by calculating the correlations within each Type separately, and averaging these values. The overall distribution of correlation coefficients is illustrated in Fig. 2. Averaged across Types, the mean pairwise correlation coefficient in the occurrence vs. non-occurrence of insertion sequences equals 0.171 ± 0.044. It is this residual correlation which it is proposed is explained by the model presented in the next section.

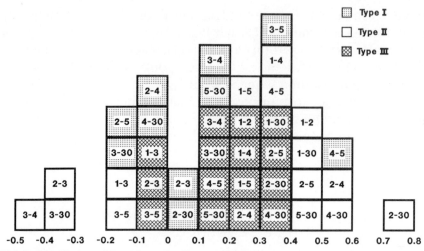

Fig. 2. Distribution of Pearson correlation coefficients for presence or absence of pairs of insertion sequences in strains of the ECOR collection, classified according to Type I, Type II and Type III strains (shading). Numbers separated by hyphens within the boxes are the numerical designations of the insertion sequences having a correlation in the indicated range, namely IS*1*, IS*2*, IS*3*, IS*4*, IS*5* or IS*30*.

INFECTION MODEL WITH CORRELATIONS

The small but persistent and highly significant correlation in the presence or absence of pairs of insertion sequences was unanticipated because biological evidence suggests that the elements are independent in their processes of infection and transposition. Attempts have been made to account for the correlation based on a subtle effect of the mechanism of dissemination of insertion sequences, thus obviating the need to postulate a previously unsuspected interaction between unrelated insertion sequences.

In brief, it has been demonstrated that a correlation in the presence or absence of transposable elements disseminated by plasmids results from the possibility of simultaneous infection by more than one type of insertion sequence, even when the elements occur independently in the plasmid pools (Hartl & Sawyer, 1987).

In order to quantify this effect, consider a pair of unrelated insertion sequences, A and B. With respect to A and B, the chromosome in a bacterial cell in the population may be in any one of four possible states: (1) infected by neither A nor B, (2) infected by A but not B, (3) infected by B but not A, or (4) infected by both A and B. The four possible states are assumed to have fitnesses R, $R - s_1$, $R - s_2$, and $R - s_3$, respectively, where the $s_i > 0$ measures the deleterious effects of the insertion sequences. In this model the selective effect of an insertion sequence is independent of the copy number, so s_1 and s_2 correspond to the values of D in the appropriate CC or HC models in Sawyer et al. (1987) and in Table 2.

In addition, the chromosomes are assumed to be subject to infection by the first IS element (but not the second) at the rate u_1, by the second IS element (but not the first) at the rate u_2, and by both IS elements simultaneously at the rate u_3.

With these stipulations, it can be shown (Hartl & Sawyer, 1987) that the predicted Pearson correlation coefficient of the presence vs. absence of the two IS elements in chromosomes equals

$$\rho = \frac{u_1 u_2 (s_1 + s_2 - s_3) + u_3 s_1 s_2}{\sqrt{\Delta_1 \Delta_2}} \tag{1}$$

where

$$\Delta_1 = \frac{s_1}{s_1 - u_1} \left(u_1 u_2 (s_1 - s_3) + u_1 s_2 (s_3 - u_1) + u_3 s_2 (s_1 - u_1) \right)$$

and

$$\Delta_2 = \frac{s_2}{s_2 - u_2} \left(u_1 u_2 (s_2 - s_3) + u_2 s_1 (s_3 - u_2) + u_3 s_1 (s_2 - u_2) \right)$$

The case of greatest immediate interest is $s_3 = s_1 + s_2$, corresponding to additive or independent effects of the two IS elements on fitness. Equation (1) implies that there is nevertheless a correlation between the elements A and B, which results from doubly infected plasmids.

Estimates of the s_i's and u_i's were obtained using the data of Sawyer et al. (1987) in order to determine whether the observed correlations might be explained as substantially due to the plasmid-mediated mode of transmission and the occurrence of multiply infected plasmids. Accordingly, for each pair of insertion sequences we assume additivity in the effects on fitness ($s_3 = s_1 + s_2$). Details of the estimation procedure are described in Hartl and Sawyer (1987).

AGREEMENT BETWEEN OBSERVED AND PREDICTED CORRELATIONS

For each pair of insertion sequences, Equation (1) was used to obtain the expected correlation coefficient within strains of Type I, Type II and Type III. Averaged across Type, the observed and expected correlations are summarised in Table 7. (For each pair of numbers in the table, the upper number is the observed correlation, the lower

Table 7. *Pairwise correlations averaged across type*

	IS2	IS3	IS4	IS5	IS30
IS1	0.378 0.367	−0.101 0.121	0.318 0.256	0.230 0.240	0.429 0.214
IS2		−0.110 0.096	0.240 0.233	0.215 0.032	0.355 0.204
IS3			−0.045 0.069	0.014 0.075	−0.117 0.071
IS4				0.345 0.052	0.293 0.087
IS5					0.259 0.181

Table 8. *Averages of pairwise correlations*

	Observed ± SEM	Expected
Type I	0.094 ± 0.072	0.107
Type II	0.206 ± 0.102	0.159
Type III	0.188 ± 0.037	0.149
Overall	0.171 ± 0.044	0.142

number the expected correlation.) While there is great variation in the observed pairwise correlations, at least some of the variation certainly results from the small sample sizes. Except for correlations involving IS3, in which four of five are negative, all the rest are positive. Many of the correlations agree with the predicted values very well (e.g. IS1 vs. IS2, IS1 vs. IS4, IS1 vs. IS5, IS2 vs. IS4, IS5 vs. IS30), whereas others fit not so well. However, the sampling variances are still quite large.

Comparison of the observed and expected correlations within the three Types of strains is given in Table 8. Within each Type, the expected correlation is essentially within one standard error of the observed. Pooling across all pairs of insertion sequences and all three Types of strains results in a correlation of 0.171 ± 0.044, as compared with the expected value of 0.142. It appears that the branching-process model of infectious transmission accounts satisfactorily for the correlation in the presence vs. absence of pairs (and higher multiples) of insertion sequences among natural isolates of *E. coli*.

DISCUSSION

The results of Sawyer *et al.* (1987) on the distribution of individual insertion sequences among strains, when combined with the results of the present analysis and that of Hartl and Sawyer (1987) on the correlations among unrelated elements, provides a reasonably complete picture of the evolutionary forces which determine the short-term fate of insertion sequences in bacterial populations. The occurrence of elements among possible hosts is determined by the rate of plasmid-mediated infection. Among infected hosts, the distribution of numbers is determined by the rates of transposition $T(n)$ and the effects on fitness $D(n)$, both of which are functions of copy number. In some cases, for example IS2, IS4 and IS30, the distribution of elements is well described by the CC model in which $T(n) = T$ constant and $D(n) = D$ constant. This model implies regulation of transposition in that the probability of transposition *per element* decreases with copy number $(T(n)/n = T/n)$, but fitness is independent of copy number. In the cases of IS1 and IS5, a better model is the LL model in which $T(n) = Tn$ and $D(n) = Dn$, implying weak regulation of transposition $(T(n)/n = T)$ and a fitness that decreases linearly with copy number. However, the CC model also gives a satisfactory fit with IS1 and IS5. Insertion sequence IS3 is unusual in that both the CC and LL models can be rejected, and the HC model fits the best. This result implies that IS3 has a very strong regulation of transposition $(T(n)/n = 1/n^2)$.

Beyond the distribution of individual insertion sequence isolates, there exists a positive correlation between pairs and higher multiples of elements: isolates that contain one insertion sequence tend to contain one or more additional unrelated insertion sequences. It has been shown that such a positive correlation is expected as a consequence of the plasmid-mediated mode of transmission of insertion sequences among hosts.

That plasmid transmission should induce a postive correlation can be justified intuitively by considering a case in which infection occurs at a very low rate but in which plasmids are long lived. Under these conditions, plasmids will accumulate insertion elements through time, and when they are transferred, will simultaneously infect the host with two or more unrelated insertion sequences. Occurring in the population as a whole, the possibility of simultaneous infection will induce a positive correlation in the presence or absence of the insertion sequences. This effect occurs even when insertion sequences transpose into plasmids independently.

Although undoubtedly oversimplified somewhat, the plasmid-transmission model is important in demonstrating that the small positive correlations observed between unrelated insertion sequences can be explained by the fact that they are transmitted in plasmids, even in the absence of any biological interactions. Biological interactions include cross regulation in rates of transposition, or nonlinear interactions affecting fitness ($s_3 \neq s_1 + s_2$).

There is a feedback between insertion sequences in the chromosome and plasmids that the theoretical model does not take into account explicitly. On the one hand, the abundance of insertion sequences in plasmids determines the rate at which uninfected strains become infected. On the other hand, as plasmids circulate among different hosts, insertion sequences from the host chromosomes transpose into them, and thus the abundance of insertion sequences in the chromosomes also affects the abundance of insertion sequences in the plasmid pools. The net effect of this feedback is that the presence of multiple insertion sequences in plasmids, while inducing a positive correlation in the presence or absence of insertion sequences in chromosomes, should also, because of the chromosomal correlation, induce secondarily a positive correlation in the presence or absence of insertion sequences in plasmids. This view of a feedback process is supported by the finding that, among 15 pairwise correlations for presence vs. absence of insertion sequences within the plasmid pools of 62 ECOR strains containing plasmids, 11 are positive ($P \simeq 0.07$), and the average pairwise Pearson correlation coefficient equals 0.104 ± 0.042. In applying the plasmid-transfer model to the data, this effect is taken into account by assuming equilibrium and using the observed values of singly and doubly infected plasmids to estimate the infection rates.

Although the plasmid-transmission model seems to fit the observations rather well, the fit may be less good than meets the eye. If correlations involving IS3 are excluded, which may be justified because there are so many negative correlations involving IS3, then the overall mean pairwise correlation among the remaining IS sequences is 0.301 ± 0.041, which is to be compared with a theoretically expected mean of 0.174. The fit is obviously not as good as before, and only about half of the correlation can be accounted for by the plasmid transmission model. This discrepancy, if it is real, might result from any of several simplifications in the model. For example, the model assumes that the plasmid pools are shared freely among the three identified Types of strains. However, it is possible that plasmids do not occur at random with regard to host Type. Non-random occurrence of plasmids among serotypes has been found by Hartl *et al.* (1986), but the situation has not been studied in detail. The plasmid-transmission model is also oversimplified somewhat in assuming that all plasmids bearing insertion sequences have the same transmission rates. Finally, the fact that the observed correlations may be somewhat greater than expected may result from unrecognised stratification in the sample, that is, from associations between IS elements and strains possessing particular characteristics other than those which define the three identified Types.

REFERENCES

BOTSTEIN, D. & SHORTLE, D. (1985). Strategies and applications of in vitro mutagenesis. *Science*, **229**, 1193–201.

CALOS, M. P. & MILLER, J. H. (1980). Transposable elements. *Cell*, **20**, 579–95.

CAMPBELL, A. (1981). Evolutionary significance of accessory DNA elements in bacteria. *Annual Review of Microbiology*, **35**, 55–83.

CHAO, L., VARGAS, C., SPEAR, B. B. & COX, E. C. (1983). Transposable elements as mutator genes in evolution. *Nature*, **303**, 633–5.

CHARLIER, D., PIETTE, J. & GLANSDORF, N. (1982). IS*3* can function as a mobile promoter in *E. coli. Nucleic Acids Research*, **10**, 5935–48.

CIAMPI, M. S., SCHMID, M. B. & ROTH, J. R. (1982). Transposon Tn*10* provides a promoter for transcription of adjacent genes. *Proceedings of the National Academy of Sciences, USA*, **79**, 5016–20.

DOOLITTLE, F. W. & SAPIENZA, C. (1980). Selfish DNA, the phenotype paradigm and genome evolution. *Nature*, **284**, 601–3.

DYKHUIZEN, D. E., SAWYER, S. A., GREEN, L., MILLER, R. D. & HARTL, D. L. (1985). Joint distribution of insertion elements IS*4* and IS*5* in natural isolates of *Escherichia coli. Genetics*, **111**, 219–31.

GAFFNEY, T. D. & LESSIE, T. G. (1987). Insertion-sequence-dependent rearrangements of *Pseudomonas cepacia* plasmid pTGL1. *Journal of Bacteriology*, **169**, 224–30.

GEORGOPOULOS, C., MCKITTERICK, N., HERRICK, G. & EISEN, H. (1982). An IS*4* transposition causes a 13-bp duplication of phage lambda DNA and results in constitutive expression of the *cI* and *cro* gene products. *Gene*, **20**, 83–90.

GREEN, L., MILLER, R. D., DYKHUIZEN, D. E. & HARTL, D. L. (1984). Distribution of DNA insertion element IS*5* in natural isolates of *Escherichia coli. Proceedings of the National Academy of Sciences, USA*, **81**, 4500–4.

HARTL, D. L., DYKHUIZEN, D. E., MILLER, R. D., GREEN, L. & FRAMOND, J. DE (1983). Transposable element IS50 improves growth rate of E. coli cells without transposition. Cell, 35, 503–10.

HARTL, D. L., DYKHUIZEN, D. E. & BERG, D. E. (1984). Accessory DNAs in the bacterial gene pool: playground for coevolution. pp. 233–45. In D. Evered and G. M. Collins (eds), Ciba Symposium 102: Origins and Development of Adaptations. Pitman Books, London.

HARTL, D. L., MEDHORA, M., GREEN, L. & DYKHUIZEN, D. E. (1986). The evolution of DNA sequences in Escherichia coli. Philosophical Transactions of the Royal Society, London, B312, 191–204.

HARTL, D. L. & SAWYER, S. A. (1987). Why do unrelated insertion sequences occur together in the genome of Escherichia coli? Genetics, in press.

HU, M. & DEONIER, R. C. (1981). Comparison of IS1, IS2 and IS3 copy number in Escherichia coli strains K-12, B and C. Gene, 16, 161–70.

JAURIN, B. & NORMARK, S. (1983). Insertion of IS2 creates a novel ampC promoter in Escherichia coli. Cell, 32, 809–16.

KLECKNER, N. (1981). Transposable elements in prokaryotes. Annual Review of Genetics, 15, 341–404.

MILLER, R. D., DYKHUIZEN, D. E., GREEN, L. & HARTL, D. L. (1984). Specific deletion occurring in the directed evolution of 6-phosphogluconate dehydrogenase in Escherichia coli. Genetics, 108, 765–72.

NYMAN, K. H., OHTSUBO, H., DAVISON, D. & OHTSUBO, E. (1983). Distribution of insertion element IS1 in natural isolates of Escherichia coli. Molecular and General Genetics, 189, 516–18.

OCHMAN, H. & SELANDER, R. K. (1984), Standard reference strains of Escherichia coli from natural populations. Journal of Bacteriology, 157, 690–3.

ORGEL, L. E. & CRICK, F. H. C. (1980). Selfish DNA: The ultimate parasite. Nature, 284, 604–7.

REYNOLDS, A. E., FELTON, J. & WRIGHT, A. (1981). Insertion of DNA activates the cryptic bgl operon in E. coli. Nature, 293, 625–9.

SAEDLER, H., CORNELIS, G., CULLUM, J., SCHUMACHER, B. & SOMMER, H. (1980). IS1-mediated rearrangements. Cold Spring Harbor Symposium on Quantitative Biology, 45, 93–8.

SAEDLER, H., REIF, H. J., HU, S. & DAVIDSON, N. (1974). IS2, a genetic element for turn-off and turn-on of gene activity in E. coli. Molecular and General Genetics, 132, 265–89.

SAWYER, S. A., DYKHUIZEN, D. E., BUBOSE, R. F., GREEN, L., MUTANGADURA-MHLANGA, T., WOLCZYK, D. F. & HARTL, D. L. (1987). Distribution and abundance of insertion sequences among natural isolates of Escherichia coli. Genetics 115, 51–63.

SAWYER, S. A. & HARTL, D. L. (1986). Distribution of transposable elements in prokaryotes. Theoretical Population Biology, 30, 1–16.

SCORDILIS, G. E., REE, H. & LESSIE, T. G. (1987). Identification of transposable elements which activate gene expression in Pseudomonas cepacia. Journal of Bacteriology, 169, 8–13.

SILVER, R. P., AARONSON, W., SUTTON, A. & SCHEERSON, R. (1980). Comparative analysis of plasmids and some metabolic characteristics of Escherichia coli K12 from diseased and healthy individuals. Infection and Immunity, 29, 200–6.

WHITTAM, T. S., OCHMAN, H. & SELANDER, R. K. (1983). Multilocus genetic structure in natural populations of Escherichia coli. Proceedings of the National Academy of Sciences, USA, 80, 1751–5.

ZAFARULLAH, M., CHARLIER, D. & GLANSDORFF, N. (1981). Insertion of IS3 can 'turn on' a silent gene in Escherichia coli. Journal of Bacteriology, 146, 415–17.

MECHANISM AND REGULATION OF TRANSPOSITION

DOUGLAS E. BERG, TONI KAZIC, SUHAS H. PHADNIS, KAREN W. DODSON and JENNIFER K. LODGE

Departments of Microbiology and Immunology, and of Genetics, Washington University School of Medicine, St Louis, MO. 63110, USA

OVERVIEW

Transposable elements are specialised DNA segments which move to new genomic locations without extensive sequence homology. They cause insertion mutations and chromosome rearrangements, alter the expression of genes near their sites of insertion, and speed the spread of antibiotic resistance among unrelated bacterial species. In recent years they have gained a special technical importance as tools for molecular genetic research (for reviews, see Berg & Berg, 1983; 1987; Grindley & Reed, 1985; Craig & Kleckner, 1987).

Transposition depends on element-encoded proteins called transposases, on host-encoded factors, and on the sequences at the element's ends upon which these proteins act. It can occur with or without replication, depending on the element. Most elements exhibit some preference for particular insertion sites. Transposition is regulated in a variety of complementary ways. Here we review our current understanding of the mechanisms and regulation of transposition in bacteria. Our examples are drawn primarily from Tn5 (Fig. 1), with additional instances from Tn3, Tn9, Tn10 and phage Mu.

FUNCTIONAL ORGANISATION

Two groups of prokaryotic elements, the insertion (IS) elements and transposons, can be distinguished by structure and gene content. The IS elements contain only the genes and sites needed for their own regulated transposition, and are generally less than 2 kb in size. At least seven types of IS elements are resident in the genome of *Escherichia coli* K-12. They differ in size, the transposition proteins they encode, the sequences at their ends, and the structures they form by transposition (Iida, Meyer & Arber, 1983). Transposons

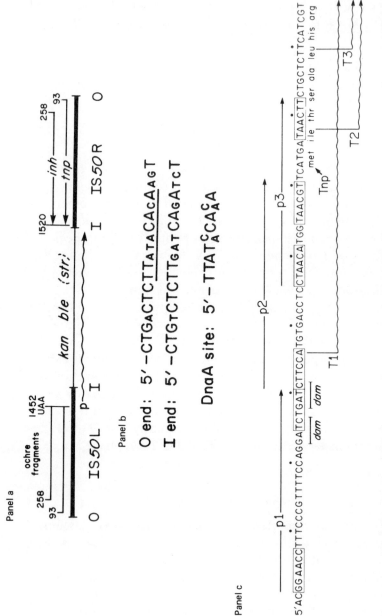

Fig. 1. The structure of Tn5. Panel a: general arrangement. The transposon is indicated with the transposase (*tnp*) and inhibitor (*inh*) reading frames above the line and the drug-resistance transcript below it. The positions of the outside (O) and inside (I) ends are indicated. The ochre protein fragments from IS50L are also shown, and the nucleotide coordinates at which proteins begin and end are indicated. *kan*, *ble*, and *str* are the genes encoding resistance to kanamycin, bleomycin, and streptomycin, respectively. Adapted from Johnson *et al.* (1982) and Mazodier *et al.* (1985). Panel b: the outside and inside ends compared. Identical bases are in large type. The consensus DnaA binding site is underneath, and the DnaA site in the O end is underlined. Panel c: The promoter region of IS50R, beginning at base 30. The T1, T2, and T3 transcripts (wavy lines) are initiated from the p1, p2, and p3 promoters, respectively. For each promoter, the −35 and −10 regions are boxed, and the extent of the promoter is indicated by the arrow. The first eight amino acids of the transposase (Tnp) are shown, and the sites of methylation by the Dam protein are marked (•).

contain auxiliary genes, such as those encoding virulence or antibiotic resistance, in addition to transposition genes. Many have terminal repeats of intact IS elements, and the mobility of these composites is due simply to the pair of IS elements transposing in unison. The 5.8 kb Tn5 element (Fig. 1), from an R factor plasmid of *Klebsiella* (Berg *et al.*, 1975), is a representative example. The IS elements in Tn5, called IS50 (Berg *et al.*, 1981; Isberg & Syvanen, 1981; Berg *et al.*, 1982*a*), are not related to the IS elements resident in *E. coli* K-12 (Berg & Drummond, 1978). Two of the other three resistance transposons considered here are also composites: Tn10, conferring tetracycline resistance (Tet^r) contains inverted repeats of IS10, and Tn9, conferring chloramphenicol resistance (Cam^r) contains direct repeats of IS1, whereas Tn3, conferring ampicillin resistance (Amp^r), does not contain a separate IS element (Kleckner, 1983; Iida *et al.*, 1983; Heffron, 1983).

The central region of Tn5 contains genes encoding resistance to kanamycin, bleomycin, and streptomycin (Mazodier *et al.*, 1985), which are transcribed from the promoter near the end of IS50L (Rothstein *et al.*, 1980; Rothstein & Reznikoff, 1981). The *str^r* gene is usually cryptic in *E. coli,* but is expressed in non-enteric bacteria (Mazodier, Giraud & Gasser, 1982). The *kan^r* gene encodes an aminoglycoside phosphotransferase which is related to aminoglycoside phosphotransferases found in many Gram-negative and Gram-positive pathogens and in several antibiotic producing organisms (Gray & Fitch, 1983; see also Chater *et al.*, this volume).

Below we discuss: (i) transposition mechanisms; (ii) ınsertion specificity; (iii) transposable element ends; and (iv) the regulation of IS50 transposition.

MECHANISMS OF TRANSPOSITION

There are two groups of transposition models: conservative (cut-and-paste) and replicative. Tn5 and Tn10 seem to move by conservative mechanisms, while Tn3 moves by a replicative process. Mu and IS1 (Tn9) transpose by both replicative and conservative mechanisms, depending on the circumstances. Nearly all transposable elements generate short direct repeats of target sequences when they transpose (Iida *et al.*, 1983): for example, Tn5 and Tn10 each make nine base pair (bp) target duplications (Schaller, 1979; Kleckner, 1979). Current models invoke a staggered break and the filling of gaps formed

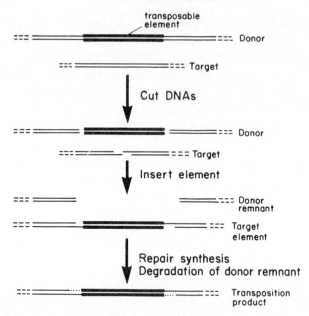

Fig. 2. A model of conservative transposition. Adapted from Berg (1977, 1983).

by the joining of element and target DNAs. Certain mutations in the *polA* gene result in a 10- to 20-fold decreased transposition activity (Sasakawa, Uno & Yoshikawa, 1981; Syvanen, Hopkins & Clements, 1982), which may reflect a need for DNA polymerase I to fill the gaps.

In conservative transposition (Fig. 2) double-stranded breaks are made between the element and its vector, and the element and target DNAs are joined together without replication. The ends of donor DNA are not rejoined and the remnant is exonucleolytically degraded. Loss of the donor DNA species from the cell lineage is avoided when cells contain duplicates of the donor sequence. Although one copy is consumed, the other survives, and the resulting lineage contains one element at the original site and a second at a new site. The number of elements is increased as surely as if transposition had been replicative (Berg, 1983; Berg, Berg & Sasakawa, 1984*a*).

In symmetric replicative models (Fig. 3; Shapiro, 1979; Craigie *et al.* and Hatfull *et al.*, this volume) cuts on just one of the two strands are made at each end of the element. The free ends of element DNA are joined to the target, and the remaining 3′ ends prime replication across the element. The product in the case shown in

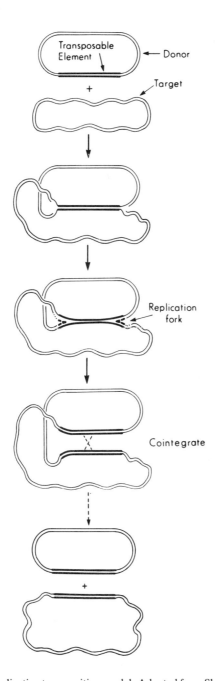

Fig. 3. A symmetric replicative transposition model. Adapted from Shapiro (1979).

Fig. 3 (intermolecular transposition) consists of complete vector and target DNAs joined by direct repeats of the element (a 'cointegrate'). Site-specific recombination in these repeats separates the target (containing the transposed element) and donor moieties ('resolution'). Asymmetric replicative models have also been proposed, primarily to account for instances in which neither cointegrates nor resolution functions are found (Harshey & Bukhari, 1981; Galas & Chandler, 1981).

Three results indicate that Tn*5* and IS*50* transposition is conservative. First, the Kan^r products of transposition from populations of pBR322::Tn*5* plasmids (kept uniformly monomeric in *recA*$^-$ cells) contain simple insertions of Tn*5* but no pBR322 sequences. Many of the Kan^r products of transposition from uniformly dimeric pBR322::Tn*5* plasmids contain larger DNA segments, e.g. a continuous segment with one copy of the pBR322 moiety as well as two IS*50* or Tn*5* sequences. These larger inserts also consist of just a fragment of the donor molecule (Berg, 1983). Many transposition products formed in recombination-proficient (*recA*$^+$) cells contain pBR322 as well as Tn*5* sequences; they were called cointegrates, and were interpreted as evidence for replicative transposition (Hirschel *et al.*, 1982*a*; Ahmed, 1986). However, our reconstruction experiments indicate that dimeric and larger forms of pBR322-related plasmids accumulate during the growth of *recA*$^+$ cells. Thus, these 'cointegrates' probably also arise by conservative transposition from concatemeric donors in the heterogeneous plasmid populations, and not by replicative transposition from monomeric donors (Berg, 1983; Hirschel *et al.*, 1982*b*). Second, Tn*5* does not encode a resolvase that recombines direct repeats of Tn*5* or IS*50* elements (Berg, 1983; Isberg & Syvanen, 1985a; Phadnis & Berg, 1985). The instance of plasmid instability originally interpreted as evidence for a resolvase (Zupancic *et al.*, 1983) actually reflects selection for recombinants superimposed on the normal low levels of *recA*-independent homologous recombination (Phadnis & Berg, 1986). Finally, the destruction of the donor remnant is implied during conservative transposition because it is not recovered genetically (Berg, 1977). Although an event that has been termed transposon excision is readily detected by reversion of Tn*5*- and Tn*10*-induced insertion mutations, it is independent of transposase functions and unrelated to transposition *per se*. It appears to result from normal deletion processes (Egner & Berg, 1981; Foster *et al.*, 1981).

Two recent results show that the transposition of Tn*10* (IS*10*) is also conservative. First, transposase overproduction results in the cleavage of Tn*10* from vector DNA; the ends of the vector are not rejoined, and the vector DNA is degraded. The free Tn*10* DNA is found primarily in a circular form (Morisato & Kleckner, 1984) that could be a transposition intermediate. Second, products of transposition from heteroduplex λ::Tn*10-lac* DNAs in which one strand is *lac*$^+$ and the other is *lac*$^-$ contained just one copy of the Tn*10-lac* element but formed sectored colonies. Thus the transposition product contained both the *lacZ*$^+$ and the *lacZ*$^-$ DNA strands (Bender & Kleckner, 1986). In contrast, Tn*3* transposition is replicative: mutants which form relatively stable cointegrates were shown to be defective either in the gene for a site-specific recombination protein (resolvase), or the site at which resolvase acts (Gill *et al.*, 1978; Arthur & Sherratt, 1979; Kostriken, Morita & Heffron, 1981).

Phage Mu transposes repeatedly during lytic growth, and analyses of intracellular Mu DNAs revealed cointegrates with direct repeats of Mu sequences (Chaconas *et al.*, 1981). Detailed *in vitro* analyses have shown that replicative transposition starts at each Mu end as in Fig. 3, and is primed by the free 3′ ends of target DNA (Mizuuchi & Craigie, 1986; Craigie *et al.*, this volume). In contrast, the first insertion of Mu DNA during infection is conservative (Liebart, Ghelardini & Paolozzi, 1982; Akroyd & Symonds, 1983; Chaconas, Kennedy & Evans, 1983; Harshey, 1984). The injected Mu DNA is flanked by sequences from its former host, and the ends are held together by the Mu N protein (Harshey & Bukhari, 1983; Gloor & Chaconas, 1986). Conservative insertion may also begin with strand transfers as in replicative transposition (Fig. 3), but breakage or digestion of the flanking DNA from the former host would destroy one arm of each potential replication fork, producing conservative insertions (see Mizuuchi & Craigie, 1986).

Some products of intermolecular Tn*9* (IS*1*) transposition are cointegrates that must have arisen by replicative transposition (Galas & Chandler, 1982). However, others are simple insertions of just a portion of the donor DNA. Because IS*1* does not encode a resolution function equivalent to that of Tn*3*, it is less clear how these simple insertions are formed. In one model, replicative transposition is asymmetric, initiating at one end and terminating at the other, and the pattern of strand joining at termination determines whether simple insertions or cointegrates are formed (Galas & Chandler,

1981). However, when dimeric pBR322::Tn9 (IS*1*) was used as the transposition donor (Biel & Berg, 1984), the products obtained differed from those predicted by the asymmetric model. This indicates that the simple insertions arise either by conservative transposition (Fig. 2), or by breakage of an intermediate in symmetric replicative transposition (such as in Fig. 3) before or during replication (Ohtsubo *et al.*, 1981; Biel & Berg, 1984).

SPECIFICITY AND PRECISION OF INSERTION

Genetic mapping indicated that Tn5 can insert into many sites in a single gene and at many chromosomal locations (Berg, 1977; Shaw & Berg, 1979; Berg, Weiss & Crossland, 1980; Miller *et al.*, 1980). More detailed analyses using plasmid pBR322 as a target revealed a complex distribution of frequently and infrequently used sites: 28 different sites of insertion were found among 75 independent Tet[s] insertion mutations, with most sites used just once or twice. However one-fourth and one-tenth of all insertions in the 1.3 kb *tet* locus were at just two sites ('hotspots' I and II, respectively). These two sites were also used repeatedly for insertion involving a pair of IS50 inside (I) ends (Nag *et al.*, 1985).

G/C pairs are present at the first and ninth base pairs of the target duplications at five Tn5 insertion hotspots and at 10 of 11 other hotspots found in transpositions involving pairs of I ends, but no other common features were evident. The importance of these G/C pairs at one hotspot was shown directly by making base substitution mutations. G/C to A/T substitutions reduced the frequencies of Tn5 insertion into the hotspot from about 25% to less than 5%, whereas G/C to C/G transversions resulted in a frequency of about 16%. About half of the sites used infrequently for Tn5 insertion contain G/C pairs at both ends, and most of the rest contain G/C at one end and A/T at the other (Lodge *et al.*, 1984; Berg *et al.*, 1984*b*; Lodge & Berg, unpublished data). Given the hundreds of G/C pairs with a nine bp spacing in pBR322 that are not used as Tn5 insertion hotspots, additional features must also help guide Tn5 to its preferred sites. These might consist of members of a highly degenerate family of DNA sequences or particular topological structures near the insertion site. The hotspot described above coincides with the −10 region of the *tet* promoter (P_{tet}), and is near a second promoter ($P_{anti-tet}$) in the opposite orientation (Stueber & Bujard,

1981). A variety of substitutions within 20 bp of hotspot I have been found to decrease but not eliminate its use during Tn5 insertion, and there is an intriguing correlation between the transcriptional activity of P_{tet} and $P_{anti-tet}$ and the use of hotspot I in transposition (Lodge & Berg, unpublished data).

Studies of Tn9 (IS1) insertions in F' episomal and chromosomal targets indicated that G/C pairs are found preferentially at the ends of the nine bp direct repeats, and showed that one preferred region is A + T rich and contains sequences matching 7 bp at IS1 ends (Galas, Calos & Miller, 1980). In pBR322, two IS1-based transposons exhibited a striking preference for an A + T rich region near the beginning of the *amp* gene (80% of all insertions in a 200 bp segment). These transposons also tended to insert near a short poly (dA.dT) segment added to pBR322 (Zerbib *et al.*, 1985). Strong hotspots for Tn10 insertion occur at a density of about one per kilobase, with relatively fewer insertions at other sites (Foster, 1977; Kleckner *et al.*,1979; Miller *et al.*, 1980). DNA sequence analyses of eleven hotspots revealed a fairly close match to a consensus sequence, 5'-nGCTnAGCn, and suggested that positions ♯2–4 and ♯6–8 of the nine bp duplicated by insertion are recognised (Halling & Kleckner, 1982).

Neither extensive DNA sequence homology nor generalised recombination functions are needed for transposition. Two lines of evidence suggest that the topology of target DNA is important for Tn5 transposition. First, the frequency of transposition is reduced in strains that lack DNA gyrase or topoisomerase (Isberg & Syvanen, 1982; Sternglanz *et al.*, 1981). Second, when an infecting λ phage is used as a transposition target, the addition of the gyrase inhibitor coumermycin before infection reduces transposition about eightfold, compared to controls which receive the drug shortly after infection. Adding coumermycin before, but not after, infection inhibits supercoiling of the λ DNA target; the supercoiling of the donor DNA is independent of the time of addition. Coumermycin thus seems to inhibit transposition by altering the conformation of target DNA (Isberg & Syvanen, 1982). These results are consistent with the reduced transposition frequencies seen in *rho* strains (Datta & Rosner, 1987), which also show lower negative supercoiling (Fassler, Ferstandig Arnold & Tessman, 1986). The insertion specificities of many elements may reflect base sequence recognition by their cognate transposition proteins, and this recognition may in turn depend on local DNA conformation.

TRANSPOSABLE ELEMENT ENDS

Mutational analyses identified sites needed for transposition

The ends of elements are essential for transposition, probably as the sites upon which the cognate transposase and associated proteins act. The sizes of these sites have been estimated experimentally by making deletions that encroach on the termini and then assaying the mutants for transposition activity. In IS*50* the critical sites were found to begin at the first base pair (Sasakawa *et al.*, 1985), and to extend about 19 bp (Sasakawa, Carle & Berg, 1983; Johnson & Reznikoff, 1983; Fig. 1b). The comparable sites extend 21 to 25 bp in IS*1* (Gamas, Galas & Chandler, 1985), 35 to 38 bp in Tn*3* (Huang *et al.*, 1986), and 13 to 27 bp in IS*10* (although sequences extending another 40 bp are also involved in transposition; Way & Kleckner, 1984). The ends of IS*1*, Tn*3* and IS*10* consist of relatively long inverted repeats (23 bp, 38 bp and 22 bp, respectively). The correspondence between essential sites and inverted repeats is consistent with models in which each end of a given element is recognised similarly by a protein or set of proteins.

It is interesting that the two ends of IS*50* ('outer,' O, and 'inner,' I) are well matched in their left, but not in their right halves (8/9 v. 4/10 bp; Fig. 1b). To analyse the functional organisation of these substrate ends, one or more substitution mutations were generated at each of the 19 positions in the O end of IS*50*. Twenty-seven of the 31 mutations obtained caused significant reductions in transposition that ranged from three- to 100-fold, depending on the position and nature of the substitution. At only one position (#12) was there no significant effect of base substitution on transposition activity (Phadnis & Berg, 1987). Thus, all but possibly one of the 19 positions of the IS*50* O end are important at some stage in transposition. Some will be specifically recognised by transposition proteins; others may only affect local DNA conformation and so indirectly influence binding. The pattern of sequence conservation in IS*50* ends and the importance of nearly every position in the O end suggests to us that each end consists of two domains. The conserved region would be bound by a common factor such as transposase, and the divergent region by other factors, one specific for the O end and another for the I end.

A host factor for the IS50 O end

It was suggested that the host-encoded DnaA protein participates in Tn5 transposition (Johnson & Reznikoff, 1983) because the O end contains a good match to the consensus sequence of the DnaA binding site (Zyskind et al., 1983; Fig. 1b) and DnaA binds to the O end in vitro (Fuller, Funnell & Kornberg, 1984). Our mutational analysis provides direct evidence for DnaA protein involvement in transposition: substitution mutations at the seven positions that correspond to invariant residues among DnaA binding sites severely reduced transposition, whereas changes at the two variable positions caused no or only slight reductions. Moreover, the frequency of Tn5 (O end-mediated) transposition from an infecting λ::Tn5 phage was reduced in a $dnaA^-$ null mutant to about 1% of that in an isogenic $dnaA^+$ control (Phadnis & Berg, 1987).

The DnaA protein is needed to initiate replication of the E. coli chromosome (McMacken, Silver & Georgopoulos, 1987), and the possible involvement of DnaA protein in Tn5 transposition was initially interpreted in the context of a replicative transposition model (Johnson & Reznikoff, 1983). We prefer a different interpretation because the structures of transposition products indicate that Tn5 transposition is probably conservative (see above). Recent work has shown that DnaA protein does not directly catalyse DNA or RNA primer synthesis, and suggests that it is needed for the proper assembly of a large complex of replication proteins (the 'replisome'; Baker et al., 1986). We propose that in the case of Tn5, DnaA helps assemble transposase (and perhaps other proteins) on the IS50 O end, and that the ensuing transposition is conservative.

Since new protein synthesis is required at the initiation of each round of replication (McMacken et al., 1987), the activities of DnaA or associated proteins may fluctuate during the cell cycle and in response to environmental changes. A requirement for a host protein that is available or active only during periods of rapid growth could be used to restrict Tn5 transposition to cells that are likely to contain multiple copies of the chromosome. This is advantageous when transposition consumes the donor DNA molecule (Fig. 2; Berg & Berg, 1983; Berg et al., 1984a).

There is additional evidence suggesting transposition and cell growth are coupled. Tn5 transposition is reduced in lon^- mutants, but normal in lon^- $sulA^-$ double mutants or in lon^- mutants grown with DL-pantoyl lactone (Sasakawa, Uno & Yoshikawa, 1987). The *sulA* gene encodes a cell division inhibitor that is highly expressed after DNA damage. The *lon* gene encodes a protease that cleaves SulA protein, and DL-pantoyl lactone induces septation directly, thereby bypassing the *sulA* inhibition (Donachie & Robinson, 1987; Gottesman, 1987). Assuming a basal level of SulA protein, septation may be slightly inhibited in *lon⁻* mutants. Since the *lon* gene encodes a protease, it might also be needed to activate a protein of the transposition complex or to inactivate an inhibitor.

The activity of the IS50 I ends

Tests with a variety of IS*50*-based transposons had shown that pairs of I ends are about 100- to 1000-fold less active in transposition than pairs of O ends (Guarente *et al.*, 1980; Isberg & Syvanen, 1981; Sasakawa & Berg, 1982; Berg *et al.*, 1982a). Recent evidence has shown that the low activity of I ends is not due to an inherent defect (below) and that it may reflect their position in IS*50* and adenine methylation. Transcription directly inhibits the use of an O end in transposition (Sasakawa *et al.*, 1982), and by analogy is probably also inhibitory at I ends. One transcript crosses the I end of IS*50*L, and another enters and probably crosses the I end of IS*50*R (which contains the last two codons of the *tnp* gene; Rothstein & Reznikoff, 1981). Thus both I ends could be inhibited by transcription. Further, the I end contains two 5'GATC methylation sites, whereas the O end does not (Fig. 1): interactions of proteins with DNA substrates that contain this sequence are often affected by methylation (Sternberg, 1985).

The intrinsic activities of IS*50* ends were estimated directly using a set of synthetic transposons that differed only in the combinations of O and I ends that bracketed the *tnp* and *tet* genes. In *dam⁺* (adenine methylation proficient) cells, a pair of O ends was 10-fold more active than a pair of I ends. The situation is reversed in isogenic *dam⁻* cells: the I end pair was about sevenfold more active than an O end pair. Although the transposition of each type of element was stimulated by the lack of methylation, the magnitude differs fourfold for an O end pair but 300-fold for an *I* end pair (Dodson & Berg, unpublished data). The I end of IS*10*, which also has a 5'GATC sequence, is inhibited by methylation of both strands, but one of the two hemi-methylated species is as active as the non-methylated species (Roberts *et al.*, 1985).

The high activity of I ends unencumbered by transcription or by adenine methylation implies that they too are recognised efficiently by transposition proteins. Given the sequence divergence between O and I ends and the importance of each position in the O end, it is likely that complexes of transposition proteins specific to each end differ in one or more components. The I end does not contain a DnaA consensus sequence (Fig. 1b). An as yet unknown host protein may help assemble the transposition apparatus at the I end, mimicking the postulated action of DnaA protein at the O end. Since replication generates hemi-methylated DNA (Lyons & Schendel, 1984), the sensitivity of I end activity to methylation may be as effective as the O end dependence on DnaA protein in limiting transposition to cells containing at least two copies of the donor sequence.

REGULATION OF IS50 TRANSPOSITION

The proteins of IS50

Two proteins are encoded in each IS50 element (Fig. 1a). The proteins of IS50R are the transposase (also called protein 1 or p58, 58 kd) and the inhibitor (protein 2 or p54, 54 kD; Rothstein et al., 1980). These are translated in the same frame and are identical in primary structure except that the N-terminal 55 amino acids of transposase are missing from the inhibitor (Johnson, Yin & Reznikoff, 1982; Krebs & Reznikoff, 1986). The proteins of IS50L are truncated by an ochre (UAA) nonsense codon, but are otherwise identical to those of IS50R. They seem to play no role in transposition or inhibition in non-suppressing strains (Johnson et al., 1982; Isberg, Lazaar & Syvanen, 1982). Transposase is absolutely required for transposition and complements very poorly in trans (Rothstein et al., 1980; Isberg & Syvanen, 1981). A protein of 30 kD which precipitates with anti-transposase/inhibitor antibody has also been seen, but its significance is unclear (Isberg & Syvanen, 1985b).

Joint control of transposition by transposase and inhibitor

The finding that an established Tn5 element inhibits transposition of entering Tn5 showed that transposition is regulated (Biek & Roth, 1980). Studies with tnp-lacZ gene fusions showed that the inhibitor does not regulate the synthesis of transposase. It may inhibit transposition by interacting with transposase, with the IS50 ends, or with a host component (Johnson et al., 1982; Isberg et al., 1982). Each of these possible mechanisms differs from those used for transposition control by other elements, for example classical repression of transcription of transposase genes in the cases of Tn*3* and Mu (Heffron, 1983; Toussaint & Resibois, 1983), and inhibition of translation of the transposase message by anti-sense RNA in the case of IS*10* (Simons & Kleckner, 1983).

Analyses of IS*50*-proteins on SDS gels indicated there is two- to four-fold more inhibitor than transposase, and their amounts are directly proportional to IS*50*R copy number (Johnson *et al.*, 1982; Isberg *et al.*, 1982; Krebs & Reznikoff, 1986). About 500 molecules of IS*50*R encoded proteins are found per cell when Tn*5* is present on a multicopy colE1 plasmid or about 5 molecules of transposase and 20 of inhibitor per copy of IS*50*R (Johnson & Reznikoff, 1984*b*). In contrast about 0.15 molecules of IS*10* transposase are found per element per cell (Raleigh & Kleckner, 1986). Cell fractionation experiments using strains that overproduce the IS*50*R proteins suggested that the transposase is localised in the membrane but the inhibitor is cytoplasmic. This is consistent with their *cis* and *trans* actions, respectively (Isberg & Syvanen, 1985*b*).

The extent to which inhibitor decreases the transposition of an entering element depends on the copy number of the resident IS*50*R, and ranges from about 2-fold to about 100-fold (Isberg *et al.*, 1982; Johnson *et al.*, 1982; Johnson & Reznikoff, 1984*a,b*; Krebs & Reznikoff, 1986). Consistent with this, the chance that any given element will transpose in the steady state is inversely related to the IS*50*R copy number, and the total transposition frequency per cell is essentially constant (Johnson & Reznikoff, 1984*a*). More detailed studies revealed situations in which changes in transposase and inhibitor levels do not cause simple proportionate changes in transposition frequency in the steady state. For example, estimated reductions of the transposase and inhibitor levels to approximately 6% of those of colE1::Tn*5*, or an increase in the inhibitor:transposase ratio from 4:1 to about 30:1, reduced the transposition frequency to just about half of the control values (Johnson & Reznikoff, 1984*b*). These nonlinear responses led to the suggestion that transposition frequencies are determined primarily by the levels of a limiting host factor (Johnson & Reznikoff, 1984*a,b*). The results can also be explained by hypothesising that the proteins are normally made in excess, so that little additional effect is observed when they are overproduced.

Transcripts of IS50R

The three promoters shown in Fig. 1c are thought to contribute differently to transposase and inhibitor synthesis. Only the T1 transcript specifies transposase since the T2 and T3 transcripts start within the *tnp* reading frame, and IS*50* elements from which the promoter for T1 was deleted (p1) do not make transposase. Most inhibitor synthesis is thought to depend on the T2 transcript because deletions which remove the p1 but not p2 or p3 promoters do not significantly affect inhibitor concentrations (Krebs & Reznikoff, 1986), and the T2 transcript is about 100-fold more abundant than T3. The T2 transcript is also 100-fold more abundant and slightly more stable than T1. There is thus a striking disparity between the relative amounts of the T1 and T2 transcripts (1:100) and the transposase and inhibitor proteins in the steady state (1:2 or 1:4) (Krebs & Reznikoff, 1986; McCommas & Syvanen, personal communication).

The conclusion that the T2 transcript is responsible for most inhibitor synthesis seems weakened by studies in which efficient promoter and translation initiation signals were inserted at or near the first codon of the *tnp* gene. These fusions resulted in elevated levels of both inhibitor and transposase (Isberg & Syvanen, 1985*b*; Krebs & Reznikoff, 1986). One hypothesis to reconcile these results is to suppose that the 5'-untranslated region of T1, which is missing in these fusions, affects translational initiation at the inhibitor start sites. Clarification of this question is of practical as well as conceptual importance since purification of inhibitor-free transposase would greatly facilitate the development of an *in vitro* transposition system.

Methylation and the control of tnp expression

The frequencies of Tn*5* transposition are four- to 18-fold higher in *dam⁻* than in *dam⁺* strains (Roberts *et al.*, 1985; Dodson & Berg, unpublished data). Although there are no 5'GATC sequences in the O end, the p1 promoter region contains two 5'GATC sites (Fig. 1c, labelled 'dam'), and this promoter is more active in *dam⁻* than in *dam⁺* cells (McCommas & Syvanen, personal communication). It is tempting to think that the level of transposase normally limits the frequency of Tn*5* transposition, despite the results described above. Circumstantial support for this idea comes from studies of Tn*10*. Its transposition is stimulated in *dam⁻* cells, its *tnp* promoter is sensitive to methylation, and its transposase levels clearly limit the frequency of transposition. In addition, the transposition of Tn*3* and Tn*9*, whose *tnp* promoters do not contain dam sites, is not sensitive to *dam* methylation. These results led to the suggestion that transient hemimethylation during the cell cycle triggers a burst of Tn*10* transposition

(Roberts *et al.*, 1985). This mechanism could also be involved in the control of Tn*5* transposition. Nevertheless, alternative explanations for Tn*5*, such as methylation control of a limiting host factor or of target DNA availability, remain to be tested.

Effects of external promoters on transposition

The *tnp* gene of IS*50* starts less than 100 bp from flanking host sequences, and consequently an element in a highly transcribed region might be expected to make high levels of transposase. However, little if any transposase protein is made from transcripts extending into the *tnp* reading frame from external promoters, and it has been suggested that the translational start site is sequestered from ribosomes in a region of mRNA secondary structure (Krebs & Reznikoff, 1986). In addition, transcription across an IS*50* O end interferes directly with the ability of that end to participate in transposition (Sasakawa *et al.*, 1982). Thus, even if the transposition frequency does reflect transposase levels, cells are protected from deleterious context effects by sequestration and transcriptional inhibition. IS*10* transposase levels are also insensitive to transcription from outside promoters (Davis, Simons & Kleckner, 1985), and the use of IS*10* and IS*1* ends is also inhibited by transcription from external promoters (Machida *et al.*, 1983; Chandler & Galas, 1983; Biel, Adelt & Berg, 1984; Davis *et al.*, 1985).

Transposase action is restricted to nearby ends

A *tnp*⁺ allele does not significantly complement *tnp*⁻ IS*50*-related elements in *trans*. Even in *cis*, complementation is efficient only when the ends are relatively close to the active *tnp*⁺ gene (Isberg & Syvanen, 1981; Isberg, *et al.*, 1982; Berg *et al.*, 1982; Johnson *et al.*, 1982; Phadnis, Sasakawa & Berg, 1986). The transposase proteins of IS*1* and IS*10* also act preferentially on cognate ends near their sites of synthesis (Machida *et al.*, 1982; Chandler, Clerget & Galas, 1982; Morisato *et al.*, 1983; Way & Kleckner, 1984). Such spatial limitation could reflect the sum of several factors. If nascent transposase binds DNA, then ends adjacent to actively transcribed and translated *tnp* genes would be preferentially bound. Transposase may be constrained from moving freely along a DNA molecule; for example, the Tn*5* transposase is probably membrane-bound (Isberg & Syvanen, 1985*b*). More than one transposase molecule may be bound to an element, and the concentrations sufficient for simultaneous binding at distant sites may rarely be achieved. Synthesis of transposase may be co-ordinated with transient availability of the ends, or the protein may be unstable.

CONCLUSIONS AND PROSPECTS

Transposition requires the interaction of element-encoded transposases and of host factors with sequences at transposable element ends. Beyond this common theme, transposable elements are diverse. They transpose by a variety of mechanisms, some conservative (Tn5, Tn10), some replicative (Tn3), and some mixed (Tn9, Mu). Each element exhibits insertion specificity which may reflect recognition of particular sequences or conformations in target DNAs. Recognition of the elements themselves is at least in part due to the cognate transposase proteins, although host factors are probably also involved. In the case of IS50 (Tn5) all but perhaps one of the 19 bp at the O end are needed as a site for transposition, and the host DnaA protein also acts there.

Transposition is sensitively regulated in numerous ways. IS50, like IS10 and IS1, makes a transposase that acts preferentially on the element that encodes it. Tn5 (IS50) transposes most efficiently when it first enters a cell; this favours its establishment in the cell lineage even when the vector is unstable, but minimises the frequency of subsequent mutations. The interference with transposition of an entering element by resident IS50 entails a *trans*-acting inhibitor that may bind to, or compete with, the *cis*-acting transposase. Finally, IS50 may use the DnaA requirement at the O end, and *dam* methylation sites of the transposase promoter and the I end to couple transposition to cell division, thereby ensuring that transposition occurs primarily in cells with multiple copies of the donor chromosome.

Our discussion has indicated some remaining major questions. There is a need for thorough understanding of the mechanisms and regulation of transposition processes at the biochemical level. Important insights into replicative transposition have emerged from *in vitro* studies of Mu insertion, and there have been some significant steps in the development of an *in vitro* conservative transposition system using Tn10. The following questions are among the important issues that need to be resolved. Are element or host encoded protein(s) limiting for transposition? Is transposition regulated by normal physiological changes during the cell cycle? The coming years should clarify how Tn5 and other elements transpose.

ACKNOWLEDGEMENTS

We thank Dr C. M. Berg for discussion and critical reading of the manuscript. T. Kazic is supported by training grant A107172 from the United States Public Health Service. This work was supported by research grants GM-37138 and DMB-8608193 to D. Berg from the United States Public Health Service and the National Science Foundation, respectively.

REFERENCES

AHMED, A. (1986). Evidence for replicative transposition of Tn5 and Tn9. *Journal of Molecular Biology*, **191**, 75–84.

AKROYD, J. E. & SYMONDS, N. (1983). Evidence for a conservative pathway of transposition of bacteriophage Mu. *Nature*, **303**, 84–6.

ARTHUR, A. & SHERRATT, D. (1979). Dissection of the transposition process: a transposon-encoded site-specific recombination system. *Molecular and General Genetics*, **175**, 267–74.

BAKER, T. A., SEKIMIZU, K., FUNNELL, B. E. & KORNBERG, A. (1986). Extensive unwinding of the plasmid template during staged enzymatic initiation of DNA replication from the origin of the *Escherichia coli* chromosome. *Cell*, **45**, 53–64.

BENDER, J. & KLECKNER, N. (1986). Genetic evidence that Tn10 transposes by a nonreplicative mechanism. *Cell*, **45**, 801–15.

BERG, D. E. (1977). Insertion and excision of the transposable kanamycin resistance determinant Tn5. In *DNA Insertion Elements, Plasmids, and Episomes*. ed. A. I. Bukhari, J. A. Shapiro & S. L. Adhya, pp. 205–12. Cold Spring Harbor, NY: Cold Spring Harbor Laboratory.

BERG, D. E. (1983). Structural requirement for IS50-mediated gene transposition. *Proceedings of the National Academy of Sciences, USA*, **80**, 792–96.

BERG, D. E. & BERG, C. M. (1983). The prokaryotic transposable element Tn5. *Biotechnology*, **1**, 417–35.

BERG, C. M. & BERG, D. E. (1987). Uses of transposable elements and maps of known insertions. In *Escherichia coli and Salmonella typhimurium. Cellular and Molecular Biology*. Vol. 2. ed. F. C. Neidhardt, J. L. Ingraham, K. B. Low, B. Magasanik, M. Schaechter & H. E. Umbarger, pp. 1071–109. Washington, D.C.: American Society for Microbiology.

BERG, D. E., BERG, C. M. & SASAKAWA, C. (1984a). The bacterial transposon Tn5: evolutionary inferences. *Molecular Biology and Evolution* **1**, 411–22.

BERG, D. E., DAVIES, J., ALLET, B. & ROCHAIX, J.-D. (1975). Transposition of R factor genes to bacteriophage λ. *Proceedings of the National Academy of Sciences, USA*, **72**, 3628–32.

BERG, D. E. & DRUMMOND, M. (1978). Absence of DNA sequences homologous to transposable element Tn5 (Kan) in the chromosome of *Escherichia coli* K-12. *Journal of Bacteriology*, **136**, 419–22.

BERG, D. E., EGNER, C., HIRSCHEL, B. J., HOWARD, J., JOHNSRUD, L., JORGENSEN, R. A. & TLSTY, T. D. (1981). Insertion, excision, and inversion of Tn5. Cold Spring Harbor Symposium on Quantitative Biology, vol. 45, pp. 115–23. Cold Spring Harbor, NY: Cold Spring Harbor Laboratory.

BERG, D. E., JOHNSRUD, L., McDIVITT, L., RAMABHADRAN, R. & HIRSCHEL, B. J. (1982a). Inverted repeats of Tn5 are transposable elements. *Proceedings of the National Academy of Sciences, USA*, **79**, 2632–5.

BERG, D. E., LODGE, J., SASAKAWA, C., NAG, D. K., PHADNIS, S. H., WESTON-HAFER, K. & CARLE, G. F. (1984b). Transposon Tn5: specific sequence recognition and conservative transposition. In Cold Spring Harbor Symposium on Quantitative Biology, vol. 49, pp. 215–26. Cold Spring Harbor, NY: Cold Spring Harbor Laboratory.

BERG, D. E., LOWE, J. B., SASAKAWA, C. & McDIVITT, L. (1982b). The mechanism and control of Tn5 transposition. In *Stadler Symposium*, vol. 14, pp. 5–28. Columbia, Mo.: University of Missouri Press.

BERG, D. E., SCHMANDT, M. A. & LOWE, J. B. (1983). Specificity of transposon Tn5 insertion. *Genetics*, **105**, 813–28.

BERG, D. E., WEISS, A. & CROSSLAND, L. (1980). Polarity of Tn5 insertion mutations in *Escherichia coli*. *Journal of Bacteriology*, **142**, 439–46.

BIEK, D. & ROTH, J. R. (1980). Regulation of Tn5 transposition in *Salmonella typhimurium*. *Proceedings of the National Academy of Sciences, USA*, **77**, 6047–51.

BIEL, S. W., ADELT, G. & BERG, D. E. (1984). Transcriptional control of IS1 transposition in *Escherichia coli*. *Journal of Molecular Biology* **174**, 251–64.

BIEL, S. W. & BERG, D. E. (1984). Mechanism of IS1 transposition in *E. coli*: choice between simple insertion and cointegration. *Genetics*, **108**, 319–30.

CHACONAS, G., HARSHEY, R. M., SARVETNICK, N. & BUKHARI, A. I. (1981). Predominant end-products of prophage Mu DNA transposition during the lytic cycle are replicon fusions. *Journal of Molecular Biology*, **150**, 341–59.

CHACONAS, G., KENNEDY, D. L. & EVANS, D. (1983). Predominant integration end products of infecting bacteriophage Mu DNA are simple insertions with no preference for integration of either Mu DNA strand. *Virology* **128**, 48–59.

CHANDLER, M., CLERGET, M. & GALAS, D. J. (1982). The transposition frequency of IS1-flanked transposons is a function of their size. *Journal of Molecular Biology*, **54**, 229–43.

CHANDLER, M. & GALAS, D. J. (1983). Cointegrate formation mediated by Tn9. II. Activity of IS1 is modulated by external DNA sequences. *Journal of Molecular Biology*, **170**, 61–91.

CRAIG, N. L. & KLECKNER, N. (1987). Transposition and site-specific recombination. In *Escherichia coli and Salmonella typhimurium. Molecular and Cellular Biology*, Vol. 2. ed. F. C. Neidhardt, J. L. Ingraham, K. B. Low, B. Magasanik, M. Schaechter, & H. E. Umbarger, pp. 1054–70. Washington, D.C.: American Society for Microbiology.

DATTA, A. R. & ROSNER, J. L. (1987). Reduced transposition in *rho* mutants of *Escherichia coli* K-12. *Journal of Bacteriology*, **169**, 888–90.

DAVIS, M. A., SIMONS, R. W. & KLECKNER, N. (1985). Tn10 protects itself at two levels from fortuitous activation by external promoters. *Cell*, **43**, 379–87.

DONACHIE, W. D. & ROBINSON, A. C. (1987). Cell division: parameter values and the process. In *Escherichia coli and Salmonella typhimurium. Cellular and Molecular Biology*, vol. 2, ed. F. C. Neidhardt, J. L. Ingraham, K. B. Low, M. Schaechter & H. E. Umbarger, pp. 1578–93. Washington, D. C.: American Society for Microbiology.

EGNER, C. & BERG, D. E. (1981). Excision of transposon Tn5 is dependent on the inverted repeats but not on the transposase function of Tn5. *Proceedings of the National Academy of Sciences, USA*, **78**, 459–63.

FASSLER, J. S., FERSTANDIG ARNOLD, G. & TESSMAN, I. (1986). Reduced super-helicity of plasmid DNA produced by the *rho-15* mutation in *Escherichia coli*. *Molecular and General Genetics*, **204**, 424–9.

FOSTER, T. J. (1977). Insertion of the tetracycline resistance translocation unit Tn*10* in the *lac* operon of *Escherichia coli* K12. *Molecular and General Genetics*. **154**, 305–9.

FOSTER, T. J., LUNDBLAD, V., HANLEY-WAY, S., HALLING, S. M. & KLECKNER, N. (1981). Three Tn*10*-associated excision events: relationship to transposition and role of direct and inverted repeats. *Cell*, **23**, 215–27.

FULLER, R. S., FUNNELL, B. E. & KORNBERG, A. (1984). The *dnaA* protein complex with the *E. coli* chromosomal replication origin (*oriC*) and other DNA sites. *Cell*, **38**, 889–900.

GALAS, D. J., CALOS, M. P. & MILLER, J. H. (1980). Sequence analysis of Tn*9* insertions in the *lacZ* gene. *Journal of Molecular Biology*, **144**, 19–41.

GALAS, D. J. & CHANDLER, M. (1981). On the molecular mechanisms of transposition. *Proceedings of the National Academy of Sciences, USA*, **78**, 4858–62.

GALAS, D. J. & CHANDLER, M. (1982). Structure and stability of Tn*9*-mediated cointegrates. Evidence for two pathways of transposition. *Journal of Molecular Biology*, **154**, 245–72.

GAMAS, P., GALAS, D. & CHANDLER, M. (1985). DNA sequence at the end of IS*1* required for transposition. *Nature*, **317**, 458–60.

GILL, R., HEFFRON, F., DOUGAN, G. & FALKOW, S. (1978). Analysis of sequences transposed by complementation of two classes of transposition-deficient mutants of Tn*3*. *Journal of Bacteriology*, **136**, 742–56.

GLOOR, G. & CHACONAS, G. (1986). The bacteriophage Mu *N* gene encodes the 64–kDa virion protein which is injected with, and circularizes, infecting Mu DNA. *Journal of Biological Chemistry*, **261**, 16682–8.

GOTTESMAN, S. (1987). Regulation by proteolysis. In *Escherichia coli and Salmonella typhimurium. Cellular and Molecular Biology*. Vol. 2. ed. F. C. Neidhardt, J. L. Ingraham, K. B. Low, B. Magasanik, M. Schaechter, & H. E. Umbarger, pp. 1308–12. Washington, D.C.: American Society for Microbiology.

GRAY, G. S. & FITCH, W. M. (1983). Evolution of antibiotic resistance genes: the DNA sequence of a kanamycin resistance gene from *Staphylococcus aureus*. *Molecular Biology and Evolution*, **1**, 57–66.

GRINDLEY, N. D. F. & REED, R. R. (1985). Transpositional recombination in prokaryotes. *Annual Review of Biochemistry*, **54**, 893–96.

GUARENTE, L. P., ISBERG, R. R., SYVANEN, M. & SILHAVY, T. J. (1980). Conferral of transposable properties to a chromosomal gene in *Escherichia coli*. *Journal of Molecular Biology*, **141**, 235–48.

HALLING, S. M. & KLECKNER, N. (1982). A symmetrical six-base-pair target site sequence determines Tn*10* insertion specificity. *Cell*, **28**, 153–63.

HARSHEY, R. M. (1984). Transposition without duplication of infecting bacteriophage Mu DNA. *Nature*, **311**, 580–1.

HARSHEY, R. M. & BUKHARI, A. I. (1981). A mechanism of DNA transposition. *Proceedings of the National Academy of Sciences, USA*, **78**, 1090–4.

HARSHEY, R. M. & BUKHARI, A. I. (1983). Infecting bacteriophage Mu DNA forms a circular DNA-protein complex. *Journal of Molecular Biology*, **167**, 427–41.

HEFFRON, F. (1983). Tn*3* and its relatives. In *Mobile Genetic Elements*. ed. J. A. Shapiro, pp. 223–60. New York: Academic Press.

HIRSCHEL, B. J., GALAS, D. J., BERG, D. E. & CHANDLER, M. (1982a). Structure and stability of transposon 5-mediated cointegrates. *Journal of Molecular Biology*, **159**, 557–80.

HIRSCHEL, B. J., GALAS, D. J. & CHANDLER, M. (1982b). Cointegrate formation by Tn5, but not transposition, is dependent on recA. *Proceedings of the National Academy of Sciences, USA*, **79**, 4530–4.

HUANG, C.-J., HEFFRON, F., TWU, J.-S., SCHLOEMER, R. H. & LEE, C.-H. (1986). Analysis of Tn3 sequences required for transposition and immunity. *Gene*, **41**, 23–31.

IIDA, S., MEYER, J. & ARBER, W. (1983). Prokaryotic IS elements. In *Mobile Genetic Elements*, ed. J. A. Shapiro, pp. 159–221. New York: Academic Press.

ISBERG, R. R., LAZAAR, A. L. & SYVANEN, M. (1982). Regulation of Tn5 by the right-repeat proteins: control at the level of the transposition reaction? *Cell*, **30**, 883–92.

ISBERG, R. R. & SYVANEN, M. (1981). Replicon fusions promoted by the inverted repeats of Tn5. The right repeat is an insertion sequence. *Journal of Molecular Biology*, **150**, 15–32.

ISBERG, R. R. & SYVANEN, M. (1982). DNA gyrase is a host factor required for transposition of Tn5. *Cell*, **30**, 9–18.

ISBERG, R. R. & SYVANEN, M. (1985a). Tn5 transposes independently of cointegrate resolution. Evidence for an alternative model for transposition. *Journal of Molecular Biology*, **182**, 69–78.

ISBERG, R. R. & SYVANEN, M. (1985b). Compartmentalization of the proteins encoded by IS50R. *Journal of Biological Chemistry*, **260**, 3645–51.

JOHNSON, R. C. & REZNIKOFF, W. S. (1983). DNA sequences at the ends of transposon Tn5 required for transposition. *Nature*, **304**, 280–2.

JOHNSON, R. C. & REZNIKOFF, W. S. (1984a). Copy number control of Tn5 transposition. *Genetics*, **107**, 9–18.

JOHNSON, R. C. & REZNIKOFF, W. S. (1984b). Role of the IS50R proteins in the promotion and control of Tn5 transposition. *Journal of Molecular Biology*, **177**, 645–61.

JOHNSON, R. C., YIN, J. C. P. & REZNIKOFF, W. S. (1982). Control of Tn5 transposition in *Escherichia coli* is mediated by protein from the right repeat. *Cell*, **30**, 873–82.

KLECKNER, N. (1979). DNA sequence analysis of Tn10 insertions: origin and role of 9 bp flanking repetitions during Tn10 translocation. *Cell*, **16**, 711–20.

KLECKNER, N. (1983). Transposon Tn10. In *Mobile Genetic Elements*. ed. J. A. Shapiro, pp. 261–98. New York: Academic Press.

KLECKNER, N., STEELE, D. A., REICHARDT, K. & BOTSTEIN, D. (1979). Specificity of insertion by the translocatable tetracycline-resistance element Tn10. *Genetics*, **92**, 1023–40.

KOSTRIKEN, R., MORITA, C. & HEFFRON, F. (1981). Transposon Tn3 encodes a site-specific recombination system: identification of essential sequences, genes, and actual site of recombination. *Proceedings of the National Academy of Sciences, USA*, **78**, 4041–5.

KREBS, M. P. & REZNIKOFF, W. S. (1986). Transcriptional and translational initiation sites of IS50. Control of transposase and inhibitor expression. *Journal of Molecular Biology*, **192**, 781–91.

LIEBART, J. C., GHELARDINI, P. & PAOLOZZI, L. (1982). Conservative integration of bacteriophage Mu DNA into pBR322 plasmid. *Proceedings of the National Academy of Sciences, USA*, **79**, 4362–6.

LODGE, J. K., WESTON-HAFER, K., LOWE, J. B., DODSON, K. W. & BERG, D. E. (1984). Determinants of Tn5 insertion specificity. In *Genome Rearrangement*. UCLA Symposium on Molecular and Cellular Biology, new series. vol. 20, pp. 37–42.

LYONS, S. M. & SCHENDEL, P. F. (1984). Kinetics of methylation in *Escherichia coli* K-12. *Journal of Bacteriology*, **159**, 421–3.

McMACKEN, R., SILVER, L. & GEORGOPOULOS, C. (1987). DNA replication. In *Escherichia coli and Salmonella typhimurium. Cellular and Molecular Biology*. Vol. 1. ed. F. C. Neidhardt, J. L. Ingraham, K. B. Low, B. Magasanik, M. Schaechter & H. E. Umbarger, pp. 564–612. Washington, D.C.: American Society for Microbiology.

MACHIDA, C., MACHIDA, Y., WANG, H.-C., ISHIZAKI, K. & OHTSUBO, E. (1983). Repression of cointegration ability of insertion element IS*1* by transcriptional readthrough from flanking regions. *Cell* **34**, 135–42.

MACHIDA, Y. MACHIDA, C., OHTSUBO, H. & OHTSUBO, E. (1982). Factors determining frequency of plasmid cointegration mediated by insertion sequence IS*1*. *Proceedings of the National Academy of Sciences, USA*, **79**, 277–81.

MAZODIER, P., COSSART, P., GIRAUD, E. & GASSER, F. (1985). Completion of the nucleotide sequence of the central region of Tn*5* confirms the presence of three resistance genes. *Nucleic Acids Research*, **13**, 195–205.

MAZODIER, P., GIRAUD, E. & GASSER, F. (1982). Tn*5* dependent streptomycin resistance in *Methylobacterium organophilum. FEMS Microbiology Letters*, **13**, 27–30.

MILLER, J. H., CALOS, M. P., GALAS, D., HOFER, M., BUECHEL, D. E. & MUELLER-HILL, B. (1980). Genetic analysis of transpositions in the *lac* region of *Escherichia coli. Journal of Molecular Biology*, **144**, 1–18.

MIZUUCHI, K. & CRAIGIE, R. (1986). Mechanism of bacteriophage Mu transposition. *Annual Review of Genetics*, **20**, 385–429.

MORISATO, D. & KLECKNER, N. (1984). Transposase promotes double strand breaks and single strand joints at Tn*10* termini *in vivo. Cell*, **39**, 181–90.

MORISATO, D., WAY, J. C., KIM, H.-J. & KLECKNER, N. (1983). Tn*10* transposase acts preferentially on nearby transposon ends *in vivo. Cell*, **32**, 799–807.

NAG, D. K., DASGUPTA, U., ADELT, G. & BERG, D. E. (1985). IS*50*-mediated inverse transposition: specificity and precision. *Gene*, **34**, 17–26.

OHTSUBO, E., ZENILMAN, M., OHTSUBO, H., McCORMICK, M., MACHIDA, C. & MACHIDA, Y. (1981). Mechanism of insertion and cointegration mediated by IS*1* and Tn*3*. Cold Spring Harbor Symposium on Quantitative Biology, vol. 45, pp. 283–95. Cold Spring Harbor, NY: Cold Spring Harbor Laboratory.

PHADNIS, S. H. & BERG, D. E. (1985). *recA*-independent recombination between repeated IS*50* elements is not caused by an IS*50*-encoded function. *Journal of Bacteriology*, **161**, 928–32.

PHADNIS, S. H., SASAKAWA, C. & BERG, D. E. (1986). Localization of action of the IS*50*-encoded transposase protein. *Genetics*, **112**, 421–7.

RALEIGH, E. A. & KLECKNER, N. (1986). Quantitation of insertion sequence IS*10* transposase gene expression by a method generally applicable to any rarely expressed gene. *Proceedings of the National Academy of Sciences, USA*, **83**, 1787–91.

ROBERTS, D., HOOPES, B. C., McCLURE, W. R. & KLECKNER, N. (1985). IS*10* transposition is regulated by DNA adenine methylation. *Cell*, **43**, 117–30.

ROTHSTEIN, S. J., JORGENSEN, R. A., POSTLE, K. & REZNIKOFF, W. S. (1980). The inverted repeats of Tn*5* are functionally different. *Cell*, **19**, 795–805.

ROTHSTEIN, S. J. & REZNIKOFF, W. S. (1981). The functional differences in the inverted repeats of Tn*5* are caused by a single base pair nonhomology. *Cell*, **23**, 191–9.

SASAKAWA, C. & BERG, D. E. (1982). IS*50*-mediated inverse transposition. Discrimination between the two ends of an IS element. *Journal of Molecular Biology*, **159**, 257–71.

SASAKAWA, C., CARLE, G. F. & BERG, D. E. (1983). Sequences essential for transposition at the termini of IS50. *Proceedings of the National Academy of Sciences, USA*, **80**, 7293–7.

SASAKAWA, C., LOWE, J. B., McDIVITT, L. & BERG, D. E. (1982). Control of transposon Tn5 transposition in *Escherichia coli*. *Proceedings of the National Academy of Sciences, USA*, **79**, 7450–4.

SASAKAWA, C., PHADNIS, S. H., CARLE, G. F. & BERG, D. E. (1985). Sequences essential for IS50 transposition. The first base-pair. *Journal of Molecular Biology*, **182**, 487–93.

SASAKAWA, C., UNO, Y. & YOSHIKAWA, M. (1981). The requirement for both DNA polymerase and 5' to 3' exonuclease activities of DNA polymerase I during Tn5 transposition. *Molecular and General Genetics*, **182**, 19–24.

SASAKAWA, C., UNO, Y. & YOSHIKAWA, M. (1987). Lon-SulA regulatory function affects the efficiency of transposition of Tn5 from λ *b221 cI857* Pam *O*am to the chromosome. *Biochemical and Biophysical Research Communications*, **142**, 879–84.

SCHALLER, H. (1979). The intergenic region and the origins for filamentous phage DNA replication. In *Cold Spring Harbor Symposium on Quantitative Biology*, vol. 43, pp. 401–8. Cold Spring Harbor, NY: Cold Spring Harbor Laboratory.

SHAPIRO, J. A. (1979). Molecular model for the transposition and replication of bacteriophage Mu and other transposable elements. *Proceedings of the National Academy of Sciences, USA*, **76**, 1933–7.

SHAW, K. J. & BERG, C. M. (1979). *Escherichia coli* K-12 auxotrophs induced by insertion of the transposable element Tn5. *Genetics*, **92**, 741–7.

SIMONS, R. W. & KLECKNER, N. (1983). Translational control of IS10 transposition. *Cell*, **34**, 683–91.

STERNBERG, N. (1985). Evidence that adenine methylation influences DNA-protein interactions in *Escherichia coli*. *Journal of Bacteriology*, **164**, 490–3.

STERNGLANZ, R., DiNARDO, S., VOELKEL, K. A., NISHIMURA, Y., HIROTA, Y., BECHERER, K., ZUMSTEIN, L. & WANG, J. C. (1981). Mutations in the gene coding for *Escherichia coli* DNA topoisomerase I affect transcription and transposition. *Proceedings of the National Academy of Sciences, USA*, **78**, 2747–51.

STUEBER, D. & BUJARD, H. (1981). Organization of transcriptional signals in plasmids pBR322 and pACYC184. *Proceedings of the National Academy of Sciences, USA*, **78**, 167–71.

SYVANEN, M., HOPKINS, J. D. & CLEMENTS, M. (1982). A new class of mutants in DNA polymerase I that affects gene transposition. *Journal of Molecular Biology*, **158**, 203–12.

TOUSSAINT, A. & RESIBOIS, A. (1983). Phage Mu: transposition as a life-style. In *Mobile Genetic Elements*. ed. J. A. Shapiro, pp. 105-58. New York: Academic Press.

WAY, J. C. & KLECKNER, N. (1984). Essential sites at transposon Tn10 termini. *Proceedings of the National Academy of Sciences, USA*, **81**, 3452–6.

ZERBIB, D., GAMAS, P., CHANDLER, M., PRENTKI, P., BASS, S. & GALAS, D. (1985). Specificity of insertion of IS1. *Journal of Molecular Biology*, **185**, 517–24.

ZUPANCIC, T. J., MARVO, S. L., CHUNG, J. H., PERALTA, E. G. & JASKUNAS, S. R. (1983). *RecA*-independent recombination between direct repeats of IS50. *Cell*, **33**, 629–37.

ZYSKIND, J. W., CLEARY, J. M., BRUSILOW, W. S. A., HARDING, N. E. & SMITH, D. W. (1983). Chromosomal replication origin from the marine bacterium *Vibrio harveyi* functions in *Escherichia coli: oriC* consensus sequence. *Proceedings of the National Academy of Sciences, USA*, **80**, 1164–8.

MECHANISM OF THE DNA STRAND TRANSFER STEP IN TRANSPOSITION OF Mu DNA

ROBERT CRAIGIE, MICHIYO MIZUUCHI, KENJI ADZUMA and KIYOSHI MIZUUCHI

Laboratory of Molecular Biology, National Institute of Diabetes, and Digestive and Kidney Diseases, National Institutes of Health, Bethesda, Maryland 20892, USA

Transposable elements are common in both prokaryotes and eukaryotes. These elements are segments of DNA, bounded by specific terminal sequences, that can move to new locations in the host genome. Each type of element encodes its own transposase protein; in some cases the transposase functions are divided among more than one polypeptide. Transposase recognises the ends of the element and is an essential factor for transposition (for reviews, see Kleckner, 1981; Grindley & Reed, 1985; Berg and Hatfull *et al.*, this volume). In addition to insertion at new sites in DNA, the transposition process can generate many types of genomic rearrangements, including inversions and deletions. Transposable elements are therefore likely to have played an important role in the evolution of both prokaryotic and eukaryotic genomes. Some prokaryotic transposons, in addition to their transposase gene, also carry genes that can confer a direct selective advantage on the host cell; resistance to antibiotics is the best known example.

Most of our knowledge of transposition mechanisms has come from studies on prokaryotic elements, where genetic approaches laid a foundation for the more recent biochemical analysis of the reaction mechanism. Early clues concerning the mechanism of transposition came from analysis of the structure of transposition products. Two types of intermolecular product are observed: one type is the simple insertion of a transposable element, from the original (donor) DNA molecule, into another (target) DNA molecule; the other type of product, which is commonly called a cointegrate, comprises the entire donor molecule joined with the target DNA. In the cointegrate structure, directly repeated copies of the transposable element are found at the junctions between the nontransposon part of the donor molecule and the target DNA; the presence of two copies of the element in a cointegrate implies that the element is replicated in this transposition pathway. In both kinds of product a short sequence (typically less than 10 bp and of a length characteristic for each type of element) is duplicated at the site of integration into the target DNA. This target-site duplication has been explained in terms of staggered cleavage of the target DNA (Grindley & Sherratt, 1978); subsequent single strand gap repair makes a directly repeated target-sequence duplication. These early discoveries led investigators to propose a variety of models for transposition mechanisms (reviewed in Mizuuchi & Craigie, 1986).

BACTERIOPHAGE Mu AS A TRANSPOSABLE ELEMENT

Bacteriophage Mu is a temperate phage of *E. coli*, and some other species of enterobacteria, that utilises the replicative cointegration transposition pathway to make multiple copies of its genome during lytic growth; Mu also uses the other transposition pathway, non-replicative simple insertion, to integrate its genome into the host chromosome after infection of a sensitive cell (for a general review of Mu as both a transposable element and a phage, see Toussaint & Résibois, 1983). Here, we only consider the features of Mu that are directly involved in transposition.

Only two Mu gene products are directly involved in the transposition reaction: the Mu A protein (transposase), which is essential, and the Mu B protein; several other Mu genes are involved in the regulation of transposition. The Mu A protein recognises the ends of Mu DNA and is expected to play the central role in the DNA cleavage and joining steps. The Mu B protein is not essential for transposition, but it greatly enhances the efficiency of the intermolecular reaction. The left (L) and right (R) ends of Mu are the only essential DNA sequence requirements.

AN OVERVIEW OF THE Mu TRANSPOSITION REACTION

The availability of a cell-free system for Mu transposition (Mizuuchi, 1983) has enabled the reaction mechanism to be studied at the molecular level. The reaction utilises, as the transposon donor, a mini-Mu plasmid containing the Mu left and right ends in their normal inverted orientation; another circular DNA molecule serves as the target.

Both the Mu A and B genes have been cloned on an inducible expression vector (Craigie & Mizuuchi, 1985a). Extracts, made from *E. coli* cells induced for expression of these proteins, provide the source of Mu A and B proteins. Cell extract also supplies the necessary host proteins. Both simple insert and cointegrate transposition products are made in this cell-free reaction.

Under certain conditions, simple insert and cointegrate products are not formed in the cell-free reaction, but a new DNA structure accumulates (Craigie & Mizuuchi, 1985b). Analysis of this structure revealed that it results from joining the 3' ends of the Mu DNA sequence, that were connected to flanking DNA in the donor molecule, to the 5' ends of a staggered cut made in the target DNA (Fig. 1C). This product, which has two Y-shaped junctions, is the expected intermediate in the cointegration mechanism proposed by Shapiro (1979) and by Arthur & Sherratt (1979). Replication of the mini-Mu part of the structure, using the free 3' ends of the target DNA as primers for leading strand DNA synthesis, generates a cointegrate product (Fig. 1D). It has been shown that this branched DNA structure, made in a cell-free reaction and subsequently deproteinised, can be resolved to make a cointegrate transposition product upon incubation with an *E. coli* cell extract (Craigie & Mizuuchi, 1985b). Therefore, the Shapiro structure is indeed an intermediate in the Mu cointegration reaction. How does Mu make simple insertion transposition products? Mu simple insertions made *in vivo* (Harshey, 1984) and *in vitro* (Mizuuchi, 1984) are not newly replicated along their entire length. This eliminates the pathway, involving a primary cointegrate product, that is used by Tn3 and some other transposons (for a review, see Heffron, 1983). Tn3 and its relatives first make a cointegrate and then recombination between the two transposon copies in the cointegrate regenerates the donor molecule together with an insertion of the transposon into the target DNA. Instead, we have proposed (Mizuuchi, 1984; Craigie & Mizuuchi, 1985b) that the Mu simple insertion reaction occurs by an alternative resolution pathway of the same Shapiro structure that is an intermediate in cointegration. Degradation of the DNA that flanked the Mu part of the donor molecule in this structure (symbolised by wavy lines in Fig. 1C), results in a simple insert product with single strand gaps at the junctions between the Mu DNA and the target DNA. A gap repair process would then complete the simple insertion (Fig. 1E). Incubation of the purified Shapiro structure with an *E. coli* cell extract does indeed generate simple insert as well as cointegrate

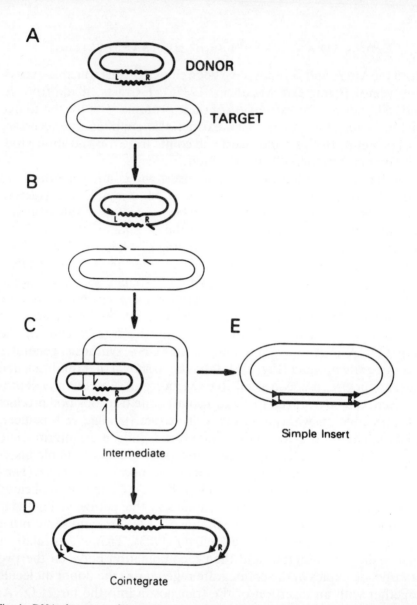

Fig. 1. *DNA cleavage and joining events involved in Mu transposition.* A The transposon donor molecule is represented by bold lines. The mini-Mu DNA segment of the donor is depicted by smooth lines, and the DNA flanking this segment by wavy lines. The left (L) and right (R) ends of the mini-Mu DNA segment are labeled. The target DNA is represented by thin lines. B Nicks are made at each 3′ end of the Mu sequence in the donor molecule, and a staggered cut with a 5′-protruding overhang of 5 bases is made at an essentially random site in the target DNA. Half-arrows depict the 3′ ends of DNA strands. C Each 3′ end of the cleaved Mu DNA is joined to one of the 5′ ends of the cleaved target DNA, to generate a Shapiro structure. D Leading and lagging strand DNA synthesis, using the free 3′ ends of the target DNA in the Shapiro structure as primers for synthesis of the leading strand, results in a cointegrate product. Arrowheads represent the 5 bp directly repeated target-sequence duplication that results from staggered cleavage. E Alternatively, the Shapiro intermediate may be resolved by nucleolytic cleavage and gap repair to make a simple insert product. The nucleolytic cleavage step removes the old flanking DNA (wavy lines in C), to form a simple insert with single strand gaps at the junctions between the Mu and target DNA. These gaps can be filled by a gap repair process to complete the simple insertion.

products, demonstrating that it is also an intermediate in the Mu simple insertion pathway (Craigie & Mizuuchi, 1985*b*).

THE Mu DNA STRAND TRANSFER REACTION

Transfer of the 3′ ends of Mu DNA to the 5′ ends of a staggered cut in the target DNA is a central step in the Mu transposition reaction. We shall now focus on this DNA strand transfer reaction (for a recent review, see Mizuuchi & Craigie, 1986). This part of the transposition reaction can be carried out in a defined system with purified proteins. An efficient intermolecular reaction requires only the Mu A protein, Mu B protein, and the *E. coli* histone-like protein HU as protein factors, together with ATP and Mg^{2+} (Craigie *et al.*, 1985). Strand transfer still occurs in the absence of Mu B protein and/or ATP. However, this reaction is much less efficient and the products are mainly intramolecular (target site of strand transfer on the same molecule as the participating pair of Mu ends), in contrast to the predominantly intermolecular reaction that occurs in the presence of both Mu B protein and ATP. The donor molecule must be supercoiled and carry the two Mu ends in inverted orientation. The target DNA need not be supercoiled and, to a first approximation, the sites of integration into the target DNA are nonspecific with respect to DNA sequence.

The DNA strand transfer reaction can be dissected into several steps based on our current knowledge (see Fig. 2). Synapsis of the ends of Mu DNA is followed by single-strand cleavage at the Mu 3′ ends. This cleaved donor DNA, which exists as a noncovalent nucleoprotein complex, is the active species in capturing the target DNA. After capture, the target DNA is cut with a 5 bp stagger and the precut 3′ ends of the Mu DNA are joined to the protruding 5′ ends of this staggered cut. We now discuss these steps in more detail.

Synapsis

We use the term 'synaptic complex' for a nucleoprotein complex, involving a pair of Mu ends and protein factors, that precedes cleavage at the ends of the Mu DNA sequence. There may well be more than one such metastable complex along the pathway leading

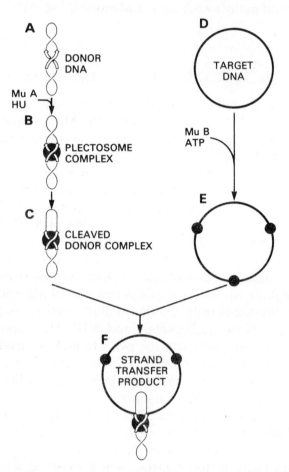

Fig. 2. *Steps involved in the Mu DNA strand transfer reaction.* A A negatively supercoiled mini-Mu donor DNA with the Mu ends (open arrows) in inverted orientation. B The Mu end sequences in the donor DNA associate with Mu A protein and protein HU, to form a plectosome complex. This synaptic complex requires, for its stabilisation, a negatively super- coiled donor molecule with the Mu ends in inverted orientation. The shaded area depicts protein factors noncovalently associated with the Mu end DNA sequences. C Single strand cleavages are made at each 3′ end of the Mu DNA, allowing the flanking DNA segment of the donor molecule to relax. In this cleaved donor complex, the superhelicity of the mini-Mu DNA segment is maintained because the Mu end DNA sequences are stably bridged by noncovalently associated protein. The geometry of the Mu end DNA sequences in this complex is likely to differ from their spatial organisation in the plectosome complex, because superheli- city is no longer required to complete strand transfer once the cleaved donor complex is formed. D A target DNA molecule. E Mu B protein molecules (represented by an arbitrary number of solid circles) interact primarily with the target DNA, increasing the efficiency with which it may be captured by a cleaved donor complex (Adzuma & Mizuuchi, in prep- aration). This association of Mu B protein with the target DNA requires a nucleotide cofactor. F Transfer of the cleaved 3′ ends of the Mu DNA, to a staggered cut made in the target DNA, completes strand transfer.

L-END

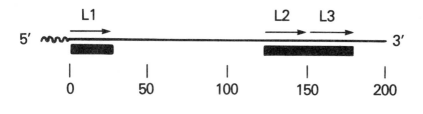

R-END

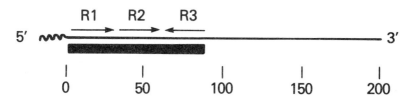

Fig. 3. *Binding of Mu A protein to the ends of Mu DNA.* Sequences at the left (L) and right (R) ends of the Mu DNA that are protected from nuclease attack by Mu A protein are represented as solid bars. The distance from the junction between the Mu DNA (straight lines) and flanking DNA (wavy lines) is indicated in base pairs. Mu A protein binding sites within the protected sequences, which conform to the consensus 5'-TGNTTCAPyTNPuAA-PuPyPuCGAAAPu-3', are represented by arrows. These sites are numbered L1, L2, L3, and R1, R2, and R3, in order of their distance from the termini of the Mu left and right ends, respectively.

to cleavage at the 3' ends of Mu DNA. The restricted term 'plecto-some complex' is reserved for the specific hypothetical synaptic struc-ture, described below, that can be stabilised by the free energy of negative supercoiling only when the Mu ends are in inverted orien-tation in the donor DNA molecule.

An early step in the DNA strand transfer reaction is synapsis of a pair of Mu ends. Mu A protein is expected to play the key role because it specifically binds to sequences at each end of Mu DNA (Craigie *et al.*, 1984), whereas only non-specific DNA binding is observed with Mu B protein (Chaconas *et al.*, 1985) and protein HU (Rouvière-Yaniv & Gros, 1975; Holck & Kleppe, 1985). Each Mu end has three binding sites for Mu A protein, designated L1, L2, and L3 at the left end and R1, R2, and R3 at the right end (Fig. 3). Sites L1 and R1 are identically positioned with respect to the terminal TG dinucleotide of the left and right ends of Mu DNA, respectively. The other sites differ in their location between the left and right ends. The DNA sequences of these binding sites conform to a consensus, but the individual sites differ from one another in their exact sequence and affinity for Mu A protein. Deletion and mutation analysis demonstrates that all these sites, with the possible exception of R3, are required for optimal transposition efficiency *in vivo* (Groenen *et al.*, 1985; Groenen & van de Putte, 1986). This complex arrangement of binding sites for Mu A protein at the ends of Mu DNA is likely to reflect a similarly complex three-dimensional organization of the Mu end DNA segments in a synaptic complex. Protein HU, in addition to Mu A protein, must be involved at some stage prior to cleavage of the Mu ends, because not even a partial reaction is detected in its absence. Binding of Mu A protein to individual binding sites on linear DNA induces bending of the DNA (Adzuma & Mizuuchi, 1987), and this bending is expected to contribute to the structure of the synaptic complex. However, no direct structural studies have yet been attempted with a supercoiled donor molecule, or with Mu A protein together with protein HU. As described below, we expect that negative supercoiling is necessary to achieve the full synaptic structure that is required for the reaction to proceed.

Although the detailed structure of the synaptic complex remains to be studied, some information has been deduced from the requirements for a specific relative orientation of the Mu ends in a negatively supercoiled donor molecule (Craigie & Mizuuchi, 1986). These two requirements are thought to be necessary for the formation of a specific synaptic structure, a prerequisite for the reaction to proceed. Figure 4 illustrates how a particular synaptic complex can be favoured by negative supercoiling only when the Mu ends are in inverted orientation in the donor DNA. Figure 4A depicts such a negatively supercoiled donor molecule with the Mu ends in inverted orientation; it forms an interwound right-handed superhelix due to its negative superhelicity. Let us suppose, as the simplest possible example, that the Mu ends also form the same right-handed helix in a synaptic complex. Then, negative supercoiling will energetically favour this synaptic structure if the Mu ends are in inverted orientation (as shown in Fig. 4A); we call this hypothetical synaptic structure a 'plectosome complex'. Conversely, if the ends are in direct repeat orientation (as shown in Fig. 4B), the synaptic structure depicted in Fig. 4A can only be formed at the energetic cost of disrupting the interwound superhelix of the donor molecule. The actual path of the Mu end DNA sequences in the plectosome complex is unlikely

A   B

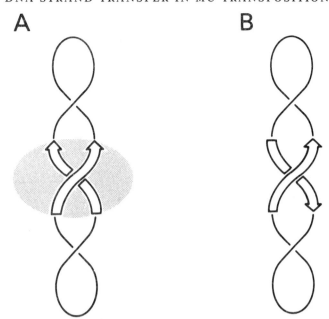

Fig. 4. *Stabilisation of a synaptic complex by negative supercoiling.* A A negatively supercoiled mini-Mu donor molecule, with the Mu end DNA sequences (open arrows) in inverted orientation along the linear path of the DNA. The molecule forms a right-handed interwound superhelix as a consequence of its negative superhelicity. If the Mu end DNA sequences form such a right-handed superhelix in a synaptic complex with protein factors, represented by the shaded area, the negative superhelicity of the donor DNA will stabilise this complex. B Conversely, if the Mu end DNA sequences are in direct repeat orientation in the donor molecule, the same geometric configuration of these sequences in a synaptic complex can only be achieved at the expense of disrupting the right-handed superhelix of the donor molecule. This is energetically unfavourable, hence the synaptic complex shown in A cannot be formed efficiently.

The actual configuration of the Mu end DNA sequences in this hypothetical synaptic complex, which we call a plectosome complex, is unlikely to be a simple right-handed superhelix; a significant component of this superhelical geometry is all that is required in this model.

to be such a simple right-handed superhelix, but there is expected to be a significant component of this superhelical geometry. This model for explaining the requirement for the Mu ends to be in inverted repeat orientation in a negatively supercoiled donor molecule is supported by the results of tests with specially constructed artificial donor substrates (Craigie & Mizuuchi, 1986).

Cleavage at the 3' ends of Mu DNA

When supercoiled mini-Mu donor DNA is incubated with Mu A protein and protein HU, a conversion to the open circle form occurs. This donor DNA cleavage reaction, like the complete DNA strand transfer reaction, requires a negatively supercoiled mini-Mu molecule carrying the two Mu ends in inverted repeat orientation. Mu B Protein, ATP, or target DNA are not required.

Analysis of the cleaved donor DNA reveals that the single strand cleavages are located precisely at the 3' termini of the Mu left and right end sequences (Craigie & Mizuuchi, submitted). They can be sealed by T4 DNA ligase and are therefore nicks with 3' hydroxyl and 5' phosphoryl termini. The donor DNA remains noncovalently associated with protein after cleavage. The Mu ends are bridged by protein in this cleaved donor complex, maintaining the negative superhelicity of the Mu DNA segment of the donor molecule, but allowing the flanking DNA to relax (see Fig. 2C).

The cleaved donor complex can go on to complete strand transfer if Mu B protein, ATP, and a target DNA are added to the reaction mixture after cleavage has occurred. The resulting strand transfer product is identical to that formed under the normal reaction conditions when all the components necessary for strand transfer are present from the start of the reaction. Cleavage of the 3' ends of Mu DNA, to generate a nucleoprotein complex in which the Mu DNA segment of the donor molecule remains supercoiled, is therefore an intermediate step in the Mu DNA strand transfer reaction. The requirement for negative superhelicity of the donor molecule together with an inverted orientation of the Mu ends is lifted once the cleaved donor complex is formed; both the Mu and flanking DNA segments in the cleaved donor complex can be cut by restriction enzymes without reducing the efficiency of subsequent DNA strand transfer.

Capture of the target DNA by a cleaved donor complex

The reaction steps discussed above can occur in the absence of Mu B protein, ATP and a target DNA. It is therefore likely that these events also precede interaction with the target DNA in the normal Mu DNA strand transfer reaction. Interaction between a cleaved donor complex and potential target DNA appears to be mediated through the Mu B protein. Mu B protein binds to potential target DNA, thereby increasing the efficiency with which it may be captured by a cleaved donor complex (Adzuma & Mizuuchi, in preparation). The carboxyl-terminal domain of Mu A protein may be involved in this interaction between a cleaved donor complex and Mu B protein that is already associated with a potential target DNA. An interaction between the carboxyl-terminus of Mu A protein and Mu B protein is suggested by the finding that deletion mutants for the carboxyl-terminus of Mu A protein are phenotypically also Mu B⁻; this can be simply explained if the truncated Mu A protein has lost its ability to recognise Mu B protein (Harshey & Cuneo, 1986).

Mu B protein requires a nucleotide cofactor for its activity. ATP, which is hydrolysed by Mu B protein (Maxwell *et al.*, 1987), satisfies this requirement (Craigie *et al.*, 1985). However, strand transfer still occurs efficiently when the non-hydrolysable analogue ATPγS is substituted for ATP in the reaction (Adzuma & Mizuuchi, in preparation). Therefore, ATP hydrolysis does not seem to be directly involved in the role of Mu B protein in the DNA strand transfer reaction. Preincubation experiments (Adzuma & Mizuuchi, in preparation) indicate that Mu B protein becomes tightly associated with DNA in the presence of ATPγS, so that this DNA becomes committed as the potential target, even when it is subsequently challenged with an excess of competitor DNA. In contrast, with ATP as the cofactor, Mu B protein is able to redistribute when challenged with a competitor DNA. It appears that ATP hydrolysis is involved in allowing Mu B protein to dissociate from DNA, rather than playing a direct role in the capture step.

On the macroscopic level there is little sequence specificity for the DNA that serves as the target for strand transfer. However, the results of sequencing a large number of junctions between mini-Mu insertions and adjacent DNA have revealed that the sequence 5′ −Py C/G Pu − 3′, aligned with nucleotide positions 2 to 4 of the 5 bp target-site duplication, is a strongly preferred target site (M. Mizuuchi & K. Mizuuchi, unpublished). Since this specificity is limited to three base pairs, with a degeneracy at each position, it is observed only at the nucleotide level. This target site preference, which was initially observed with *in vivo* Mu transpositions, is also seen when the strand transfer reaction is carried out with purified proteins and the Shapiro intermediate is subsequently converted to simple insert and cointegrate products by incubation with an *E. coli* cell extract; this result is obtained even when the DNA strand transfer step is performed in the absence of Mu B protein. Therefore, only Mu A and HU proteins seem to be involved in recognising this sequence. In addition to this 3 bp site, there is also a preference for the utilisation of multiple sites in certain stretches of DNA as the target for strand transfer. These regional hotspots are observed even when strand transfer is carried out with purified proteins (M.

Mizuuchi & K. Mizuuchi, unpublished); the molecular basis of this phenomenon remains to be determined.

Transfer of the Mu 3' ends to the target DNA

The Mu DNA strand transfer reaction does not require a high energy cofactor (ATP is required only for the stimulatory activity of the Mu B protein), so the energy required to form the new phosphodiester bonds between the Mu and target DNA must come from the DNA or protein substrates. Other known site-specific recombination reactions (for a mini-review, see Sadowski, 1986) store the energy of a cleaved phosphodiester bond as a covalent DNA-protein linkage. The DNA end involved is then transferred, from the protein, to a free end of DNA. The cleavage and joining steps are therefore coupled through a DNA-protein intermediate. Does the Mu DNA strand transfer reaction utilise a similar mechanism? Four phosphodiester bonds are broken during the course of the reaction, but only two new phosphodiester bonds are made. The joining of Mu to target DNA could, in principle, be coupled with either cleavage of the donor or the target DNA. There is also the alternative possibility that Mu A protein may be in an activated state prior to strand transfer and is consumed during the reaction. Although we cannot eliminate the possibility of an activated Mu A protein, we currently favour the alternative hypothesis that the joining step is energetically coupled with DNA cleavage.

Cleavage at the 3' ends of Mu DNA, to yield free 3' hydroxyl termini, occurs before interaction with the target DNA, as discussed above. The energy derived from cleavage of the phosphodiester bonds at the 3' ends of the Mu sequence in the donor DNA therefore cannot be utilised for subsequent joining of the Mu ends to the target DNA. In fact, cleavage of the donor DNA may be performed by a restriction endonuclease, instead of by Mu A and HU proteins (Craigie & Mizuuchi, submitted). Cleavage by a restriction enzyme at the correct site in a pair of Mu and DNA sequences has been accomplished by engineering a *Hin*dIII restriction site, overlapping with the terminal base pair of each Mu end sequence, such that *Hin*dIII cuts precisely at each 3' end of the mini-Mu DNA; the cleavages on the complementary strands are located four nucleotides into the flanking DNA. This linear mini-Mu substrate, which has free 3' hydroxyl ends produced by *Hin*dIII cleavage, can efficiently complete strand transfer in the presence of Mu A protein, Mu B protein and ATP. Protein HU is not required for this reaction with precut mini-Mu DNA.

Since the energy of hydrolysis of the phosphodiester bonds at the ends of the Mu sequence in the donor DNA is not conserved, it is likely that the joining step is energetically coupled with cleavage of the target DNA. This could be achieved through a covalent intermediate between the 5′ end of the staggered cut in the target DNA and Mu A protein; the reaction mechanism would then be similar to that of the other known site-specific recombination reactions, mentioned above. However, unlike the reactions that are known to involve a covalent intermediate, cleavage at the ends of Mu DNA, to yield free 3′ hydroxyl termini, precedes cleavage of the target DNA to which the Mu ends are to be joined. An alternative reaction mechanism is therefore feasible. The precut ends of Mu DNA may be located in the active site of Mu A protein in the cleaved donor complex, such that the 3′ hydroxyl groups at the ends of the Mu DNA act as nucleophiles to attack the target DNA. In this mechanism the 5′ ends of the cut target DNA would be transferred directly to the recipient Mu 3′ ends, instead of indirectly through a covalent DNA-protein intermediate. Further work is required to test these models.

DO OTHER PROKARYOTIC ELEMENTS TRANSPOSE BY A SIMILAR MECHANISM?

Detailed information on the mechanism of transposition is so far available only for Mu. Progress towards establishing cell-free reaction systems for other transposons should soon enable the similarities and differences in transposition mechanism between elements to be clarified. Transposition of Tn10 has been detected in vitro, but the efficiency of the reaction is rather low at present (D. Morisato & N. Kleckner, personal communication); another, as yet poorly understood, reaction of Tn10, excision and self-circularisation (Morisato & Kleckner, 1984), occurs efficiently in a cell-free system (D. Morisato & N. Kleckner, personal communication). However, for most transposons, available data are still limited to structural analysis of the end products of transposition.

The major products of intermolecular transposition are either non-replicative simple inserts or replicative cointegrates, as discussed earlier. Some transposons, such as Tn3 (for reviews, see Heffron,

1983 and Hatfull *et al.*, this volume) make cointegrates; others, like Tn*10* (Bender & Kleckner, 1986) and Tn*5* (Berg, 1983), make non-replicative simple inserts (see Berg *et al.*, this volume). Many elements, including Mu and some IS elements, can make both types of product at an easily detectable frequency (see Kleckner, 1981). Even transposons that appear to exclusively make only one of these types of product may also make the alternative product, but at a frequency below the detection limit of many experimental systems; for example, a low frequency of non-replicative simple insertion has been reported with Tn*3* (Bennett *et al.*, 1983), which was thought to produce only cointegrates as primary products.

If we make the assumption that transposition of all the elements involves an essentially similar DNA strand transfer reaction, simple variations in how the products are processed can account for the different frequencies with which each type of transposon makes simple inserts or cointegrates. Cointegration can be favoured by factors that stabilise the Y-shaped junctions in the Shapiro intermediate against nucleolytic attack, or facilitate the assembly of a replication complex at these junctions. The transposase proteins of some elements may play such a role, in addition to their primary function of mediating DNA strand transfer. It may be significant in this respect that, after completion of Mu DNA strand transfer, the product remains associated with proteins as a noncovalent nucleoprotein complex (Surette *et al.*, 1987). Similarly, secondary activities of other transposase proteins may lead to predominantly simple insertion products. Any activity that increases the probability of the old flanking DNA being degraded, before DNA replication can initiate in the Shapiro structure, favours simple insertion over cointegration. In fact, if the transposase can make an additional pair of cleavages on the strands complementary to those directly involved in strand transfer, this flanking DNA could be severed before strand transfer occurs. This is essentially the 'cut and paste' mechanism, involving a double strand break at the ends of the transposon, that has been proposed for Tn*10* (Morisato & Kleckner, 1984). Even in cases where the Shapiro intermediate is formed, some transposases may promote its resolution to a simple insert. This could be achieved directly, by making additional transposase-mediated cleavages after strand transfer has occurred, or indirectly by perturbing local DNA structure so as to make the old flanking DNA adjacent to the transposon ends hypersensitive to endonucleolytic attack. Alternatively, inhibiting, or simply not facilitating, the assembly of a replication complex at the Y-shaped junctions in the Shapiro structure may be sufficient to favour simple insertion over cointegration *in vivo*.

The simple insertion of Mu DNA into the host chromosome, after infection of a sensitive cell, is a special case. Although the infecting Mu DNA is linear, the ends, which consist of old host flanking DNA, are reported to become non-convalently bridged by protein after infection (Harshey & Bukhari, 1983; Ljungquist & Bukhari, 1979; Ljungquist *et al.*, 1978). This DNA can be supercoiled (Harshey & Bukhari, 1983). The strand transfer product with such a donor would have a non-covalent protein linkage in the old flanking DNA. This linkage may be sufficiently labile that the structure is readily resolved to a simple insert. Even if replication of the product is successfully initiated, an attempt to make a cointegrate from such an intermediate would lead to a double strand break, and likely consequent loss of the chromosome copy into which Mu integrated.

In addition to transposition by non-replicative simple insertion, Tn*10* also makes a transposon circle product (Morisato & Kleckner, 1984), as mentioned above. In this structure, the 3' and 5' ends of a pair of IS*10* termini are covalently joined on one of the two DNA strands; the other strand, which remains unjoined, frequently contains an interruption. The absolute requirement for transposase and the high frequency of this reaction suggests that it is related to normal transposition. Excision of Tn*10* from the donor molecule, by a double strand break, has been suggested to be a common step in the generation of both types of product (Morisato & Kleckner, 1984). The pair of pre-cut ends could then attack a target DNA to form a simple insert or, alternatively, become joined to make a transposon circle. It is not known if joining of the transposon ends is mediated by transposase itself or by a host DNA ligase. In either case, it is surprising and intriguing that Tn*10* promotes this apparently non-productive reaction with such efficiency.

PERSPECTIVES

We have attempted to briefly summarise current knowledge of the Mu DNA strand transfer reaction mechanism and the role of this reaction in the transposition of Mu DNA. The Mu strand transfer reaction is a novel recombination reaction that differs from other known site-specific recombination reactions in several fundamental respects. Three DNA sites are involved: the two specific sequences at the ends of Mu DNA, and a target site that is essentially non-specific with respect to DNA sequence. Four phosphodiester bonds are broken during the course of the reaction, but only two new bonds are formed.

Many systems other than Mu are likely to utilise a similar DNA strand transfer reaction to join the ends of a mobile genetic element to a target DNA. In fact, all integration reactions that result in a short sequence duplication at the site of insertion into the target DNA are potential candidates. These duplications are the hallmarks of staggered cleavage of the target DNA and subsequent filling in of the single strand gaps by a DNA polymerase; the DNA strand transfer reaction leaves the 3' hydroxyl ends of the target DNA unjoined, and therefore available to prime synthesis. In addition to other prokaryotic transposons, many eukaryotic transposable elements are likely to join their ends to a target DNA by means of a DNA strand transfer reaction that is fundamentally similar to the Mu reaction. Retrotransposons, such as the yeast Ty and Drosophila *copia* elements, and retroviruses, are especially strong candidates. These elements (see Kingsman *et al.*, and Kuff & Lueders, this volume; for brief reviews see also Baltimore 1985; Panganiban, 1985) first make an RNA transcript, which is then reverse-transcribed to make a DNA copy which is the precursor for integration. Joining of this DNA copy to a target DNA is likely to involve a pair of DNA strand transfers to make single strand connections between each end of the element and the target DNA. Completion of integration would then simply require a nucleolytic cleavage step to remove a few extra base pairs that flank the LTR sequences at each end of the DNA copy, followed by gap repair.

Since the Mu DNA strand transfer reaction can be carried out with purified components, it is possible to study the molecular mechanism of this reaction in detail. In addition to increasing our understanding of the Mu transposition reaction, it is anticipated that the information gained will provide a framework for the study of related systems which may not be as readily amenable to biochemical analysis.

REFERENCES

ADZUMA, K. & MIZUUCHI, K. (1987). Mu A protein-induced bending of the Mu end DNA. In *Proceedings of the Fifth Conversation in Biomolecular Stereodynamics*. Adenine Press, in press.

ARTHUR, A. & SHERRATT, D. J. (1979). Dissection of the transposition process:

a transposon-encoded site-specific recombination system. *Molecular and General Genetics*, **175**, 267–74.

BALTIMORE, D. (1985). Retroviruses and retrotransposons: the role of reverse transcription in shaping the eukaryotic genome. *Cell*, **40**, 481–2.

BENDER, J. & KLECKNER, N. (1986). Genetic evidence that Tn*10* transposes by a nonreplicative mechanism. *Cell*, **45**, 801–15.

BENNETT, P. M., DE LA CRUZ, F. & GRINSTED, J. (1983). Cointegrates are not obligatory intermediates in transposition of Tn*3* and Tn*21*. *Nature*, **305**, 743–4.

BERG, D. E. (1983). Structural requirement for IS*50*-mediated gene transposition. *Proceedings of the National Academy of Sciences, USA*, **80**, 792–6.

CHACONAS, G., GLOOR, G. & MILLER, J. L. (1985). Amplification and purification of the bacteriophage Mu encoded B transposition protein. *Journal of Biological Chemistry*, **260**, 2662–9.

CRAIGIE, R. & MIZUUCHI, K. (1985a). Cloning of the A gene of bacteriophage Mu and purification of its product, the Mu transposase. *Journal of Biological Chemistry*, **260**, 1832–5.

CRAIGIE, R. & MIZUUCHI, K. (1985b). Mechanism of transposition of bacteriophage Mu: structure of a transposition intermediate. *Cell*, **41**, 867–76.

CRAIGIE, R. & MIZUUCHI, K. (1986). Role of DNA topology in Mu transposition: mechanism of sensing the relative orientation of two DNA segments. *Cell*, **45**, 793–800.

CRAIGIE, R., ARNDT-JOVIN, D. J. & MIZUUCHI, K. (1985). A defined system for the DNA strand transfer reaction at the initiation of bacteriophage Mu transposition: protein and DNA substrate requirements. *Proceedings of the National Academy of Sciences, USA*, **82**, 7570–4.

CRAIGIE, R., MIZUUCHI, M. & MIZUUCHI, K. (1984). Site-specific recognition of the bacteriophage Mu ends by the Mu A protein. *Cell*, 39, 387–94.

GRINDLEY, N. D. F. & REED, R. R. (1985). Transpositional recombination in prokaryotes. *Annual Review of Biochemistry*, **54**, 863–96.

GRINDLEY, N. D. F. & SHERRATT, D. J. (1978). Sequence analysis at IS*1* insertion sites: models for transposition. *Cold Spring Harbor Symposia on Quantitative Biology*, **43**, 1257–61.

GROENEN, M. A. M. & VAN DE PUTTE, P. (1986). Analysis of the ends of bacteriophage Mu using site-directed mutagenesis. *Journal of Molecular Biology*, **189**, 597–602.

GROENEN, M. A. M., TIMMERS, E. & VAN DE PUTTE, P. (1985). DNA sequences at the ends of the genome of bacteriophage Mu essential for transposition. *Proceedings of the National Academy of Sciences, USA*, **82**, 2087–91.

HARSHEY, R. M. (1984). Transposition without duplication of infecting bacteriophage Mu DNA. *Nature*, **311**, 580–1.

HARSHEY, R. M. & BUKHARI, A. I. (1983). Infecting bacteriophage Mu DNA forms a circular DNA-protein complex. *Journal of Molecular Biology*, **167**, 427–41.

HARSHEY, R. M. & CUNEO, S. D. (1986). Carboxyl-terminal mutants of phage Mu transposase. *Journal of Genetics*, **65**, 159–74.

HEFFRON, F. (1983). Tn*3* and its relatives. In *Mobile Genetic Elements*, ed. J. A. Shapiro, pp. 223–60. New York: Academic Press.

HOLCK, A. & KLEPPE, K. (1985). Affinity of protein HU for different nucleic acids. *FEBS Letters*, **185**, 121–4.

KLECKNER, N. (1981). Transposable elements in prokaryotes. *Annual Review of Genetics*, **15**, 341–404.

LJUNGQUIST, E. & BUKHARI, A. I. (1979). Behavior of bacteriophage Mu DNA upon infection of *Escherichia coli* cells. *Journal of Molecular Biology*, **133**, 339–57.

148   R. CRAIGIE *ET AL*

LJUNGQUIST, E., KHATOON, H., DUBOW, M., AMBROSIO, L., DE BRUIJN, F. & BUK-
    HARI, A. I. (1978). Integration of bacteriophage Mu DNA. *Cold Spring Harbor
    Symposia on Quantitative Biology*, **43**, 1151–8.
MAXWELL, A., CRAIGIE, R. & MIZUUCHI, K. (1987). B protein of bacteriophage
    Mu is an ATPase that preferentially stimulates intermoleular DNA strand
    transfer. *Proceedings of the National Academy of Sciences, USA*, **84**, 699–703.
MIZUUCHI, K. (1983). *In vitro* transposition of bacteriophage Mu: a biochemical
    approach to a novel replication reaction. *Cell*, **35**, 785–94.
MIZUUCHI, K. (1984). Mechanism of transposition of bacteriophage Mu: polarity
    of the DNA strand transfer reaction at the initiation of transposition. *Cell*, **39**,
    395–404.
MIZUUCHI, K. & CRAIGIE, R. (1986). Mechanism of bacteriophage Mu transposition.
    Annual Review of Genetics, **20**, 385–429.
MORISATO, D. & KLECKNER, N. (1984). Transposase promotes double strand breaks
    and single strand joints at Tn*10* termini *in vivo*. *Cell*, **39**, 181–90.
PANGANIBAN, A. T. (1985). Retroviral DNA integration. *Cell*, **42**, 5–6.
ROUVIÈRE-YANIV, J. & GROS, F. (1975). Characterization of a novel, low-molecular-
    weight DNA-binding protein from *Escherichia coli*. *Proceedings of the National
    Academy of Sciences, USA*, **72**, 3428–32.
SADOWSKI, P. (1986). Site-specific recombinases: changing partners and doing the
    twist. *Journal of Bacteriology*, **165**, 341–7.
SHAPIRO, J. A. (1979). Molecular model for the transposition and replication of
    bacteriophage Mu and other transposable elements. *Proceedings of the National
    Academy of Sciences, USA*, **76**, 1933–7.
SURETTE, M. G., BUCH, S. J. & CHACONAS, G. (1987). Transpososomes: stable
    protein–DNA complexes involved in the *in vitro* transposition of bacteriophage
    Mu DNA. *Cell*, **49**, 253–62.
TOUSSAINT, A. & RÉSIBOIS, A. (1983). Phage Mu: transposition as a lifestyle. In
    Mobile Genetic Elements, ed. J. A. Shapiro, pp. 105–58. New York: Academic
    Press.

SITE-SPECIFIC RECOMBINATION BY THE γδ RESOLVASE

G. F. HATFULL, J. J. SALVO, E. E. FALVEY,
V. RIMPHANITCHAYAKIT and N. D. F. GRINDLEY*

Department of Molecular Biophysics and Biochemistry, Yale University, 333 Cedar St., New Haven, CT 06510, USA

INTRODUCTION

The scope of this article

Transposons of the Tn*3* family encode two specialised recombination systems (for reviews see Grindley, 1983; Heffron, 1983; Grindley & Reed, 1985). The first, which consists of the transposase protein and the two ends of the transposon, is responsible for insertion of the entire donor molecule into the target (Fig. 1(i)). During the process the transposon is duplicated, and the product with copies of the transposon at both donor–target junctions is called a cointegrate. The second transposon-encoded recombination system reduces a cointegrate into the two final products: the regenerated donor molecule and the target DNA now with a simple insertion of the transposon (Fig. 1(ii)). It is this second process – cointegrate resolution – that is the topic of this article. Our research has been with the resolution system of the γδ transposon. This system is virtually identical to the resolution system of Tn*3* and components of these two systems can be readily interchanged. We have used a multi-faceted attack to explore the details of cointegrate resolution, including biochemistry using purified components *in vitro*, molecular genetics of both the resolvase protein and its DNA substrate, and, in collaboration with the research group of Thomas Steitz, crystallography. These studies have been complemented by topological and other studies performed by the research groups of Nicholas Cozzarelli & David Sherratt (Krasnow & Cozzarelli, 1983; Cozzarelli *et al.*, 1984; Benjamin *et al.*, 1985; Wasserman & Cozzarelli, 1985; Wasserman *et al.*, 1985; Wasserman & Cozzarelli, 1986; Boocock *et al.*, 1986, 1987).

* To whom correspondence should be sent

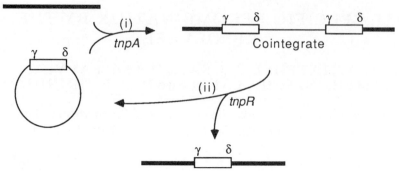

Fig. 1. γδ Transposition: formation and resolution of a cointegrate. The heavy line represents target DNA, the open box represents the transposon γδ. (i) Formation of a cointegrate requires the *tnpA* gene product, transposase. The cointegrate contains two identically oriented copies of γδ. (ii) Resolution of the cointegrate requires action of the *tnpR* gene product, resolvase, at an internal site, *res*.

Overview of cointegrate resolution

Cointegrate resolution is a site-specific recombination which is carried out by resolvase, product of the *tnpR* gene, acting at the duplicated *res* site which is located just upstream of *tnpR*. The biochemical requirements of the recombination are very simple. *In vitro*, a superhelical DNA molecule with two *res* sites in the same orientation is converted by purified resolvase protein into two singly catenated circles, each with a single *res* (Reed, 1981; Krasnow & Cozzarelli, 1983). Mg^{2+} is required for the efficient reaction. In the absence of Mg^{2+} an intermediate is formed in which the two resolvase-bound *res* sites are paired together in a potentially active form (Fig. 2). Treatment of this complex with proteinase yields linear forms of the normal circular products, in which each *res* is cleaved at the crossover point. The cleavages result in two-base 3' extensions (see Fig. 3B) with free 3' hydroxyls and with the recessed 5' ends covalently attached to the remains of the proteolysed resolvase by a phospho–serine linkage (Reed & Grindley, 1981; Reed & Moser, 1984). This covalent association of resolvase and *res*, together with the fact that recombination shows no requirement for a high energy nucleotide cofactor, indicates that resolvase acts like a topoisomerase, conserving the energy of the broken phosphodiester bond. Indeed, topoisomerase activity of resolvase can readily be demonstrated (Krasnow & Cozzarelli, 1983). The topoisomerase and Mg^{2+}-independent cleavage reactions only occur with a proper recombination substrate, that is, one with two correctly oriented

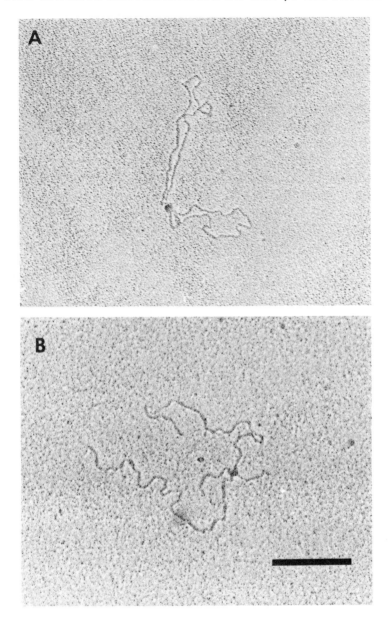

Fig. 2. Electron micrographs of paired resolvase–*res* complexes (Salvo, 1987). The cointegrate plasmid pRR51 (Reed, 1981) was incubated with resolvase in the absence of Mg^{2+}. Complexes were fixed by addition of glutaraldehyde and formaldehyde. The protein-free spreading procedure of Williams (1977) was used. (A) A molecule showing the typical segregation of the two domains and exclusion of interdomainal linkages. (B) After fixing, the DNA was digested with *Eco*RI before spreading. Measurement of the arms clearly indicates pairing of the two *res* sites. Bar marker $1\,\mu$m.

res sites, indicating that the recombinational potential requires assembly of a two *res*–site synaptic complex.

The coupled cutting of both DNA strands at the crossover sites of both *res* sites that is observed in the absence of Mg^{2+} suggests that recombination normally occurs by a mechanism that proceeds in three sequential steps: (i) the formation of double stranded breaks, (ii) the switching of recombinational partners (in a topologically constrained rotational process), and (iii) the resealing of the recombinant duplex products. Such a mechanism would contrast with that of the site-specific recombination associated with integration and excision of bacteriophage λ. In the case of λ, there is apparently an ordered and sequential nicking and joining of single strands, the first pair of cuts and joins forming a reciprocal exchange of single strands, that is, a Holliday structure (see Nunes–Düby *et al.*, 1987 and Nash *et al.*, 1987). From their demonstration of the type 1 topoisomerase activity of resolvase, Krasnow & Cozzarelli (1983) concluded that resolvase transiently interrupts only one strand during relaxation, apparently in conflict with a recombinational mechanism that involves concerted double strand breaks. There are three possible resolutions to this conflict. First, recombination occurs by concerted double-strand breaks but topoisomerase activity only involves nicking of a single strand. Secondly, both recombination and topoisomerase activity occur by a double-strand break process: in this case topoisomerase activity would involve breakage of both strands at a single *res*, a 360° rotation of one duplex end relative to the other, and eventual religation (such a process would change the topological linking number in steps of one, characteristic of a type 1 topoisomerase). Thirdly, both recombination and topoisomerase activities occur by separate single strand scissions with recombination involving a Holiday intermediate; in this case the double strand breaks seen in the absence of Mg^{2+} would suggest that the process has become uncoupled, the second cut occurring even though the strands cut first had not been rejoined.

THE *res* SITE

The DNA substrate for cointegrate resolution, the *res* site, is a complex region of 120 bp. In the case of γδ and Tn*3*, *res* lies between the divergently transcribed *tnpA* and *tnpR* genes of these transposons (Fig. 3) and in addition to its recombinational function, resolvase acts as a repressor of both genes. Within *res* are three binding sites

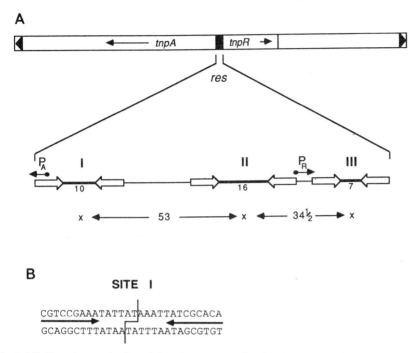

Fig. 3.(A) Genetic organisation of the γδ transposon showing the *res* site. The upper portion shows the entire γδ with *tnpA* and *tnpR* genes transcribed in the direction of the arrows. Below is a schematic of *res*. I, II and III indicate the three resolvase binding sites, each of which consists of inverted copies of a 9 bp imperfectly conserved sequence (open arrows) separated by a short spacer (heavy line). Numerals indicate the length (bp) of the spacers within each site and also the distances between the centres of adjacent sites. P_A and P_R label the start of the transcripts of the *tnpA* and *tnpR* genes respectively. (B) DNA sequence of site I of *res* showing the crossover site (marked ⌐⌐).

for the resolvase protein (Grindley *et al.*, 1982). The site farthest from the *tnpR* gene, site I, contains the recombinational crossover point, but the other two sites are also required for efficient recombination (Reed & Grindley, 1981; Wells & Grindley, 1984). The organisation of the three resolvase binding sites is unusual and irregular (see Fig. 3). Each site consists of inverted repeats of a conserved 9 bp segment (consensus sequence $TGTCYR_T^ATA$ where Y = pyrimidine, R = purine), but these 9 bp segments are separated by different-sized spacers within each site – 10 bp in site I, 16 bp in site II and 7 bp in site III. If each site is bound by a dimer of resolvase, one consequence of intra-site spacer variability is that the different geometries of each binding site can only be accommodated either if resolvase has a highly extended and flexible structure, or if the DNA within each site is distorted. In fact, we have evidence that both alternatives are true (see below). In addition to the different spacings within each site, the spacings between sites vary: 53 bp separate the centres of sites I and II, $34\frac{1}{2}$ bp separate the centres of II and III. These inter-site spacings are not arbitrary, however. Additions of an integral number of helical turns (10 bp, 21 bp) between sites I and II have little or no effect on cointegrate resolution, but additions of partial helical turns (for example, 6 bp or 17 bp) are highly inhibitory (see below). Similarly, alterations to the length of the II–III connecting segment have inhibitory effects. It is clear that an appropriate juxtaposition of the three binding sites is essential for the assembly of a resolvase–*res* complex that can subsequently participate in recombination.

THE RESOLVASE PROTEIN

Properties of wild-type resolvase

The $\gamma\delta$ resolvase is a 183-residue (21,000 mol. wt) polypeptide with two structural domains (Abdel-Meguid *et al.*, 1984). Mild proteolysis cleaves the protein into two fragments. The larger, consisting of the 140 N-terminal residues, contains the recombinational active site and is responsible for dimer formation; it has no detectable DNA binding activity. The smaller fragment, consisting of the 43 C-terminal residues, binds to *res* DNA, interacting specifically with each half of the three binding sites. Since each half-site is protected independently and with its own characteristic affinity by the small C-terminal domain, the uniform protection of each binding site afforded by the intact resolvase is clearly a result of a cooperative binding of two monomeric units to the two halves of a binding site. This cooperativity (and thus, resolvase dimerisation) is dependent on the large domain. The C-terminal domain contains a region with amino acid sequence homology to the helix–turn–helix structural element that is found in several other sequence-specific DNA-binding proteins (Pabo & Sauer, 1984). This structural element is known to provide the basis of protein–DNA recognition by allowing interactions within the major groove of DNA between amino acid side chains on the surface of the second α-helix and individual base pairs (Anderson *et al.*, 1987).

The two domains of resolvase may be connected by a flexible hinge. Both the intact resolvase and the large N-terminal domain can be crystallised in an isomorphous form (Abdel-Meguid *et al.*,

1984). Comparison of the diffraction patterns of these crystals shows that there are no obvious intensity differences. This similarity is surprising since the small domain constitutes almost one-quarter of the total molecule, and it suggests that the small domain is disordered relative to the large domain in the native crystals. Flexibility between the domains is one way by which the different geometries of the binding sites could be accommodated.

The $\gamma\delta$ resolvase is related to other resolvases and to DNA–invertases

The $\gamma\delta$ resolvase is just one of a large family of related site-specific recombination proteins. To date, the resolvase genes of some seven different transposons have been sequenced (Fig. 4). The corresponding resolvase proteins fall into four different complementing groups, yet all are clearly related with 13% of the amino acid residues being identical throughout the family. The resolvases are also closely related to a second group of site-specific recombination proteins, the DNA–invertases (Plasterk & van de Putte, 1984; van de Putte, this volume). These enzymes are responsible for inversion of several specific segments of DNA, resulting in alternative patterns of gene expression. All 23 of the amino acid residues conserved throughout the resolvase family are also found in the DNA invertases. Virtually all the wholly conserved residues lie within the N-terminal (catalytic) domain. The lack of homology between the C-terminal (DNA binding) domains is not surprising since different members of the family recognise different DNA sequences.

Analysis of resolvase mutants gives further insight into resolvase function

Mutants of the $\gamma\delta$ resolvase have been isolated and analysis of their properties throws some light on the functions of the protein (Newman & Grindley, 1984; Hatfull & Grindley, 1986; Hatfull *et al.*, 1987; G. Hatfull & N. Grindley, unpublished). We have obtained some 33 different amino acid substitutions at 28 different sites; the collection includes substitutions in 13 of the 23 wholly conserved residues (see Fig. 5). About 60% of the mutations that map in the N-terminal domain alter wholly conserved residues, even though these constitute only 15% of the domain. Assays *in vivo* enabled us to group mutants into two classes: those which have lost recombinational activity but retain repressor function (that is, still bind to

```
              10        20        30        40        50        60        70
              •         •         •         •         •         •         •
Tn 501   MQGHRI-GYVRVSSFDQNPERQ-----LEQTQ--VSKVFTDKASGK--DTQRPQLEALLSFVREGDTVVVHSMDRLA
Tn 21    MT.Q..-..I...T...........-----..GVK--.DRA.S.....--.VK.......I..A.T..........

Tn 3     MRIFGYARVSTSQQSLDIQIRALKDAGVK--ANRIFTDKASGS--STDREGLDLLRMKVEEGDVILVKKLDRLG
γδ       M.L......................V........--...........--.S..K..................

Tn 2501  MSRVFAYCRVSTLEQTTENQRREIEAAGFAIRPQRLIEEHISGSVAASERPGFIRLLDRMENGDVLIVTKLDRLG

Tn 917    MIFGYARVSTDDQNLSLQIDALTHYG----IDKLFQEKVTGA--KKDRPQLEEMINLLREGDSVVIYKLDRIS

Hin      MATI-GYIRVSTIDQNIDLQRNALTSAN----CDRIFEDRISGK--IANRPGLKRALKYVNKGDTLVVWKLDRLG
Gin      ML.-..V...ND..TD......VC.G----.EQ....KL..T--RTD.....RA.KRLQK..........
Pin      ML.-..V...ND..TD......NC.G----.EL....KI..T--KSE.....KL.RTLSA...V.........
Cin      ML.-..V...NE..TA......ES.G----.EL....KA..K--KAE.....KV.RMLSR.............
```

```
              80        90        100       110       120       130       140
              •         •         •         •         •         •         •
Tn 501   RNLDDLRRLVQKLTQRGVRIEFLKEGLVFTGEDSPMANLMLSVMGAFAEFERALIRERQREGITLAKQRGAYR
Tn 21    .......I..T......H...V..H.S.................................A........

Tn 3     RDTADMIQLIKEFDAQGVAVRFIDDGISTDGD---MGQMVVTILSAVAQAERRILERTNEGRQEAKLKGIKF
γδ       .................SI...........E---..K..............Q............MA..VV.

Tn 2501  RNAMDIRKTVEQLASSDIRVHCLALGGVDLTS--AAGRMTMQVISAVAEFERDLLLERTHSGIARAKATGKRF

Tn 917   RSTKHLIELSELFEELSVNFISIQDNVDTSTS---MGRFFFRVMASLAELERDIIERTNSGLKAARVRGKKG

Hin      RSVKNLVALISELHERGAHFHSLTDSIDTSS---AMGRFFFHVMSALAEMERELIVERTLAGLAAARAQGRLG
Gin      .MKH.IS.VG..RE..IN.R.........S---P.........G........I..M...AA..NK..I.
Pin      .MRH.VV.VE..RE..IN.R.........T---P.........G........V..K...ET..AQ..I.
Cin      .MRH.VV.VE..RD..IN.R.........T---P.........G........V..R...DA..AE..I.
```

```
              150       160       170       180
              •         •         •         •
Tn 501   GRKKALSDEQAATLRQRATAGEPKAQLAREFNISRETLYQYLRTDD
Tn 21    ....S..S.RI.E....VE...Q.TK.....G............Q

Tn 3     GRRRTVDRNVVLTLHQK---GTGATEIAHQLSIARSTVYKILEDERAS
γδ       ..K.KI..DA..NMW.Q---.L..SH.SKTMN........VINESN

Tn 2501  GRPSALNEEQQLTVIARINAGISISAIAREFNTTRQTILRVKAGQQSS

Tn 917   GRPSKGKLSIDLALKMYDSKEYSIRQILDASKLKNNLLPLPQ

Hin      GRPRAINKHEQEQISRLLEKGHPRQQLAIIFGIGVSTLYRYFPASSIKKRMN
Gin      ..PPKLTKAEWE.AG..LAQ.I..KQV.L.YDVAL....KKH..KRAHIENDDRIN
Pin      ..RPKLTPEQWA.AG..IAA.T..QKV.I.YDVGV....KRF..GDK
Cin      ..RPKYQEETWQ.MR..LEK.I..KQV.I.YDVAV....KKF..SSFQS
```

Fig. 4. Comparison of known resolvase and DNA-invertase protein sequences. The amino acid sequences are aligned to maximise homology; totally conserved residues are enclosed in stippled boxes. Numbering refers to the γδ residues. Chymotryptic cleavage of the γδ resolvase occurs at position 140, so that the bottom group of sequences correspond to the C-terminal DNA-binding domain. Adapted from Hatfull & Grindley (1987).

res DNA; shown above the map in Fig. 5), and those which have lost both activities (below the map in Fig. 5) (Hatfull & Grindley, 1987). Not surprisingly, mutations in the C-terminal domain (particularly within the helix–turn–helix structural element) eliminate or severely impair the ability of resolvase to bind *res*. Interestingly, binding to *res* is also eliminated by seven mutations that cluster in the N-terminal domain between residues 120 and 141 (plus an

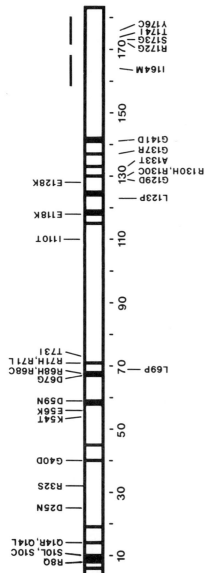

Fig. 5. Mutants of the γδ resolvase. The horizontal open bar represents the primary sequence of resolvase with the C-terminal domain shown shaded. Vertical bars within resolvase indicate totally conserved residues (see Fig. 4). Amino acid substitutions above resolvase result in loss of recombination activity but retention of DNA binding (i.e. repressor) activity, substitutions below cause loss of both activities. The position of the postulated helix–turn–helix DNA- binding super secondary structure is indicated by the two horizontal lines above resolvase. The nomenclature for mutants indicates the wild-type residue (single letter code), the position and the mutant residue; thus E118K is a change from glutamate to lysine at residue 118.

additional substitution at residue 69). It seems unlikely that these seven substitutions all prevent effective folding of the N-terminal domain (although the proline substitutions of residues 69 and 123 may well do so). This leaves three possible explanations for their phenotype. First, this region of resolvase may be involved in dimer formation. This seems unlikely because in a preliminary study some of these mutants exhibit partial dominance over wild-type suggesting that the corresponding heterodimers are formed and are inactive. Secondly, this portion of the N-terminal domain may make specific contacts with the resolvase binding sites. Comparison of the DNA contacts specified by the intact resolvase and by its C-terminal domain suggest that the N-terminal domain specifically contacts no more than a very small region of DNA (see below). Thus, it seems unlikely that all of the segment of resolvase from residues 120 to 141 is involved in such binding interactions, although a portion may be. The third possible explanation is that this region of resolvase may be involved in connecting the dimerisation and DNA binding domains in a configuration appropriate for recognising the different binding site geometries. This latter explanation is supported by the properties of the lysine substitution at residue 128. The E128K resolvase shows an abnormal sensitivity to site geometry; it fails to bind to site III (with its 7 bp spacer), but binds normally to a site III analogue in which the spacer was increased to 10 bp, as well as to sites I and II (Hatfull *et al.*, 1987 and unpublished observations).

It is likely that residue 10 of the $\gamma\delta$ resolvase is the active site serine that is linked transiently to the crossover point during recombination. Only two serines are well conserved among the family of resolvase and DNA–invertase proteins: serine-10 and serine-39 (which is a theonine in the Tn917 resolvase). Substitutions of serine-10 with either the iso-structural residue, cysteine, or the more bulky and hydrophobic leucine both destroy recombination activity (Hatfull & Grindley, 1986). Although the S10C resolvase binds site I just like the wild-type resolvase, the S10L protein has reduced affinity specifically for site I, suggesting that the bulky leucine side chain prevents close contact between the crossover point and the active site region of the mutant. Further evidence against the active involvement of the other conserved serine comes from the observation that substitution of serine-39 with cysteine has no effect on recombination activity (V. Rimphanitchayakit, G. Hatfull & N. Grindley, unpublished results).

A number of additional assays for partial activities of the mutant resolvase have been developed, or are in the process of being developed. Together with forthcoming structural information from crystallographic studies, it is hoped that they will elucidate further the functions of resolvase.

RESOLVASE–*res* INTERACTIONS

Contacts between res and the C-terminal DNA binding domain

Contacts have previously been mapped between resolvase and the *res* site by determining which DNA modifications (e.g. ethylation of phosphates, methylation of purines; see Siebenlist & Gilbert, 1980) interfered with complex formation (Falvey & Grindley, 1987). These studies confirmed that resolvase interacted specifically with the conserved 9 bp (consensus-related) segments that bracket each binding site, regardless of the different spacings between these segments. Additional contacts towards the inner portions of each site were also evident, the entire contacted segment consisting of about 14 bp per half binding site. Resolvase spanned the major groove across the outer portion of this segment (the first 7 bp of the consensus sequence) and the minor groove over the inner portion. Methylation of the N-7 position of guanines at several different positions within the outer portion of the contacted segment strongly inhibited resolvase binding, indicating major groove recognition of this region (as was expected for a helix–turn–helix recognition element).

The large size of the region (within each half of a binding site) contacted by resolvase, together with the small size of the C-terminal DNA binding domain, and the possibility that the C-terminal portion of the N-terminal domain might be directly involved in resolvase–DNA interactions (see above), all prompted us to investigate the contacts specified exclusively by the C-terminal domain. Complexes between the C-terminal domain and resolvase binding sites can be separated from uncomplexed DNA by electrophoresis on high-percentage native polyacrylamide gels. We have used this separation procedure to map the inhibitory DNA modifications (V. Rimphanit-chayakit & N. Grindley, unpublished observations). A summary of the results is shown schematically in Fig. 6. Our previous data had shown that ethylation of seven phosphates within a half-site inhibited binding of the intact resolvase. Six of these seven ethylations also inhibited binding of the C-terminal domain, indicating

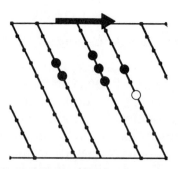

Fig. 6. Contacts between the C-terminal domain of resolvase and one half of a binding site. The upper portion shows the consensus sequence for half of a binding site (with the centre of the site to the right). The length of the segment is that necessary for maximal binding of the C-terminal domain. Heavy arrows indicate the positions of phosphates which, when ethylated, interfere with binding of both intact resolvase and the small domain. The small arrow indicates the phosphate which, when ethylated, inhibits binding only of the intact resolvase. The lower portion shows the contact points on a planar projection of the DNA helix. The open circle shows the phosphate contact that is exclusive to the intact resolvase, all other phosphate contacts are represented by solid circles. The horizontal arrow indicates the position of the 9 bp segment of conserved sequence.

that even the 43 amino acid C-terminal domain spans an adjacent major and minor groove of the DNA (about 12 bp) and covers about one-third of the circumference around the helix. The single phosphate contacted exclusively by the intact resolvase lies towards the centre of each binding site (see Fig. 6). Moreover, our data indicate that this phosphate is not an important contact at all sites (in particular, ethylation of this phosphate at site II–R does not inhibit formation of the resolvase–site II complex).

Using a complementary procedure we have shown that all the DNA contacts required for binding of the C-terminal domain are contained within a specific 15 bp DNA segment. By primer extension reactions in the presence of dideoxynucleotides, a set of DNA substrates was prepared in which synthesis of the duplex was randomly terminated at all possible positions throughout the protein binding site. Following separation of complexed from uncomplexed DNA, the extent of DNA required for complex formation was determined simply by measuring the minimum length of the extended primer on a DNA sequencing gel (Gronostajski *et al.*, 1985). The data, obtained with resolvase binding site I, show that the segment of DNA required for efficient binding of the small domain of resolvase includes the 9 bp conserved (consensus-related) segment, plus 2 bp outside it (relative to the centre of each complete binding site), and 4 bp inside it (see Fig. 6); absence of one base pair at either end reduces but does not eliminate binding. In agreement with the ethylation interference data, the 15 bp segment does not include the phosphate that is contacted exclusively by intact resolvase. From these and other studies we infer that most of the contacts required for binding of intact resolvase are made by the C-terminal domain. In particular we have identified just one phosphate contact that is made only by the intact protein, and even in this case the N-terminal domain may be acting indirectly on the positioning or folding of the C-terminal domain rather than directly contacting the phosphate.

Resolvase-induced bending of the res *site*

By electron microscopy the resolvase–*res* complex appears to be a compact structure in which the DNA is highly condensed: 120 bp (i.e. 40 nm) of DNA is contained in an approximately spherical (or cylindrical) structure of about 15 nm diameter (Salvo, 1987). To obtain this degree of compaction the DNA must be tightly wrapped. Indeed, analysis of the topological effects of a single *res*–resolvase complex indicates that between half and three-quarters of a negative superhelical turn is buried in the structure (H. Benjamin & N. Cozzarelli, personal communication). We have shown that this wrapping is a product of (i) bending of the DNA at each individual binding site, (ii) phasing of these bends in an appropriate fashion, which in turn allows (iii) bending of the DNA between sites I and II through protein–protein interactions that are dependent upon the presence of site III. The complex nucleoprotein structure formed by the interaction of resolvase with *res* may play an important role in the remarkable precision of the recombination process that is seen in the site selection, the directional specificity and the topological invariance of the reaction (see Echols, 1986).

A characteristic of bent DNA is that it migrates anomalously during electrophoresis in polyacrylamide gels; a DNA fragment with a bend close to its centre migrates more slowly than the same length of fragment with a terminal bend or with no bend at all. This

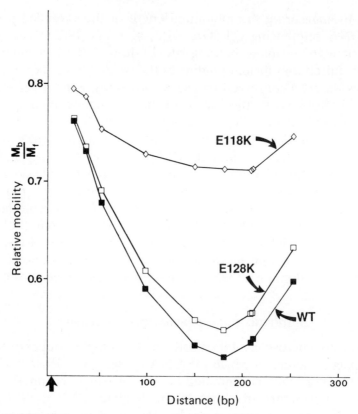

Fig. 7. Graph illustrating resolvase-dependent bending of binding site I. Complexes between resolvase (=WT) or two mutants of resolvase (E118K and E128K) and a series of 320 bp fragments of DNA containing site I were analysed by polyacrylamide gel electrophoresis. The relative mobility is the mobility of the bound DNA fragments (M_b) divided by the mobility of the free DNA fragment (M_f). The fragments are a circularly permuted set which differ from one another only in the location of site I relative to the fragment ends. The distance shown on the abscissa is that between the centre of site I (↑) through site I–R (the half normally proximal to site II) to the end of each fragment.

behaviour is typical both of DNA with intrinsic, sequence-induced bends and of protein–DNA complexes in which bending of the DNA is induced (or fixed) by the protein interactions. The dependency of retardation upon bend position can be used to map the centre of bending. The data in Fig. 7 show the position-dependence of the mobility of resolvase complexes with DNA containing site I (G. Hatfull, unpublished results). The DNA fragments are a 320 bp circularly permuted set in which the only variation is the position of the end of the fragment (and thus the location of site I relative to the ends). Similar data have been obtained with sites II and III separately, and with *res* (Salvo, 1987). That this behaviour is caused by a DNA structural alteration and not simply by varying the position of a blob of basic protein was demonstrated by the resolvase mutant E118K. This mutant when bound at site I exhibits only a modest position-dependent effect on electrophoretic mobility (see Fig. 7) even though another mutant, E128K, with exactly the same charge change, shows the same position dependency as the wild-type resolvase (Hatfull *et al.*, 1987). Perhaps more significantly, the E118K resolvase does bend site II normally, its complexes with site II exhibiting the same position dependency as those of the wild-type or E128K resolvases. Thus, the abnormal mobility of E118K–site I complexes is site-specific and can be explained easily only by a site-specific alteration in the DNA structure.

Correct phasing between the resolvase binding sites (and, therefore, between the resolvase-induced bends) is essential for assembly of an active resolvase—*res* complex. The normal distance between the centres of sites I and II is 53 bp, five helical turns of B DNA. We have tested the effects of varying this distance both on recombination and on the formation of complexes with resolvase (see Fig. 8A and B). The only alterations in the separation between sites I and II that allowed efficient recombination were insertions that approximated to integral numbers of helical turns. When we examined the complexes formed between resolvase and these altered *res* sites by polyacrylamide gel electrophoresis, we found that the final form of the wild-type complex, and the pattern of filling the binding site were only retained with the insertions of 10 bp and 21 bp (one and two helical turns of DNA). The relatively fast migration of the final (fully occupied) *res* complex (and of those with the +10 and +21 insertions) suggested that it is a compact structure that can be formed only when sites I and II are appropriately phased, with the centres of these two sites on the same face of the DNA helix, even though the linear separation between the sites is relatively unimportant.

Looping of the DNA between sites I and II

The requirement for an integral number of helical turns between the centres of site I and II might reflect not only a requirement for phasing the individually bent sites, but also a requirement for interaction between resolvase dimers on neighbouring sites. Because variations in the linear separation are acceptable, essential neigh-

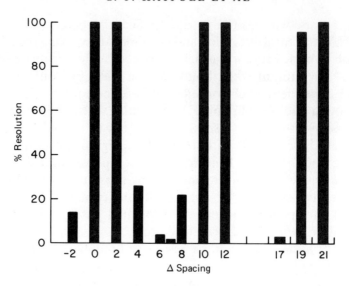

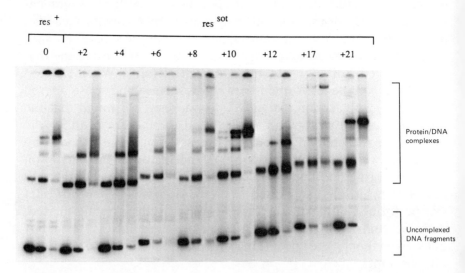

Fig. 8.(A) The effect on cointegrate resolution of changing the distance between site I and II. The bar graph shows the efficiency of resolution (using a genetic assay *in vivo*) of a set of cointegrates in which one *res* site contains small alterations to the length of the spacer between sites I and II as shown on the abscissa. (B) The effect on resolvase–*res* complex formation of changing the distance between sites I and II. DNA fragments of 123–144 bp containing the entire *res* sequence (from 4 bp to the left of site I to 5 bp to the right of site III) were incubated with three different concentrations of resolvase (10 nM, 50 nM, and 250 nM) and the complexes analysed by polyacrylamide gel electrophoresis. The numbers above each group of three lanes indicate the length of extra DNA inserted between sites I and II.

bouring interactions must be able to accommodate the extra lengths of DNA; one way to do this is to bend or loop the intervening DNA as has been shown in other systems (Hoschild & Ptashne, 1986; Kramer et al., 1987).

A characteristic of a DNA loop is that the portion of the phosphodiester backbone exposed on the outside of the loop is more easily cleaved by DNase I whereas the inside portion is resistant to cleavage (Drew & Travers, 1985). As can be seen in Fig. 9A, the DNA between sites I and II in a res–resolvase complex exhibited an alternating pattern of enhanced and suppressed cleavages by DNase I that is typical of a loop. This pattern was most clearly seen with the 21 bp and the 10 bp insertions. In res^+ only a single short stretch of enhanced cleavages was seen and in the 17 bp insertion which puts sites I and II out of phase, no enhancements or suppressions were seen within the inter-site segment. Not only did the pattern of alternating enhancements depend on correct phasing of sites I and II, it also required the presence of site III (see Fig. 9B). Thus the looping of the inter-site DNA is not simply a result of interaction of the site I complex with the adjacent site II complex. The requirement for site III could indicate either (i) that the interaction is between resolvase units at sites I and III (perhaps requiring the resolvase-induced bend at site II to allow the interaction), or (ii) that the resolvase at site I interacts with the resolvases at both of sites II and III, or (iii) that resolvase at site III modifies the site II complex (or vice versa) so that it can interact with the site I complex, or (iv) that the I–II interaction is stabilised by the site III complex. It will be of interest to see whether any of our resolvase mutants are defective in the inter-site bending. Identification of such mutants would indicate the portion of the polypeptide responsible for these protein–protein interactions and might indicate whether these interactions are essential for recombination.

The DNase I footprinting results indicate the plane of the bend between sites I and II. The face of the helix which contains the crossover site lies on the *inside* of the bend (Fig. 9C). From DNase I suceptibilities across site II (see also Grindley et al., 1982) it appears that the bend within site II is in the same plane and direction. This puts the minor groove at the centre of site II on the inside of the bend. This was not unexpected since the T6 tract that runs through the centre of site II would be expected to assist a bend into the minor groove (Drew & Travers, 1985; Satchwell et al., 1986; Burkhoff et al., 1987; Zinkel & Crothers, 1987; Salvo & Grindley, 1987).

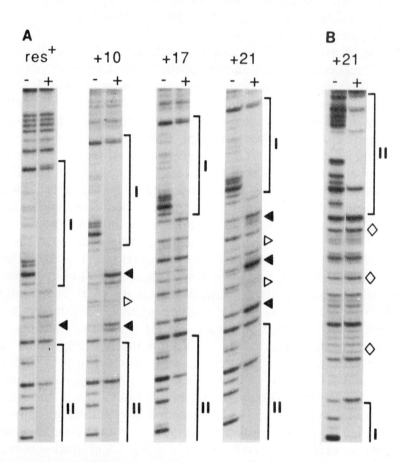

Fig. 9. Bending of the DNA between sites I and II. (A) Influence of the I–II spacing. Shown are DNase I footprints of resolvase complexed with *res*⁺ DNA or with *res* containing insertions of 10 bp, 17 bp or 21 bp between sites I and II. Alternating enhancements (◄) and suppressions (▷) of cleavage are seen with the +10 and +21 *res* derivatives but not with +17. All the fragments contained all three binding sites. (B) The inter-site bend requires site III. A *res* derivative with an insertion of 21 bp between sites I and II, and with a deletion of all of site III was constructed. A DNase I footprint of this site shows none of the enhanced and suppressed cleavages that are seen when site III is present. '◇' indicates the regions at which enhanced cleavage would have occurred if bending was independent of site III. All data are for the 'bottom' strand of DNA (as the *res* sequence is conventionally shown) but the order of the sites in B is reversed. (C) Direction of the inter-site bend. A planar projection of the DNA helix is shown, covering the region between the centres of sites I and II (shown by the diamonds) in a *res* site with a 21 bp insertion between the sites. I–R and II–L indicate the right half of site I and the left half of site II. The resolvase–DNA phosphate contacts are shown as in Fig. 6, and the crossover points in site I are indicated by the two arrow heads. The shaded sections of the sugar phosphate backbone are those with increased sensitivity to DNase I in a resolvase–*res* complex. These sections, therefore, lie on the outside of the inter-site bend.

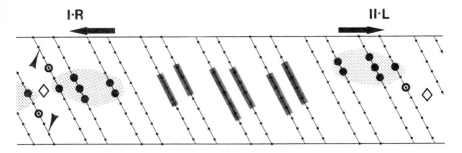

Fig. 9(C). (See opposite page for legend.)

In a similar manner, an A5 tract just to the right of site I–R also appears to help bend the inter-site segment. By contrast, as we will show below, the site I bend actually opposes the site II and inter-site bends.

Resolvase induces an unusual DNA structure at the crossover point

When the intercalating agent MPE.Fe(II) (Methidium propyl EDTA complexed with Fe(II)) is used to footprint resolvase–*res* complexes, a region of enhanced cleavage is observed at the centre of site I (Fig. 10A; Hatfull *et al.*, 1987). The enhancements are specific to site I and are not seen at either site II or site III. Three observations indicated that the enhanced MPE cleavages are a result of enhanced intercalation that in turn results from a localised distortion of the DNA structure. First, footprinting with EDTA.Fe(II) which does not intercalate showed no corresponding enhancement. Secondly, the resolvase mutant E118K, which only imparts a modest bend to site I (see Fig. 7), gave a good footprint on site I but no enhanced cleavages at its centre (Fig. 10A); thus there is a correlation between the site I bend and the MPE intercalation. Thirdly, all the enhanced cleavages could be accounted for by a single site of intercalation at the dyad axis of site I (Fig. 10B), suggesting that the two central base pairs between which the MPE intercalates are partially separated and unstacked. Since MPE intercalates into the minor groove of DNA, it was concluded that bending of site I causes (or results from) a kink which opens up the minor groove at the centre of the site. This predicts that at the centre of site I, it is the minor groove that lies on the outside of the bend (as opposed to the major groove at site II); we have shown by an independent experiment that this is indeed the case.

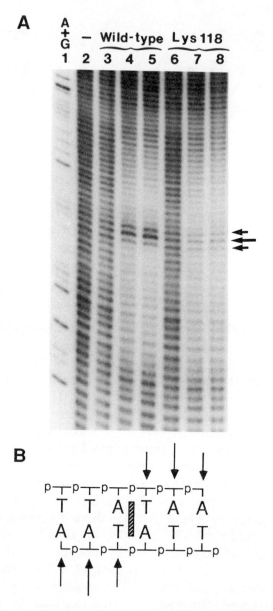

Fig. 10.(A) MPE.Fe(II) footprinting of wild-type resolvase and the E118K mutant resolvase (Lys118) at site I. The reaction in lane 2 contained no resolvase; reactions in lanes 3–5 contained 12 nM, 120 nM and 1200 nM wild-type resolvase respectively, reactions in lanes 6–8 contained 167 nM, 500 nM and 1670 nM E118K resolvase respectively. For more details see Hatfull *et al.* (1987). Arrows indicate the enhanced MPE.FeII cleavages at the centre of site I. (B) Sites of the MPE.Fe(II) cleavages within site I and the deduced position of MPE.Fe(II) intercalation. The sequence is that of the central six base pairs of site I (see Fig. 3B). The three enhanced cleavages of each strand can all be accounted for by a single site of intercalation (vertical bar) at the dyad axis of site I (see Hatfull *et al.*, 1987).

Direction of the site I bend

The effect on gel mobility of varying the length of a DNA segment that separates the resolvase-induced bend from an intrinsic bend in DNA has been analysed (Salvo & Grindley, 1988). The rationale of the experiment is that two independent bends in DNA separated by a short and variable spacer should affect electrophoretic mobility in a predictable fashion that is a function of the distance between the two centres of bending. Mobilities should be greatest whenever the bends oppose one another, and slowest when the bends reinforce one another. Because of the helical nature of DNA, mobility minima (and maxima) should occur once for every addition of 10.5 bp to the inter-bend spacer. If the direction of one bend is known, the direction of the other can be deduced from the separations at which mobility minima occur. Recently, Zinkel & Crothers (1987) have independently used the same approach to compare the directions of the bend of K–DNA (see below) and the bend induced by the *E. coli* cyclic AMP receptor protein, CAP.

For an intrinisc protein-independent bend we used a particular segment of kinetoplast DNA (K–DNA) from *Leishmania tarentolae*. This segment has been studied in detail and is bent both in solution and during polyacrylamide gel electrophoresis (Marini *et al.*, 1982; Levene & Crothers, 1983; Wu & Crothers, 1984; Hagerman, 1985); it derives its bend from a series of A tracts (A4 to A6) whose centres are separated by one turn of the DNA helix. In the constructs the distance between the centre of site I and the centre of the K segment was increased in small increments from 50 bp to 78 bp (see Fig. 11A). As is shown in Fig. 11B, the mobility of resolvase–DNA complexes varied in a cyclical manner with minima occurring every 10–12 bp. The minima occurred when the centre of site I was 63 bp and 73 bp from the centre of the K segment. Thus the bend of the K segment and the resolvase-induced site I bend reinforced each other when separated by six and seven (i.e. an integral number of) helical turns assuming 10.5 bp per turn.

Determination of the absolute direction of the site I bend depends upon knowing the direction of the K–DNA bend. From an analysis of the pattern of hydroxyl radical cleavage of A tracts, Burkhoff & Tullius (1987) concluded there was a progressive narrowing of the minor groove, consistent with a bend into the minor groove. In an experiment similar to ours and using the same K–DNA segment, Zinkel & Crothers (1987) demonstrated a similar helical phasing

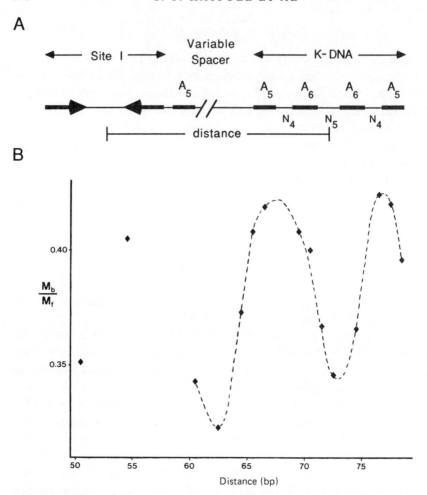

Fig. 11. Helical phasing of the resolvase–site I bend and a K–DNA bend. (A) Schematic showing the linkage of site I to a segment of K–DNA through a variable spacer. Note the alternating A tracts that constitute the K-DNA segment and one A tract just to the right of site I. (B) Relative mobility of the resolvase–DNA complexes plotted as a function of the inter-bend distance. DNA fragments of the general structure shown in Fig. 11A (but with 180 bp to the left of site I and 105 bp to the right of the K segment) were labelled, mixed with resolvase and run on an 8% polyacrylamide gel. The mobility of each resolvase-bound fragment (M_b) was divided by the mobility of the corresponding free fragment (M_f) to give the relative mobilities (data from Salvo & Grindley, 1987).

with a CAP-dependent bend and concluded that the centre of the K–DNA bend must be bent towards the major groove. Since the centre of the K segment (in their and in our experiments) was composed of four phased A tracts, its centre lay precisely between two of them. These authors therefore concluded that each A tract was bent into the minor groove. This conclusion is also predicted by several theoretical models for the A tract bend (Wu & Crothers, 1984; Koo et al., 1986; Trifonov & Sussman, 1980; Ulanovsky & Trifonov, 1987). If the effective bend of the K segment can be considered as a bend at its centre, towards the major groove, then our data indicate that the site I bend is also into the major groove. This conclusion is wholly in support of the results of MPE intercalation.

Role of the central dinucleotide of site I

The results of MPE.Fe(II) intercalation suggested there was a localised distortion of the structure of the two central base pairs of site I. These two base pairs also comprise the extent of the heteroduplex (or overlap) region (i.e. one DNA strand from each of the parental res sites) formed during recombination. Of the six base pairs across the recombination site, only the central dinucleotide is well conserved among the six sequenced res sites (ApT in five sites, GpT in Tn2501). This conservation suggested that the sequence of the central dinucleotide might play an important role in the overall recombination process. It has been found that the sequence of the central dinucleotide of site I has a profound effect on the ability of resolvase to bind to the site, to induce the structural distortion, and to perform recombination (G. Hatfull & E. Falvey, unpublished results).

Homology of the overlap region between two res sites of a cointegrate is not sufficient for recombination. During a search for mutants of site I, one was found in which the central dinucleotide was changed from ApT to CpT (E. Falvey, unpublished results). The base substitution had no effect on the binding of resolvase to site I or to res. However, the mutant site was inactive in recombination in vivo and in vitro, not only in the heterologous cross with res^+, but also in a homologous cross in a cointegrate with two copies of the mutant res site. Although no recombination in vitro was detected, changes were seen in the substrate DNA. With the [mutant $\times res^+$] cointegrate substantial relaxation of the substrate was seen with the formation of both nicked circles and topoisomers of reduced negative superhelicity. In the absence of Mg^{2+} there was formation of both linear and nicked substrate molecules, but no molecules which had been cleaved at both res sites were detected. The linearisation was a result of double-strand cleavage exclusively at the wild-type *res* site. Since a molecule with only a single *res* site is neither relaxed (in the presence of Mg^{2+}) nor cleaved (in the absence of Mg^{2+}) (Krasnow & Cozzarelli, 1983; Reed & Grindley, 1981), these results indicate that formation of an intermediate with two correctly paired *res* sites occurred as normal, but subsequent cleavage and ligation events involving the mutant *res* were impeded. With the cointegrate with two copies of the mutant *res* site, both nicking of the substrate and topoisomerase activity were seen; linearisation of the substrate was not detected. With both cointegrate substrates it has been shown that nicking occurred specifically at the mutant site I and only on the 'top' strand (at the sequence 5'-TTCT ↓ AA). Thus nicking only occurred 3' to the wild-type nucleotide (pT) and not 3' to the mutant (pG).

The accumulation of substrate molecules with a strand-specific nick at the mutant *res* site can be explained in two ways. Either (i) the top strand alone can be cleaved, or (ii) both strands can be cleaved but only the bottom strand can be re-ligated. The second hypothesis is unlikely since double strand cleavage of the mutant *res* is not seen in the absence of Mg^{2+} (conditions which inhibit all ligation steps). If the first explanation is correct then it seems that exchange of single strands in a synaptic complex does not occur (or is very inefficient or is rapidly reversed) since essentially all the substrate is converted to partially relaxed or nicked molecules. (In making this argument it is assumed that a cointegrate with a Holliday structure connecting the two *res* sites would migrate in a different position from a nicked circle.) This would imply that the strand exchanges and rotations necessary for recombination are dependent upon the formation of double-strand cuts. The topoisomerase activity seen with the doubly mutant cointegrate suggests that resolvase-catalysed relaxation can occur by a nicking-reclosing mechanism.

Mutations in the central dinucleotide affect binding of resolvase
Stimulated by the results described above, site I derivatives have been made with all 15 possible base changes in the central dinucleotide (ApT in *res*+) (G. Hatfull, unpublished results). Many of the mutations significantly reduce (30- to 100-fold) the binding of resolvase to site I when assayed *in vitro* (see Fig. 12). In addition, all effects are symmetric, that is, the binding efficiency observed with a particular central dinucleotide is always the same for its comple-

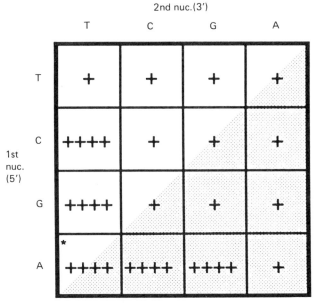

Fig. 12. Effect of the sequence of the central dinucleotide of site I on binding of resolvase. ++++ indicates normal affinity, + indicates reduced affinity (by 30-fold or more) as determined by a polyacrylamide gel electrophoresis assay for formation of resolvase–site I complexes (Hatfull & Grindley, 1986).

ment (e.g. GpT = ApC) suggesting that resolvase sees site I in a symmetric fashion. There appear to be two rules that govern the efficiency of binding. First, at least one of the 5' positions must be an A (in other words, one of the two base pairs must be wild-type). Secondly, neither of the 5' positions should be T. Context effects are dominant – the data suggest that the requirement is for an appropriate dinucleotide rather than specific base pairs at each position. A telling example of this is that the dinucleotide GpT (and its complement ApC) allows efficient complex formation; however, the 5' G and 3' C do not work in the context of a GpC dinucleotide. It is difficult to interpret these results in terms of specific resolvase–base pair contacts. In support of this, it is found (as shown in Fig. 13) that depurination of any of the central four base pairs of site I has no effect on complex formation (even though there is a contact to the phosphate immediately 5' to the central dinucleotide; Falvey & Grindley, 1987). The simplest interpretation of the data is that only a subset of dinucleotides can readily be 'kinked' into the structure that is normally observed at the crossover point in resolvase-site I complexes. One possible explanation for the inhibitory effect of

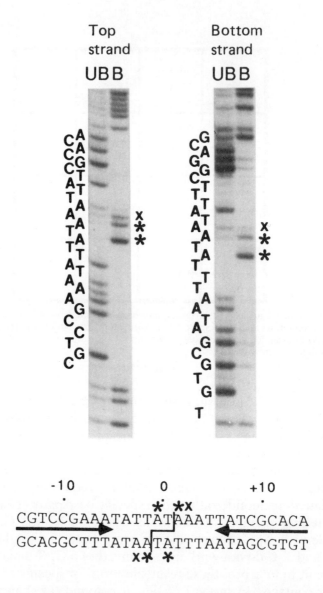

Fig. 13. Loss of purines at the centre of site I does not inhibit binding of resolvase. Resolvase complexes were formed with partially depurinated end-labelled DNA fragments that contained site I. Protein-bound and free DNA were separated by gel electrophoresis (Hatfull & Grindley, 1986), the DNAs were cleaved at the positions of depurination by heating in piperidine (Maxam & Gilbert, 1980) and were run on sequencing gels. Bands absent in the bound DNA (lanes marked B) but present in the free DNA (marked UB) correspond to purines that inhibit resolvase–site I complex formation. The marked bands at the centre of site I fail to inhibit (*) or only weakly inhibit (X) resolvase binding; these positions are indicated on the site I sequence below the autoradiograms (V. Rimphanitchayakit & N. Grindley, unpublished results).

a 5′ T is that the methyl group at the 5′ position of the pyrimidine ring (which lies in the major groove) may sterically interfere with the compression of the major groove which we propose occurs during kink formation. If this is the case, then 5′-dU may be acceptable in some dinucleotide contexts. Alternatively, introduction of a T at either 5′ position creates a run of Ts (from T3 to T5 depending on the 3′ nucleotide) which will tend to cause an intrinsic bend and as a consequence may inhibit resolvase binding or formation of the proposed kink.

To determine whether the structural deformation seen with the wild-type site is affected by the sequence of the central dinucleotide, we have examined the patterns of resolvase-dependent suppression and enhancement of MPE.Fe(II) cleavage. The general result is that there is a much increased cleavage at the centre of site I with all those sequences that reduce resolvase binding. This suggests that weakly bound sequences are deformed in a different way from the strongly bound ones. However, an alternative explanation is that weak complexes provide a greater opportunity for MPE.Fe(II) inter- calation through the association and/or dissociation of complexes during the course of the experiment (for example, the small MPE molecule may have improved access to the intercalation site in a binding intermediate, or may shift from a partially intercalated to a fully intercalated state upon resolvase dissociation).

Strong binding to site I is not sufficient for recombination. The first mutant studied, with the CpT central dinucleotide, bound resol- vase with an affinity indistinguishable from the wild-type site I and yet failed to recombine even with a homologous site I. This mutant site also showed levels of resolvase-enhanced MPE.Fe(II) cleavage similar to the wild-type site I, suggesting the formation of an approxi- mately normal structural deformation. Two possible explanations for the behaviour of this mutant are: (i) the resolvase site I complex, though structurally similar to that of the wild-type site I, is subtly different, and (ii) although specific resolvase–base pair contacts at the crossover site are not required for binding, they are required for recombination. The latter explanation is favoured because the top strand of the mutant site, which is cleaved within the wild-type dinu- cleotide 5′-TpA, is efficiently nicked (suggesting that all the structural requirements are satisfied, at least for the top strand scission) but the mutant 5′-GpA dinucleotide on the bottom strand is not cleaved.

It is notable that in the recombination sites of the related DNA- invertases, the sequence of the central dinucleotide does not appear to be similarly constrained (although there is a requirement for homology between the two recombining sites). Natural sites have ApA or GpA, both with the 3'A 'forbidden' to resolvase (see Plasterk & van de Putte, 1984), and ApT has been shown to be perfectly acceptable (Johnson & Simon, 1985). In addition secondary sites with GpT and TpT at their centres have been identified (although the effects of these dinucleotides on recombination frequency have not been investigated) (Iida & Hiestand-Nauer, 1987). There is also no evidence for a structural distortion of the DNA at the crossover point of the DNA inversion systems. MPE.Fe(II) footprinting of the complex between the *Salmonella* Hin protein (responsible for flagellar phase variation) and its recombination sites shows no sign of enhanced cleavages around the crossover point (Bruist *et al.*, 1984).

CONCLUSIONS AND IMPLICATIONS FOR RECOMBINATION

Analysis of the resolvase–*res* complex suggests that it is a compact and highly ordered structure. Condensation of the DNA occurs by bending it both within each resolvase binding site and between binding sites. It has been shown that the path of the DNA between sites I and II is curved, in a manner that is dependent upon the presence of site III. This implies a direct protein–protein connection between the resolvase molecules on site I and those on site III (and/or site II) with the intervening DNA forming a closed loop. The curve between sites I and II appears to follow the same path as the site II bend. Benjamin & Cozzarelli (personal communication) have measured the topological effect of forming a single *res*–resolvase complex and find the complex contains effectively about one half of a negative superhelical turn (this could be in the form of untwisting the DNA or wrapping the DNA). Taken together the results suggest that *res* DNA is wrapped around a resolvase protein core with the ends of the DNA emerging in approximately the same direction. Since the site I bend (or kink) appears to oppose the site II bend and the curve of the DNA between sites I and II, the portion of *res* excluding site I must be considerably wrapped. Looking down on the complex, the path of the DNA must follow a left-handed screw in order to account for the negative sign of the *res*-resolvase superhelicity.

An attractive model for the synaptic complex of two *res* sites in a cointegrate has been proposed by Boocock *et al.* (1986, 1987) which accounts for the interdomainal topology of the substrate that was deduced by Cozzarelli & coworkers (Krasnow *et al.*, 1983; Wasserman & Cozzarelli, 1985; Wasserman *et al.*, 1985). In their model Boocock *et al.*, proposed that the two *res* sites of an active complex are brought together with sites II and III aligned antiparallel and wrapped plectonemically around one another in a right-handed helix to entrap three negative interdomainal nodes. Each *res* site, when viewed along the helical axis of the interwraps, follows a right-handed screw path. If this model is correct, two changes must be made to the individual *res*–resolvase complexes when two of these sites pair to form a recombinationally active complex. First the path of each *res* must change from a left-handed to a right-handed screw (note that negative superhelicity is retained in both cases since a negative superhelical turn can adopt either a left-handed solenoidal wrap or a right-handed plectonemic wrap). Secondly, the protein–protein interactions of a single *res* complex must be broken (presumably to re-form as inter-*res* interactions) to allow the passage of DNA strands that is required for the interwrapping of the two *res* sites.

Within a *res*–resolvase complex the DNA at the crossover site is structurally altered in a manner that is presumably essential for recombination. Consistent with this, it has been shown that the recombination deficiency of the E118K mutant resolvase is a result of its inability to form a productive complex with site I (Hatfull *et al.*, 1987). The structural transformation at the crossover site is a product of the resolvase site I interaction; it does not require the presence of sites II or III, nor does it require superhelicity. However, it appears to be sensitive to the sequence of the central dinucleotide. The accessibility of the central minor groove to an intercalating agent implies that it is exposed to solvent. It will be of interest to determine whether this accessibility is retained in a synaptic complex, since loss of accessibility would imply that the two crossover sites are brought into close proximity and might explain how the recombinational potential of the *res*–resolvase complex (as determined by non-recombinational side reactions such as topoisomerase activity and *res*–site cleavage) remains dependent upon *res*–*res* pairing.

Most of the available evidence suggests that resolvase carries out recombination by making concerted double-strand breaks at each *res* site. The cutting that is observed in the absence of Mg^{2+} of both crossover points of a wild-type cointegrate, and of the single *res*$^+$

in the (mutant × res^+) cointegrate supports this view. As argued above, the nicking of the mutant *res* sites in the doubly mutant cointegrate, together with our failure to detect substantial amounts of a Holliday intermediate, also suggests that single strand exchanges do not occur. The type 1 activity of resolvase as a topoisomerase is not incompatible with a double-strand break mechanism, although we do not know whether the observed activity occurs by single- or double-strand breaks. It is of considerable interest that the total change in DNA linking number of +4 associated with resolution (Boocock *et al.*, 1987), together with the other topological parameters of the reaction (see Wasserman & Cozzarelli, 1985; Wasserman *et al.*, 1985), is totally consistent with a mechanism that involves a double strand break at each *res* followed by a single right-hand rotation of the protein-linked ends to bring them into recombinant alignment.

ACKNOWLEDGEMENTS

The research was supported by grant GM28470 from the US National Institutes of Health.

REFERENCES

ABDEL-MEGUID, S. S., GRINDLEY, N. D. F., TEMPLETON, N. S. & STEITZ, T. A. (1984). Cleavage of the site-specific recombination protein, $\gamma\delta$ resolvase; the smaller of the two fragments binds DNA specifically. *Proceedings of the National Academy of Sciences, USA*, **81**, 2001–5.

ANDERSON, J. E., PTASHNE, M. & HARRISON, S. C. (1987). Structure of the repressor–operator complex of bacteriophage 434. *Nature*, **326**, 846–52.

BENJAMIN, H. W., MATZUK, M. M., KRASNOW, M. A. & COZZARELLI, N. R. (1985). Recombination site selection by Tn3 resolvase: topological tests of a tracking mechanism. *Cell*, **40**, 147–58.

BOOCOCK, M. R., BROWN, J. L. & SHERRATT, D. J. (1986). Structural and catalytic properties of specific complexes between Tn3 resolvase and the recombination site *res*. *Biochemical Society Transactions*, **14**, 214–16.

BOOCOCK, M. R., BROWN, J. L. & SHERRATT, D. J. (1987). Topological specificity in Tn3 resolvase catalysis. In *DNA Replication and Recombination*, eds. Kelley, T. J. and McMacken, R. pp. 703–18. Alan R. Liss, Inc., New York.

BRUIST, M. F., JOHNSON, R. C., GLACCUM, M. B. & SIMON, M. I. (1984). Characterization of Hin-dependent DNA inversion and binding *in vitro*. In *Genome Rearrangement*, eds. M. Simon and I. Herskowitz. *UCLA Symposium Molecular and Cellular Biology New Series*, **20**, pp. 63–75. New York: Alan R. Liss, Inc.

BURKHOFF, A. M. & TULLIUS, T. D. (1987). The unusual conformation adopted by adenine tracts in kinetoplast DNA. *Cell*, **48**, 935–43.

COZZARELLI, N. R., KRASNOW, M. A., GERRARD, S. P. & WHITE, J. H. (1984). A topological treatment of recombination and topoisomerases. *Cold Spring Harbor Symposium in Quantitative Biology*, **49**, 383–400.

DREW, H. R. & TRAVERS, A. A. (1985). DNA bending and its relation to nucleosome positioning. *Journal of Molecular Biology*, **186**, 773–90.

ECHOLS, H. (1986). Multiple DNA–protein interactions governing high-precision DNA transactions. *Science*, **233**, 1050–6.

FALVEY, E. & GRINDLEY, N. D. F. (1987). Contacts between $\gamma\delta$ resolvase and the $\gamma\delta$ res site. *The EMBO Journal*, **6**, 815–21.

GRINDLEY, N. D. F. (1983). Transposition of Tn3 and related transposons. *Cell*, **32**, 3–5.

GRINDLEY, N. D. F., LAUTH, M. R., WELLS, R. G., WITYK, R. J., SALVO, J. J. & REED, R. R. (1982). Transposon-mediated site-specific recombination: identification of three binding sites for resolvase at the res sites of $\gamma\delta$ and Tn3. *Cell*, **30**, 19–27.

GRINDLEY, N. D. F. & REED, R. R. (1985). Transpositional recombination in prokaryotes. *Annual Reviews in Biochemistry*, **54**, 863–96.

GRONOSTAJSKI, R. M., ADHYA, S., NAGATA, K., GUGGENHEIMER, R. A. & HURWITZ, J. (1985). Site-specific DNA binding of nuclear factor I: analysis of cellular binding sites. *Molecular and Cellular Biology*, **5**, 964–71.

HAGERMAN, P. J. (1985). Sequence dependence of the curvature of DNA: a test of the phasing hypothesis. *Biochemistry*, **24**, 7033–7.

HATFULL, G. F. & GRINDLEY, N. D. F. (1986). Analysis of $\gamma\delta$ resolvase mutants *in vitro*: evidence for an interaction between serine-10 of resolvase and site I of res. *Proceedings of the National Academy of Sciences, USA*, **83**, 5429–33.

HATFULL, G. F. & GRINDLEY, N. D. F. (1987). The resolvases and DNA-invertases: a family of enzymes active in site-specific recombination. In: *Genetic Recombination*, eds. R. Kucherlapati and G. Smith. American Society for Microbiology, Washington, D.C.

HATFULL, G. F., NOBLE, S. M. & GRINDLEY, N. D. F. (1987). The $\gamma\delta$ resolvase induces an unusual DNA structure at the recombinational crossover point. *Cell*, **49**, 103–10.

HEFFRON, F. (1983). Tn3 and its relatives. In *Mobile Genetic Elements*, ed. J. A. Shapiro, pp. 223–260, Academic Press, New York.

HOSCHILD, A. & PTASHNE, M. (1986). Cooperative binding of λ repressors to sites separated by integral turns of the DNA helix. *Cell*, **44**, 681–7.

IIDA, S. & HIESTAND-NAUER, R. (1987). Role of the central dinucleotide at the crossover site for the selection of quasi sites in DNA inversion mediated by the site-specific Cin recombinase of phage P1. *Molecular and General Genetics*, **208**, 464–8.

JOHNSON, R. C. & SIMON, M. I. (1985). Hin-mediated site-specific recombination requires two 26 bp recombination sites and a 60 bp recombinational enhancer. *Cell*, **41**, 781–91.

KOO, H. S., WU., H.-M. & CROTHERS, D. M. (1986). DNA bending at adenine–thymine tracts. *Nature*, **320**, 501–6.

KRAMER, H., NIEMOLLER, M., AMOUYAL, M., REVET, B., VON WILCKEN-BERGMANN, B. & MULLER-HILL, B. (1987). *lac* repressor forms loops with linear DNA carrying to suitably spaced *lac* operators. *The EMBO Journal*, **6**, 1481–91.

KRASNOW, M. A. & COZZARELLI, N. R. (1983). Site-specific relaxation and recombination by the Tn3 resolvase: recognition of the DNA path between oriented res sites. *Cell*, **32**, 1313–24.

LEVENE, S. D. & CROTHERS, D. (1983). A computer graphics study of sequence-directed bending in DNA. *Journal of Biomolecular and Structural Dynamics*, **1**, 429–35.

MARINI, J. C., LEVENE, S. D., CROTHERS, D. & ENGLAND, P. T. (1982). Bent helical structure in kinetoplast DNA. *Proceedings of the National Academy of Sciences, USA*, **79**, 7664–8.

MAXAM, A. & GILBERT, W. (1980). Sequencing end-labelled DNA with base-specific chemical cleavages. *Methods in Enzymology*, **65**, 499–560.

NASH, H. A., BAUER, C. E. & GARDNER, J. F. (1987). Role of homology in site-specific recombination of bacteriophage λ: evidence against joining of cohesive ends. *Proceedings of the National Academy of Sciences, USA*, **84**, 4049–53.

NEWMAN, B. J. & GRINDLEY, N. D. F. (1984). Mutants of the γδ resolvase: a genetic analysis of the recombination function. *Cell*, **38**, 463–9.

NUNES-DUBY, S. E., MATSUMOTO, L. & LANDY, A. (1987). Site-specific recombination intermediates trapped with suicide substrates. *Cell*, **50**, 779–88.

PABO, C. O. & SAUER, R. T. (1984). Protein-DNA recognition. *Annual Reviews of Biochemistry*, **53**, 293–321.

PLASTERK, R. H. A. & VAN DE PUTTE, P. (1984). Genetic switches by DNA inversions in prokaryotes. *Biochemica of Biophysica Acta*, **782**, 111–19.

REED, R. R. (1981). Transposon-mediated site-specific recombination: a defined *in vitro* system. *Cell*, **25**, 713–19.

REED, R. R. & GRINDLEY, N. D. F. (1981). Transposon-mediated site-specific recombination *in vitro*: DNA cleavage and protein-DNA linkage at the recombination site. *Cell*, **25**, 721–8.

REED, R. R. & MOSER, C. D. (1984). Resolvase-mediate recombination intermediate involves a serine-DNA linkage. *Cold Spring Harbor Symposium in Quantitative Biology*, **49**, 245–9.

SALVO, J. J. (1987). Resolvase and the *res* site: binding, bending and the topological consequences. Ph.D. Thesis, Yale University, CT.

SALVO, J. J. & GRINDLEY, N. D. F. (1987). Helical phasing between DNA bends and the determination of bend direction. *Nucleic Acids Research*, in press.

SATCHWELL, S. C., DREW, H. R. & TRAVERS, A. A. (1986). Sequence periodicities in chicken nucleosome core DNA. *Journal of Molecular Biology*, **191**, 659–75.

SIEBENLIST, U. & GILBERT, W. (1980). Contacts between *Escherichia coli* RNA polymerase and an early promoter of phage T7. *Proceedings of the National Academy of Sciences, USA*, **77**, 122–6.

TRIFONOV, E. N. & SUSSMAN, J. L. (1980). The pitch of chromatin DNA is reflected in its nucleotide sequence. *Proceedings of the National Academy of Sciences*, **77**, 3816–20.

ULANOVSKY, L. E. & TRIFONOV, E. N. (1987). Estimation of wedge components in curved DNA. *Nature*, **326**, 720–2.

WASSERMAN, S. A. & COZZARELLI, N. R. (1985). Determination of the stereo structure of the product of Tn*3* resolvase by a general method. *Proceedings of the National Academy of Sciences, USA*, **82**, 1079–83.

WASSERMAN, S. A. & COZZARELLI, N. R. (1986). Biochemical topology: applications to DNA recombination and replication. *Science*, **232**, 951–60.

WASSERMAN, S. A., DUNGAN, J. M. & COZZARELLI, N. R. (1985). Discovery of a predicted DNA knot substantiates a model for site-specific recombination. *Science*, **229**, 171–4.

WELLS, R. G. & GRINDLEY, N. D. F. (1984). Analysis of the γδ *res* sites required for site-specific recombination and gene expression. *Journal of Molecular Biology*, **179**, 667–87.

WILLIAMS, R. C. (1977). Polylysine used for adsorption of nucleic acids and enzymes to electron microscope specimen films. *Proceedings of the National Academy of Sciences, USA*, **74**, 2311–15.

Wu, H.-M. & Crothers, D. (1984). The locus of sequence-directed and protein-induced DNA bending. *Nature*, **309**, 509–13.

Zinkel, S. S. & Crothers, D. M. (1987). DNA bend direction by phase sensitive detection. *Nature*, **328**, 178–81.

SITE-SPECIFIC INVERSION IN BACTERIOPHAGE MU

P. VAN DE PUTTE

Laboratory of Molecular Genetics, University of Leiden, PO Box 9505, 2300 RA Leiden, The Netherlands

Bacteriophage Mu was discovered as a phage which causes mutations in the genome of its host *E. coli* (Taylor, 1963). These mutations were due to a process of lysogenisation by the phage, whereby random sites in the host chromosomes were used as sites of integration. Mu was discovered at about the same time as IS elements and transposons, and analogies between the three systems were recognised at an early stage. That Mu in fact was a transposon became more clear when heteroduplexes of Mu–DNA were studied under the electron microscope (EM). It was found that the phage DNA had heterogeneous tails showing up as split ends under the EM. These split ends consisted of host DNA indicating that even in the phage the Mu genome was in an integrated state. Later it was shown (Ljungquist & Bukhari, 1977) that, after induction of a thermoinducible Mu prophage, the prophage was not excised but remained in the process of its replication at the same site of the chromosome. This was an important feature in the early transposition models (Shapiro, 1979).

Besides the split ends, another peculiar feature was seen in the Mu heteroduplexes under the EM: a so-called G bubble, which was due to an inversion event in the Mu DNA. The inversion fragment had a size of 3000 bp (Fig. 1) and was located near one of the ends of the phage genome. The inversion was initially thought to play a role in the transposition process. However, it turned out to have quite a different function, namely in host specificity. The inversion acts as a genetic switch, the function of which is to provide two alternative sets of tail fibres allowing adsorption to different host cells (Van de Putte, Cramer & Giphart-Gassler, 1980). The inversion reaction is under the control of a phage-coded protein called Gin. The *gin* gene is situated beside the invertible region (Kamp *et al.*, 1979). Subsequently from *in vitro* studies it was found that besides Gin a host protein (Fis or GHF: Kahman *et al.*, 1985; Kanaar, Van de Putte & Cozzarelli, 1986) was also required for the site-specific

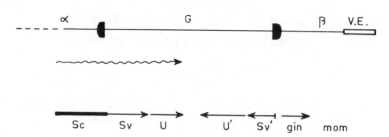

Fig. 1. Schematic representation of the organisation of the G region of Mu. The switching elements are the genes S and S' coding for different tail fibres, made up of constant (Sc) and variable (Sr, Sv') regions. The functions of U and U' are unknown. Note that the orientation of the inverted repeats (black semi-circles) on which the Gin gene product acts is inverted with respect to those in the *Salmonella* system. Other symbols: wavy arrows, mRNA; straight arrows, protein-coding regions; V. E., variable ends of Mu DNA; α, the 33.1 kb constant segment left of the G loop; β, the 17 kb constant segment right of the G loop; *mom*, gene involved in DNA modification.

recombination reaction leading to the inversion. Gin was found to act on small inverted repeats of 34 bp bordering the G region.

OTHER INVERSION SYSTEMS USING GIN-LIKE DNA INVERTASES

An invertible region with a similar function to that of Mu was found in the Mu-related phage D108 and in the phage P1. In the latter case the DNA invertase is called Cin. We have found an invertible region of 1800 bp to be present in a cryptic prophage (*e*14) present in the chromosome of *E. coli* K12 (Van de Putte, Plasterk & Kuijpers, 1984). This defective phage does not behave as a Mu-like phage as it has a specific chromosomal integration site and can be excised from the chromosome after UV irradiation. So it behaves more like a lambda-type phage. The function of the invertible region in the defective prophage is not yet known. Attempts to isolate a complete phage from the defective prophage which would allow investigation of the function of the invertible region have been unsuccessful. The DNA invertase is called Pin in this system.

A related system, but with a different function, is found in the bacterium *Salmonella typhimurium*. The invertible region is only 800 bp in this case and its function is again a genetic switch which allows the alternative expression of two different flagella proteins, H_1 and H_2, which are important antigenic determinants of the bacterium (Simon *et al.*, 1980). In this way the bacterium can escape the defences of the mammalian host. The DNA invertase is called Hin here. All the systems described above are strongly related and the DNA invertases can substitute for each other at least partially (Plasterk, Brinkman & Van de Putte, 1983).

SITE SPECIFIC INVERSION SYSTEMS THAT DO NOT INVOLVE GIN-LIKE DNA INVERTASE

An inversion system which is not related to the Mu-P1-*Salmonella* system is the FLP system present on the $2\,\mu$ plasmid in *Saccharomyces cerevisiae* (Broach, Guarascio & Javaram, 1982). The inverted repeats within which the FLP recombination acts are very long ($\pm$ 599 bp) but FLP acts only on a small segment of the actual inverted repeats at a duplication of a 13 bp sequence (Gronostajski & Sadowski, 1985).

The FLP inversion system has a completely different function from that of the Gin system in Mu. The inversion allows an amplification of the plasmid to a high copy number (Volkert & Broach, 1986). The replication of $2\,\mu$ plasmid is of a bidirectional nature and is in principle terminated at a site located opposite the origin of replication of the plasmid. When one of the replication forks is inverted by the FLP system, termination will no longer occur unless a second inversion event again inverts the direction of the replication in one of the forks. The FLP system is also different from the DNA invertase system in another aspect which is important if we consider the mechanism of inversions. The latter system is specific for inversions and catalyses deletions – when the inverted repeats are made direct by artifically inverting one of the two sites – only to a very small extent. Also recombination reactions between two sites and two different molecules are not catalysed. FLP, on the other hand, can perform all three kinds of reactions with nearly equal efficiency.

Two other site-specific recombination systems show similarities with FLP from the mechanistic point of view. Although their biological functions are not to make inversions in DNA they will be discussed briefly. The Cre-Lox system in phage P1 is responsible for site-specific recombination between prophage dimers at a specific site (loxP) on the phage genome or, rarely, between *loxP* and a site (loxB) in the host chromosome leading to the formation of monomeric circular P1 molecules (Sternberg & Hoess, 1983). The Cre-Lox system, although having a completely different function from that of FLP, is very similar in its properties (Abremski, Hoess & Stern-

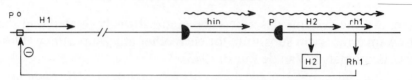

Fig. 2. Schematic representation of the organisation of the invertible region in *S. typhimurium* (see Simon *et al.*, 1980). The switching element is the promoter (P) controlling expression of H_2 and the repressor of H_1 (Rh1). Black semi-circles are inverted repeats in the DNA flanking the invertible region. Wavy arrows indicate mRNA species. Horizontal arrows are protein-coding regions. Vertical arrows indicate gene products. Rh1 acts negatively by binding at the H1 operator site (o) to repress H1 transcription.

berg, 1983). No host factors are involved, and also Cre catalyses inversions, deletions and intermolecular recombinations to the same extent (Sadowski, 1986). This is also true for the λint system, the function of which is to integrate the λ genome into the host chromosome during the process of lysogenisation (Weisberg & Landy, 1983). However, a difference from the Cre-Lox and FLP systems is that with λint a host factor (IHF) is required. This might be due to the fact that Int has to serve a dual purpose. Int is also involved in the reverse reaction: the excision of the λ-prophage from the chromosome. In the case of λ the host factor is not of crucial importance for the recombination reaction itself, since Int mutants can be isolated which no longer require the presence of IHF (Miller, Mozola & Friedman, 1980).

ORGANISATION OF INVERTIBLE REGIONS

Historically the first inversion system of which the genetic organisation was elucidated was the Hin system in *S. typhimurium* (Simon *et al.*, 1980). In this system the important component in the genetic switch is a promoter located between the inverted repeats (Fig. 2). This promoter controls the expression of the gene H_2 and the repressor of gene H_1 in one orientation and allows the expression of H_1 (because rH_1 is no longer expressed) in the other orientation. These structural genes for flagella proteins are outside the invertible region, whereas the DNA invertase gene *hin* is inside this region. The organisation of the G region in Mu, and the C (P1) and P (*e*14) regions is quite different. In these systems the structural genes themselves switch and the promoter controlling the expression of these genes is outside the invertible region. An additional feature of the Mu system is that the structural proteins which make up the tail fiber consist of a constant part and a variable part. The variable part is due to the switching and the constant part to the fact that the translation start of the genes S and S' is situated 500 bp before the invertible region. One wonders why the genetic organisation of the G, C and P regions is so different from the *Salmonella* system when inverted repeats and the DNA invertases show such a great homology to each other. In fact, the difference may be not as great as it seems to be at first sight. If the inverted repeats of *Salmonella* are compared with those of Mu they make the best fit if they are inverted. If the left IR of the Hin system is placed in the same position as the right IR of G, then a comparable genetic organisation can be obtained by transposing a stretch of DNA from the right of Hin, including the right IR, to the position of the left IR of G. As a matter of fact the systems can be made quite similar by one simple genetic rearrangement.

The genetic organisation of the C region (P1) is very similar to that of the G region, except that the *cin* gene is transcribed in a direction opposite to the direction of *gin* (Iida *et al.*, 1982). This may be due to a sequence to the right of Cin which shows homology to the inverted repeats on which Cin acts. An additional inversion reaction comprising Cin could form the basis of this difference. The organisation of the invertible P region (*e*14) looks similar to that in Mu or P1. This is based only on sequence data showing an open reading frame coming from outside the invertible region, and ending about the middle of the P region. A second open reading frame, going in the other direction, is found in the second part of the invertible region (Plasterk & Van de Putte, 1985). The sequences, however, show no homology with the S and S' genes in G of Mu and may have a different function.

The organisation of the FLP system is not comparable with the Mu or *Salmonella* systems. Its function is not on the level of expression but on that of regulation of replication termination. The information content of the piece of DNA which is inverted is irrelevant; only the presence or absence of a growing replication fork in the invertible region is important in this case.

THE DEVELOPMENT OF *IN VITRO* INVERSION SYSTEMS

The development of *in vitro* systems for DNA inversions opened up a completely new era of studies.

DNA inversion *in vitro* was first achieved using the λint system

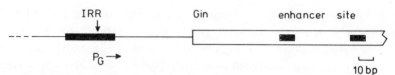

Fig. 3. Position of the enhancer site in the *gin* gene to which the host protein Fis or GHF binds (Kahmann *et al.*, 1985; Kanaar *et al.*, 1986). Symbols: IRR, right-hand inverted repeat flanking the G region; P_G, promoter of *gin*; open box, region encoding Gin.

acting on attP–attB sites which were cloned in inverted orientation on a plasmid. However, this was a rather artificial system as the normal function of Int is not to catalyse inversions. It turned out to be more difficult to achieve *in vitro* inversions with the DNA invertases Gin, Hin, etc. These inversion reactions also occur with a low frequency *in vivo*. If the inversion reaction were normally too efficient, the advantage of being in two possible states would be lost. However, the use of extracts from cells that overproduce Gin, and of a special substrate in which every inversion could be detected as a change from Lac⁻ to Lac⁺ phenotype, led to the detection of Gin promoted inversion *in vitro* (Plasterk, Kanaar & Van de Putte, 1984). This made purification of Gin and the detailed study of the inversion process possible. Now all the DNA invertases (except Pin) have been purified. The requirements for the inversion reaction have subsequently been studied in detail. A rather unexpected development was that, on further purification of the DNA invertases, it was found that an additional protein was required for the reaction. This protein has a M. wt of 12.6 kD and is called Fis (Kahmann *et al.*, 1985) for *F*actor of *i*nversion *s*timulation, or GHF (Kanaar, Van de Putte & Cozzarelli, 1986) for *G*in *H*ost *F*actor in the Mu system. This host protein binds to a sequence (*sis* site) which is situated in the *gin* gene and acts as an enhancer for the site-specific recombination reaction (Fig. 3) The enhancer contains two binding sites for the host factor. The binding sites are 4.5 helix turns apart and this distance between the sites appears to be critical (Glasgow & Kahman, personal communications). The position and orientation of the enhancer can be changed without losing its activity. The possible function of the enhancer and the host protein will be discussed in the final section. As expected, also the Cin (P1) and Hin (*Salmonella*) systems need similar host factors and enhancer sites. In contrast the FLP system in 2 μ works without host factor as does the Cre-Lox system in P1 which, however, is not a natural inversion system. The need for a host factor is reflected in a different mechanism of strand exchange as will be discussed in the next section.

The *in vitro* system also made it possible to determine the importance of the individual bases in the inverted repeats to which the DNA invertases bind and the enhancer site on which the host factor acts. Even more importantly, the availability of the purified enzymes, and the possibility of following the DNA inversion reaction *in vitro* by simple restriction enzyme analysis, allowed the study of the actual DNA strand exchange in the recombination reaction.

DNA STRAND EXCHANGE IN THE SYNAPTIC COMPLEX

Information on the strand exchange reaction in the inversion process can be obtained by following the changes in the topology of the substrate DNA when it is converted into the product of the recombination event. In principle, two parameters can be studied.

First, the topological complexity of the product can be determined, which gives information on the way the two recombining sites have encountered each other. For the inversion reaction to occur it is a prerequisite that the substrate is supercoiled. When a productive synaptic complex between the two recombining sites is formed by random collision of these sites, then it is expected that a variable number of supercoils will be entrapped in the reaction, leading to knotted structures which can be separated by gel electrophoresis. DNA inversions catalysed by Int, Cre or FLP indeed lead to just such a large variety of knotted structures (Sadowski, 1986) indicating that a random collision event is underlying these reactions. For Int and Cre this is more or less expected as these enzymes catalayse intermolecular recombination events. For Gin-mediated inversion the situation is quite different. The product of the inversion reaction is unknotted (Kanaar & Van de Putte, 1987). The only possibility for getting unknotted structures is that two or zero supercoils are entrapped in the recombination event. This suggests that the sites encounter each other in an orderly manner. This is compatible with the finding that Gin-mediated recombination is only productive when the sites are in a particular (inverted) orientation in respect to each other, and is non-productive in intermolecular reactions.

Secondly, information on the strand exchange reaction can be obtained by determining whether the recombination event leads to loss of supercoils in the substrate. This can be determined by purifying a specific topoisomer of the substrate molecules before the

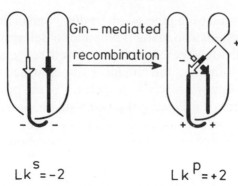

$$Lk^{S}_{=-2} \qquad\qquad Lk^{P}_{=+2}$$

Fig. 4. Model of Gin-mediated recombination explaining the observed linking-number change (ΔLk) and the absence of knotted structures in the *in vitro* reaction. According to this model the ΔLk of +4 is only due to the reversion of the sign of the overlap of the DNA strands by the inversion reaction. The negative supercoil introduced by the strand exchange is compensated by the rotation of the strands (*see also text*).

reaction and determining the change in linking-number resulting from the recombination. (The linking number specifies the number of times two DNA strands are intertwined.) If this experiment is done for Gin-mediated inversion a very specific linking number change of +4 is found (Kanaar & Van de Putte, 1987). So four supercoils are lost for every inversion event. From the combined results of the determination of the linking number change, and the absence of knotted structures in Gin-mediated inversion, a model can be put forward regarding the topological structure of a productive synaptic complex. In this complex (Fig. 4) two supercoils are entrapped, which, however, will be reversed in sign after the inversion reaction has occurred. This leads to a linking-number change of +4. Due to the recombination event itself one negative supercoil is introduced, but, in order to connect the proper 3′ ends to the 5′ ends, the strands have to rotate 180° in respect to each other leading again to the loss of one supercoil. The net balance of the reaction will then be +4, which is in agreement with the experimental findings. The strand rotation reaction presumes that a double-strand break is introduced in the strands before the actual exchange of strand occurs. For Int and Cre and also FLP it has been shown that the strand exchange reaction is initiated by a single-strand scission followed by rotation and resealing before the other pair of strands is exchanged. So in this case the DNA backbone is never entirely interrupted during the reaction. This could have its consequences for the requirements of the recombination reaction. In the category Int, Cre, FLP this could mean that every collision between the recombining sites is productive, whereas the requirements for the family of the DNA invertases could be more strict, as eight free DNA ends have to be held together in the synaptic complex. We suggest that the enhancer site and host protein play a role in establishing this special configuration.

POSSIBLE ROLE OF ENHANCER AND HOST PROTEIN

We consider three stages in the inversion process: (1) the bringing together of the sites, (2) the formation of a productive synaptic complex and (3) the actual exchange reaction. One may wonder at what stage the enhancer site and host protein act. Recent experiments suggest that the enhancer is not involved in step one. When a suitable substrate with inverted repeats is incubated with Gin without host factor and enhancer site, it can be shown by EM that a complex is formed between the two sites and Gin (R. Kanaar, unpublished results). Preliminary experiments indicate that this complex is also formed when the sites are in direct orientation in respect to each other, so an ordered process like tracking or slithering seems not to be absolutely required for the formation of this complex. However, the encounter is a non-productive one: without host protein no recombination reaction is observed. This suggests that the enhancer is acting in a later stage of the recombination reaction and we suggest that it acts in stage two. In this view the recombining sites first meet each other by a random collision event. This initial synaptic complex is stabilised only by Gin. We have seen, however, that this would lead to knotted recombination products *in vitro* which are not observed. Therefore the initial complex could be converted into a productive one in which Gin can cut the strands under the direction of the enhancer site and host protein possibly by a process of DNA wrapping. Why the complex is only stable enough to be productive when only two interdomainal supercoils are left between the two recombining sites, as we discussed in the previous section, is at present unknown and is the subject of further investigations.

REFERENCES

ABREMSKI, K., HOESS, R. & STERNBERG, N. (1983). Studies on the properties of P1 site-specific recombination: evidence for topologically unlinked products following recombination. *Cell*, **32**, 1301–11.

BROACH, J. R., GUARASCIO, V. R. & JAYARAM, M. (1982). Recombination within the yeast plasmid 2 μ circle is site-specific. *Cell*, **29**, 227–34.

GRONOSTAJSKI, R. M. & SADOWSKI, P. D. (1985). Determination of DNA sequences essential for FLP mediated recombination by a novel method. *Journal of Biological Chemistry*, **260**, 12320–7.

IIDA, S., MEYER, J., KENNEDY, R. E. & ARBER, W. (1982). A site-specific conservative recombination system carried by bacteriophage P1. Mapping the recombinase gene *cin* and the crossover sites cix for the inversion of the C segment. *The EMBO Journal*, **1**, 1445–53.

KAHMANN, R., RUDT, F., KOCH, C. & MERTENS, G. (1985). G inversion in bacteriophage Mu DNA is stimulated by a site within the invertase gene and a host factor. *Cell*, **41**, 771–80.

KAMP, D., CHOW, L. T., BROKER, T. R., KWOH, D., ZIPSER, D. & KAHMAN, R. (1979). Site-specific recombination in phage Mu. *Cold Spring Harbor Symposium on Quantitative Biology*, **13**, 1159–67.

KANAAR, R., VAN DE PUTTE, P. & COZZARELLI, N. R. (1986). Purification of the Gin recombination protein of *Escherichia coli* phage Mu and its host factor. *Biochimica et Biophysica Acta*, **866**, 170–7.

KANAAR, R. & VAN DE PUTTE P. (1987). Topological aspects of site specific DNA-inversion. *BioEssays*, in press.

LJUNGQUIST, E. & BUKHARI, A. I. (1977). State of prophage Mu DNA upon induction. *Proceedings of the National Academy of Sciences, USA*, **74**, 343–7.

MILLER, H. I., MOZOLA, M. A. & FRIEDMAN, D. I. (1980). Int-h: an Int mutation of phage λ that enhances site-specific recombination. *Cell*, **20**, 721–9.

PLASTERK, R. H. A., BRINKMAN, A. & VAN DE PUTTE, P. (1983). DNA inversion in the chromosome of *E. coli* and in bacteriophage Mu: relationship to other site-specific recombination systems. *Proceedings of the National Academy of Sciences, USA*, **80**, 5355–8.

PLASTERK, R. H. A., KANAAR, R. & VAN DE PUTTE, P. (1984). A genetic switch *in vitro*: DNA inversion by Gin protein of phage Mu. *Proceedings of the National Academy of Sciences, USA*, **81**, 2689–92.

PLASTERK, R. H. A. & VAN DE PUTTE, P. (1985). The invertible P-DNA segment in the chromosome of *Escherichia coli*. *The EMBO Journal*, **4**, 237–42.

SADOWSKI, P. (1986). Site specific recombinases: changing partners and doing the twist. *Journal of Bacteriology*, **165**, 341–7.

SHAPIRO, J. A. (1979). Molecular model for the transposition and replication of bacteriophage Mu and other transposable elements. *Proceedings of the National Academy of Sciences, USA*, **76**, 1933–7.

SIMON, M., ZIEG, J., SILVERMAN, M., MANDEL, G. & DOOLITTLE, R. (1980). Genes whose mission is to jump. *Science*, **209**, 1370–4.

STERNBERG, N. & HOESS, R. (1983). The molecular genetics of bacteriophage P1. *Annual Review of Genetics*, **12**, 123–54.

TAYLOR, A. L. (1963). Bacteriophage-induced mutation in *Escherichia coli*. *Proceedings of the National Academy of Sciences, USA*, **50**, 1043–51.

VAN DE PUTTE, P., CRAMER, S. & GIPHART-GASSLER, M. (1980). Invertible DNA determines host specificity of bacteriophage Mu. *Nature*, **286**, 218–22.

VAN DE PUTTE, P., PLASTERK, R. H. A. & KUIJPERS, A. (1984). A Mu *gin* complementing function and an invertible DNA region in *Escherichia coli* K-12 are situated on the genetic element e14. *Journal of Bacteriology*, **158**, 517–522.

VOLKERT, F. C. & BROACH, J. R. (1986). Site-specific recombination promotes plasmid amplification in yeast. *Cell*, **46**, 541–50.

WEISBERG, R. & LANDY, A. (1983) Site-specific recombination in phage lambda. in *Lambda II*, ed. R. Hendrix, J. Roberts, F. Stahl & R. Weisberg, pp. 211–50. Cold Spring Harbor: Cold Spring Harbor Laboratory.

PHASE AND ANTIGENIC VARIATION BY DNA REARRANGEMENTS IN PROCARYOTES

THOMAS F. MEYER AND RAINER HAAS

Max-Planck Institut für Biologie, Infektgenetik, Spemannstrasse 34, D-74 Tübingen, F.R.G.

INTRODUCTION

In recent years several systems for phase and antigenic variation in procaryotes have been subjected to an extensive molecular analysis. These include the flagella phase variation of *Salmonella* (Zieg *et al.*, 1977; Johnson & Simon, 1985), the control of common pilus expression in *E. coli* (Abraham *et al.*, 1985), the variation of the variable major protein (VMP) of *Borrelia hermsii* (Stoenner *et al.*, 1982; Plasterk *et al.*, 1985), and the phase and antigenic variation of pathogenic *Neisseriae* (Kellog *et al.*, 1963; Lambden & Heckels, 1979; Swanson & Barrera, 1983; Virji & Heckels, 1983). In *N. gonorrhoeae* there are in fact two systems which seem to act autonomously, one that is responsible for the control and the variation of pilin expression (Meyer *et al.*, 1982; Hagblom *et al.*, 1985; Swanson *et al.*, 1986; Haas & Meyer, 1986; Nicolson *et al.*, 1987*a*) and another that controls expression of the opacity protein (Black *et al.*, 1984; Stern *et al.*, 1984). Furthermore, a number of bacteriophages contain specific switch mechanisms allowing them to change their host ranges (Kamp *et al.*, 1978; Plasterk & van de Putte, 1984; Huber *et al.*, 1985; Kahmann *et al.*, 1985; Iida & Hiestand-Nauer, 1986).

Strictly taken, phase and antigenic switches are spontaneous events. They give rise to mixed populations, preparing the better-adapted fraction of this population to survive sudden environmental changes. Therefore, it is conceivable that phase and antigenic switches affect the expression of surface molecules which interact with specific environmental counterparts, such as antibodies or receptors on a target cell.

For some bacterial pathogens, the ability to vary their surface is an effective means to evade the immune surveillance of their infected host organisms. It also appears that the variation of surface molecules can be utilised by a pathogen to adapt to distinct microenvironments of the host during the course of an infection. In this context it is interesting that the two variable surface components of pathogenic *Neisseriae* not only represent important surface antigens, but are considered as adhesins, conferring on the bacteria the ability to bind specifically receptors on the host epithelial cells (Swanson, 1973; Lambden *et al.*, 1979; Watt & Ward, 1981; Blake & Gotschlich, 1983). Surface antigen variation in some cases may therefore affect both bacterial escape from the immune response and variability in receptor binding properties. However, in other pathogens, like *Borrelia hermsii*, antigenic variation may serve primarily as an immune escape mechanism.

It appears that different microorganisms pursue different functional goals with phase and antigenic variation. Furthermore, any particular need of a microorganism to alter its surface seems to be realised by distinct genetic mechanisms. This review therefore will stress, with special emphasis on our own work with pathogenic *Neisseriae*, the variety of mechanisms known to be used by procaryotes in order to achieve a desired degree of variability necessary for the survival of their populations.

PHASE AND ANTIGENIC VARIATION IN PATHOGENIC *NEISSERIAE*

N. gonorrhoeae, a Gram-negative coccus that causes sexually transmitted disease in humans, is the best-studied example of antigenic variation amongst the pathogenic *Neisseriae*. (Nomenclature: T1 to T4 specify the four basic colony phenotypes of gonococci as seen under a binocular microscope (Kellog *et al.*, 1963). P^+ and P^- refer to the formation of pili (irrespective of the production of pilin); O^+ and O^- refer to the production of detectable Opa, i.e., the opacity outer membrane protein or Protein II (P. II) of gonococci, or the class V protein of meningococci. The terms 'P' and 'O' therefore specify independent phases.) Both of the variable surface components of gonococci seem to play an essential role in colonisation of the human mucosal tissue by the pathogens. Pili are thought to anchor gonococci to epithelial cell surfaces in a host- and tissue-specific fashion (Swanson, 1973; Pearce & Buchanan, 1978; Lambden *et al.*, 1981). Gonococcal pili are composed of a single major protein subunit, called pilin, the primary structure of which is known (Hermodson *et al.*, 1978; Meyer *et al.*, 1984). Although this protein has been implicated with a specific receptor binding domain (Rothbard *et al.*, 1985), there exists no strict genetic evidence (compare with Norgren *et al.*, 1984, for the *E. coli* Pap-pilin system) that pilin is the adhesin molecule itself or that it is the sole structural component of gonococcal pili.

The second variable surface protein, opacity protein (Opa), also seems to be involved in the adhesion of gonococci to human epithelial cells (Lambden *et al.*, 1979; James *et al.*, 1980; Watt & Ward, 1981). It further promotes the adhesion between gonococci, thus causing the formation of bacterial aggregates (Blake & Gotschlich, 1983). Gonococcal Opa is the equivalent of the meningococcal class 5 protein (Stern & Meyer, 1987).

Two other *Neisseria* species are known to exhibit surface variation phenomena similar to the gonococcus, i.e. *N. meningitidis*, an important agent of bacterial meningitis, and *N. lactamica*, an occasional pathogen. In contrast to these latter species, antigenic variation in the gonococcus is associated with distinct colony morphology differences, which render this bacterium a most suitable subject for investigation (Kellog *et al.*, 1963; Swanson *et al.*, 1971). Four basic colony phenotypes have been described for gonococci: T1 colonies that consist of piliated (P^+) cells and produce detectable amounts of opacity protein (O^+), T2 colonies containing cells that are P^+,O^-, and T3 and T4 colonies that are typical of P^-,O^+ and P^-,O^- cells, respectively. Recently it has been found that these colony phenotypes, which are based on light microscopy, do not strictly correlate with the production of pili (Haas *et al.* submitted for publication) and/or opacity protein ('P. II') (Blake & Gotschlich, 1984; Stern *et al.*, 1984). Nevertheless, the described colony differences have proven to be extremely helpful in the isolation of spontaneous pilus and Opa variants in culture. This has allowed the elucidation of the mechanism by which the expression of pili and Opa can be turned on and off (commonly termed as phase variation) and single cells can give rise to progeny that produce structurally and antigenically distinct proteins (antigenic variation). In addition, more than one variant Opa (and perhaps pilin) can be produced at a time by a homogeneous population, and the number of variations in these proteins appears to be strikingly large (Heckels, 1981; Diaz & Heckels, 1982; Hagblom *et al.*, 1985).

THE ORGANISATION OF PILIN GENES IN *NEISSERIA GONORRHOEAE*

To assess the molecular principles underlying the phenomenon of pilus phase and antigenic variation we started, in 1981, in collaboration with So's laboratory, to isolate and characterise the genes coding for pilin of *N. gonorrhoeae* strain MS11 (Meyer *et al.*, 1982). In our genetic studies we could demonstrate that the genome of *N. gonorrhoeae* harbours multiple gene loci for pilin. Since then 5 out of a total of about 8 *pil* loci of strain MS11 have been mapped in a chromosomal segment spanning approximately 50 kb (Meyer *et al.*, 1984; Haas & Meyer, 1986). This segment includes the expression loci *pil*E1 and *pil*E2, and the major silent locus *pil*S1 which is located upstream of *pil*E1 (Fig. 1). The relative orientation of all mapped loci is identical with regard to their open reading frames and a 3'-terminal ca. 65 bp segment, bounded by *Cla*I and *Sma*I sites, that is characteristic of *pil* loci. The silent loci usually contain more than one gene copy tandemly arranged and, as in case of the large *pil*S1 locus, linked by a short repetitive sequence element, termed RS1. This repetitive element is also found in the expression loci (Fig. 1). Both expression loci harbour one complete structural gene (expression copy, EC) in addition to several silent gene copies (SC) that are located upstream in a direct orientation. The promoter sequence and the translational start signals in EC are functional and therefore promote transcription and translation of the EC gene, regardless of the phase of pilin expression (Meyer *et al.*, 1984; Swanson *et al.*, 1985; Haas *et al.*, 1987). In contrast, silent copies lack a promoter and are truncated at their 5'-ends. Most silent copies start at codon position 44, while a few other SCs begin with codons 31 or 92; the smallest silent copies are found in both expression loci and start beyond codon 141 (Fig. 1). At their 3'-termini all SCs are complete, although only some SCs carry a translational stop codon at the expected position (unpublished).

By sequence determination of RNA transcripts of several pilus variants, Hagblom & coworkers (1985) observed striking sequence changes in the expressed copies of pilin in MS11 gonococci and some fresh clinical isolates. These sequence alterations were focused onto distinct areas of the expressed pilin mRNAs with an increasing incidence towards the 3'-end of the variant expression genes, constituting a conserved 5'-region, a semi-variable central region, and a hypervariable 3'-region. Similar findings were made at the RNA

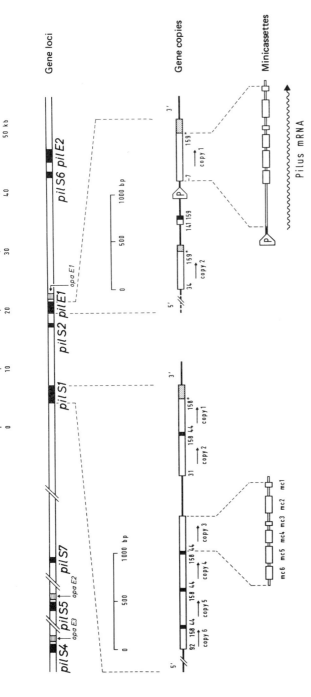

Fig. 1. Genetic map of pilus and opacity loci in *Neisseria gonorrhoeae* strain *MS11*. Besides the two expression loci (*pil*E1, *pil*E2) several silent loci (*pil*S) are located in a chromosomal segment of 50 kb. Some *pil* loci are closely linked with opacity loci (*opa*E). As seen in the central part of the figure, expression loci contain one functional pilin gene copy with a promoter (EC; i.e. copy 1 in *pil*E1). and several 5'-truncated silent copies (SC); silent loci are composed of one of several SCs; the numbering refers to the codons present in the *pil* gene copies and the asterisks refer to translational stop codons; the shaded regions represent sequences of the so-called *Sma*I/*Cla*I repeat; and black regions represent a repetitive sequence (RS1) found in *pil*S1 and *pil*E1/2 (Haas & Meyer, 1986). The pilin gene copies consist of six variable minicassettes (mc1–mc6) which are flanked by conservative sequences, as indicated in the section below.

or DNA levels by Swanson *et al.* (1986; 1987*a*) for a variant of strain MS11 carrying only one active expression site, and by Nicolson *et al.* (1987*b*) for strain P9. While in the semivariable region sequence alterations are sparsely found, the hypervariable region often undergoes complete sequence substitution. Since the latter region relates to the immunodominant region of pilin (Schoolnik *et al.*, 1983), these observations nicely explain the antigenic diversity of gonococcal pili.

Sequence analysis of SC genes reveals a pattern that is strikingly reminiscent of the variations seen in the EC genes (Haas & Meyer, 1986). Remarkably, in the silent copy genes those sequences coding for the conserved 5′-region of pilin are missing. Our observations indicate that sequence variations within both silent and expressed genes of strain MS11 are restricted to six short intragenic segments. These variable intragenic segments, designated as minicassettes (mc), are flanked by strictly conserved regions. This rule seems to be broken in a given strain only in rare instances. Recent work covering a total of 12 silent copy sequences of strain MS11 supports this view (unpublished). Furthermore, sequence data available for *N. gonorrhoeae* strain P9 (Nicolson *et al.*, 1987*b*) allow a similar assignment of variable minicassettes, although some strain-specific characteristics then become apparent. We would therefore consider the minicassette pattern as a general feature of gonococcal pilus genes.

PILIN ANTIGENIC VARIATION RESULTS FROM CONVERSION OF THE EXPRESSION COPY BY SILENT GENE SEGMENTS

After having accumulated much information on the structure and organisation of pilin genes, questions concerning the significance of the silent gene repertoire and the mechanism of antigenic variation of gonococci became more pertinent. Were the silent genes only an irrelevant remnant from previous expression loci or were they a store for variant sequence information to be used for the generation of variant expression genes? If the latter was true, how was the silent information converted to an expressed form?

The experimental key to these questions was gained by the use of synthetic oligonucleotide probes. Initially an oligonucleotide was prepared that was complementary to the most variable minicassette (mc2) of both copies 1 and 3 within the silent locus *pil*S1 (Haas

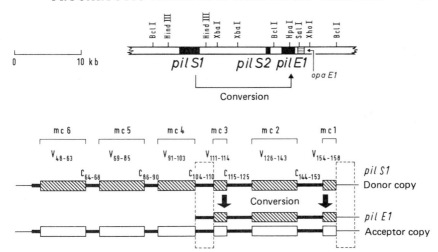

Fig. 2. Gene conversion of the expression gene in *pil*E1. In this example the *pil*E1 locus was converted by replacing three minicassettes with the corresponding ones from the silent locus *pil*S1 copy 1 located about 10 kb upstream (top panel). Conserved regions (C, solid lines in the bottom panel) are assumed to support the recombination between the donor copy and the acceptor copy. The model is based on the sequence analysis data of the gene conversion event in isogenic pilus antigenic variants (Haas & Meyer, 1986; Haas *et al.*, 1987). Numbers on top of the gene copies refer to codon positions of constant (C) and variable (V) regions in the pilus gene.

& Meyer, 1986). This oligonucleotide, which did not hybridise to any other *pil* gene, was used in Southern hybridisations for the genetic analysis of several variant gonococci, all derived from a single parental strain but producing structurally different pilin types. In some of these variants a hybridisation signal was found in the expression gene in addition to the hybridisation pattern of the parent strain. Detailed analysis showed that the expression gene in *pil*E1 had acquired sequences from the *pil*S1 locus, in a way that was reminiscent of a gene conversion process (Fig. 2).

Sequence comparisons between *pil*S1(SC1/3) and *pil*E1(EC) of the parental line and a progeny variant showed that not only the probe-homologous minicassette, but also two additional minicassettes were transferred from *pil*S1(SC1) to the expression gene of the progeny. While the corresponding minicassettes (mc1–3) in the expression locus were replaced, mc4–6 were retained (Haas *et al.*, 1987). No reciprocal changes were detectable in the silent donor locus (Fig. 2).

Following this initial finding, similar recombination events, typical of gene conversion, were detected between the EC gene and other *pil*S loci. Likewise, Swanson *et al.* (1986) implicated gene conversion from the comparison of pilin mRNA sequences of variant gonococci with the already known *pil*S1 locus. These authors also conclude that an expression gene can acquire variant gene segments of different length from a single silent copy gene (Swanson *et al.*, 1987*a*). Furthermore, gene conversion was found to take place between two silent copy genes of the *pil*S1 locus (Haas & Meyer, 1986).

All instances of gene conversion studied by us involved the transfer of one or several complete minicassettes from a silent copy gene to an expression gene. The termination points of expression gene conversion were always located in the conserved flanking sequences of the minicassettes. This led us to the conclusion that minicassettes are the basic units transferred by this mechanism (Haas & Meyer, 1986). In essence, our minicassette model implies that the conversion of *pil* genes requires flanking homology at the regions to be transferred from a donor to an acceptor gene.

Taking into account that the EC genes are constitutively transcribed and translated the general principle of pilin antigenic variation is becoming evident: the silent *pil* genes represent the store of variant minicassettes to be used for the conversion of expression genes. Depending on the combination of minicassettes in the expression copy (or copies), pilin molecules with altered primary structure are produced. Given a number of variant silent copies for each of the six minicassettes that can be reassorted in the expression gene, one can anticipate the immense potential of *N. gonorrhoeae* to produce variant pili. The total number of different immunodominant regions (mc2) of pilin alone, amongst the whole *N. gonorrhoeae* population, can be estimated on the basis of our unpublished DNA hybridisation data (S. Veit, R. Haas & T. F. Meyer) to be in the order of about one hundred.

A NOVEL PHASE OF PILIN EXPRESSION: EXCRETION OF SOLUBLE PILIN

In order to characterise the effects of gene conversions in the expression gene we have recently isolated a series of variant MS11 strains all of which had a different arrangement of variable minicassettes in their ECs (Haas *et al.*, 1987). These variants were characterised by several different criteria, including their colony morphologies and their piliation features using transmission electron microscopy (TEM). One interesting side-observation was that many variants formed colonies typical of non-piliated (T3/T4) cells, although the cells themselves were piliated by TEM-criteria. Some of the T3/T4 gonococci were indistinguishable by TEM from heavily piliated T1/T2 gonococci. This was unexpected, as hitherto the T3/T4 colony phenotypes were believed to be typical of non-piliated gonococci (Kellog et al., 1963).

A characteristic of the majority of these piliated gonococcal variants, showing the T3/T4 colony phenotype, was that their pilin was degraded to a certain extent into a smaller fragment of distinct size (Meyer et al., 1987a). Such partial degradation of pilin has been observed before by Swanson et al. (1986) with non-piliated gonococcal variants, which were referred to as 'P⁻rp⁺'.

Determination of the cellular location of the truncated pilin in our isogenic variants revealed that these truncated molecules were efficiently secreted into the extracellular environment of gonococci. This property led us to propose a novel phase of pilin expression in gonococci, the P^S phase. For some P^S variants the amount of truncated pilin (S-pilin) found in culture fluids was very high, reaching concentrations of about 2 mg/l. Such variants converted almost all of their pilin into the S-form. Other P^S variants were intermediate in terms of pilin truncation and secretion. According to the TEM analysis there was an inverse correlation between the production of S-pilin and the formation of pili on the surface of P^S cells. The more pilin that was processed into S-pilin the less that remained for the formation of pili. Rarely, P^S cells were found without any surface pili; in this case the production of S-pilin was near 100%. Therefore conversion in the EC gene affects not only the structural and antigenic properties of pilin but also the ratio of pilus formation to S-pilin production. One might assume that S-pilin production is associated with some biological function rather than being the consequence of an abortive recombination event. This point of view is supported by recent observations, gained by Swanson et al. (1987b) from studies with experimentally infected male volunteers. In that report it was noted that some of the reisolated piliated gonococci produced truncated pilin molecules which, in our view, are typical of S-pilin.

Amino acid sequence analysis of the S-pilin, prepared from one particular P^S variant, showed that it lacks the conserved amino terminal region of standard pilin and begins beyond the position 40 (Fig. 3; Haas et al., 1987). The missing region is strongly hydrophobic and has been implicated in the polymerisation of pilin (Schoolnik et al.,

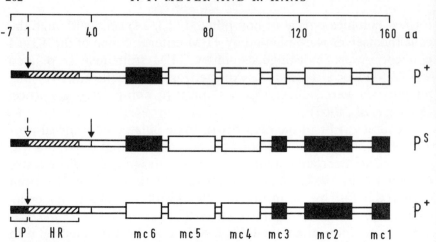

Fig. 3. Structural features of some P⁺ and Pˢ pilin molecules. Pilin is synthesised as a proform (Meyer *et al.*, 1984). After removal of a 7 amino acid leader peptide (LP), P⁺ pilin is formed. The removal of an additional 39 amino acids, which include the strongly hydrophobic region (HR) of pilin, renders the pilin in a soluble secretory form (S-pilin). (Arrows indicate processing points.) The alternative processing is thought to depend on the intrinsic structure of pilin. It is shown that a pilin, of which 95% is converted into S-pilin, can share mc1 to mc3 with a P⁺ pilin, and mc6 with another P⁺ pilin (common minicassettes black-boxed). The determinants responsible for P⁺ and P⁻ processing are as yet unknown.

1983). Its loss could therefore readily explain the solubility of S-pilin. The molecular basis for the processing of S-pilin is unknown. We would assume that the selective processing of distinct pilin variants depends on the intrinsic structural features of their pro-pilin, although the involvement of an independently controlled protease cannot be excluded.

MECHANISTIC ASPECTS OF PILUS PHASE AND ANTIGENIC VARIATION

The data discussed above suggest that the variations associated with pilin expression in gonococci are due to intragenic recombination events that affect the sequence composition in the expression copy/ies in *pil*E1 and/or *pil*E2. This is a general finding made in several laboratories (Hagblom *et al.*, 1985; Haas & Meyer, 1986; Haas *et al.*, 1987; Nicolson *et al.*, 1987*a,b*; Segal *et al.*, 1986; Swanson *et al.*, 1986, 1987*a*). Only in rare cases may point mutations play a role (Bergström *et al.*, 1986; Koomey *et al.*, 1987).

The frequent recombination seen in pilus antigenic variation is typical of gene conversion. The term 'gene conversion' was first established for non-reciprocal recombination in spore-forming fungi, where the parents and progeny were accessible to a precise genetic analysis. As used here for *N. gonorrhoeae*, this term describes the results of an apparently non-reciprocal recombination event, i.e. the duplication of a silent donor sequence followed by the precise replacement of a corresponding variant sequence in the expressed acceptor gene. This process is not connected with size changes in either of the genes involved, neglecting variations of a few triplet codons that may occur in the acceptor gene. By this definition 'gene conversion', however, does not explain the molecular mechanism by which the observed recombination phenomenon is ultimately caused.

The idea that DNA homology plays a critical role in the conversion of expression genes (Haas & Meyer, 1986; see above), is strengthened by the observation that – with essentially no exception – the third nucleotide position of triplet codons in conserved flanking regions within silent pilin genes is constant. Furthermore, a recent report on the construction of gonococcal *rec*A mutants demonstrates a strong reduction of pilin variation in such mutants (Koomey *et al.*, 1987). While conversion of the expression gene seems to be *rec*A-dependent, this does not exclude that other factors, such as site-specific DNA-binding molecules, may selectively facilitate recombination in the *pil* genes.

One other mechanism that might be intrinsically associated with antigenic variation in gonococci, and the gene conversion in particular, is transformation. It is known that gonococci, as well as some other microbial organisms, can be efficiently transformed by species-specific DNA (Sparling, 1966). Conceivably, DNA arising from cell lysis could donate variant *pil* gene DNA to a pilin expression gene by transformation of piliated gonococci. Indeed, subsequent recombination of the variant donor DNA at homologous sites of the expression gene would be in full agreement with the previous observations on gene conversion (Fig. 4).

Hitherto, one of the difficulties in accepting this point of view is that transformation is inefficient for P^- cells. Whether this applies to the P^S phase, as well, has not yet been determined. Therefore, the role of transformation in the antigenic variation of gonnococci remains an open question at this stage. The situation *in vivo* is even more poorly understood. The generation of transformation-negative mutants and an evaluation of transformation frequencies will permit discrimination between intrinsic and external EC-conversion, and provide the next step forward in the analysis of this phenomenon.

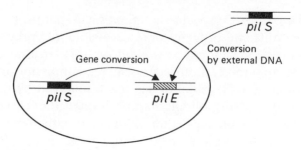

Fig. 4. Gene conversion of the expression gene. The term gene conversion is used in this work to describe the transfer of a silent gene segment (*pil*S) into an expression gene copy (*pil*E). Thereby the corresponding gene segment in the expression gene copy is replaced in a unidirectional process, while the silent donor copy remains unaltered. In this sense gene conversion may occur by different mechanisms. For the gonococcus two hypothetical models can be postulated: the first one involves conversion of the expression gene by a duplicated silent copy segment in the same cell; the second one depends on the transformation of the recipient cell by external *pil* DNA.

CHANGES IN PILIN EXPRESSION BY COARSE DNA REARRANGEMENTS

A number of pilus phase and antigenic switches observed with *N. gonorrhoeae* MS11 (and also *Neisseria meningitidis*, Perry *et al.*, 1987) do not fit into the appealing scheme of EC-conversion. While gene conversion events, as they usually occur in gonococci, are not detectable in standard Southern blots using a cross-reactive probe, other switches involve coarse DNA rearrangement in the pilin gene loci and are easily seen by Southern hybridisation. Coarse *pil* rearrangements seem to fall into at least three distinct categories, as follows.

Certain non-piliated variants have suffered deletions in their expression loci extending across the promoter and the 5'-coding sequences of the expression genes (Segal *et al.*, 1985; Bergström *et al.*, 1986). This type of rearrangement occurs with low frequency and leads to a complete loss of pilus-specific expression at both translational and transcriptional levels. The deletion formation is irreversible (Bergström *et al.*, 1986; Sparling, pers. comm., and our own unpublished data), unless a second intact expression copy is maintained in the genome. However, even in variants which carry two expression loci, such deletions may eventually occur in both copies.

Another type of rearrangement leading to an altered pilin expression, due to changes in the minicassette composition of EC, was recently detected by our laboratory (unpublished). This rearrangement resulted in a physical translocation of the expression site, presumably to a previously silent *pil* locus in the genome, but surprisingly, did not cause detectable changes in the standard *Cla*I Southern hybridisation pattern. Instead, when the genomic DNA was treated with other restriction enzymes, changes in the pattern were observed. While this type of rearrangement is not very common, one intriguing aspect is that it causes the *pil*E locus to be linked with a new opacity expression locus (*opa*E, see below).

Yet another novel *pil*E specific rearrangement has recently been observed and seems to occur quite frequently (Haas *et al.*, 1987; and further unpublished data). This type of recombination causes a reversible extension of the EC gene, while concomitant changes in other *pil* loci apparently do not occur. The extensions in EC result in elongated transcripts and translational fusions of pilin. Such pilin fusions (termed L-pilin) are over-sized and are not assembled into pili (Haas *et al.*, 1987). EC-extensions most likely are not the result of simple insertions, but may rather be caused by unequal gene conversion events replacing a segment in EC by a larger gene segment from another region in the gonococcal genome. Further DNA sequencing and hybridisation with defined oligonucleotides might provide insight into the details of this interesting P$^-$ phase, the physiological significance of which also remains to be evaluated.

Figure 5 includes the most frequent and, in our view, most relevant phases of gonococcal pilin expression. The majority of transitions in strain MS11 are phase and antigenic switches involving the P$^+$ and P^S phases (e.g. from P$^+$ to P^S to P$^+$ or, respectively, from P$_1^+$ to P$_2^+$), a direct consequence of minicassette reassortments in *pil*E(EC). The pathways leading to such reassortments are manifold, but gene conversion (or conversion via transformation) must be invoked as the most frequent mechanism. Phase transitions from P$^+$ to P$^-$ seem to occur by different mechanisms and – due to the misleading colony morphology of most P^S variants – less frequently than previously believed. It appears now that 'P$^-$rp$^+$' variants, described by Swanson *et al.* (1986), are similar to our P^S variants. We could demonstrate that P$^-$rp$^+$ variants, kindly released from Swanson's laboratory, in fact strongly secrete S-pilin (unpublished).

None the less, there exists evidence of non-piliated variants, which do not secrete S-pilin. The most significant rearrangement generating P$^-$ gonococci seems to be the one already mentioned above, which involves an extension in the expression gene and the production

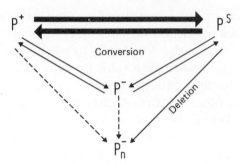

Fig. 5. Scheme for phase and antigenic variation of pilin expression in gonococci including the major phases. Conversion of the expression copy by variant gene segments from silent loci is the most frequent mechanism underlying pilin phase and antigenic variation. It gives rise to P^+ and P^S phase variants, as well as to antigenic variants of these two phases (not included in the drawing). The most frequently occurring P^- variant, included in the scheme, shows an extended expression copy that codes for an over-length pilin (L-pilin). Such phase variants may be caused by an unequal conversion in EC. Their reversion to P^+ or P^S may give rise to antigenic variants. Deletion of a unique expression copy in general generates non-revertible P_n^- variants from P^S, and probably P^+ and revertible P^-.

of L-pilin. Besides this type of P^- variant, Swanson *et al.* (1986) described a 'P^-rp^-' phase in which *pil*E1(EC) has suffered a frame-shift. In fact there seems to be a 'hot spot' for such frame-shift mutations in EC, which seem to accumulate preferentially in *rec*A strains (Koomey *et al.*, 1987). Non-reverting P^- variants, termed P_n^- by Swanson *et al.* (1985), seem to emerge from the P^S phase with low frequency, but may just as easily arise from any other phase (Fig. 5). The biological significance of both reverting and non-reverting P^- variants thus far remains speculative.

THE *opa/opr* GENE FAMILY OF PATHOGENIC *NEISSERIAE*

The organisation of the opacity genes in *N. gonorrhoeae* differs from the pilus genes, as the opacity gene family is composed of complete genes designated *opa*E (E for expression). There is no evidence that *opa*E genes are clustered in the genome as shown for the major pilin loci. Some of the *opa* loci, however, are closely linked to *pil* loci. One *opa*E locus is linked to the *pil*E1 locus (Fig. 1) (Stern *et al.*, 1984). This *pil–opa* gene linkage exibits conserved structural characteristics such as the conserved distance and downstream location of the *opa* loci relative to the *pil* loci. The genetic relevance of the *pil–opa* linkage is not clearly understood. One might speculate on a role in some as yet unknown coordinated mechanism of expression and/or recombination control of the two gene families.

As determined by sequence comparison most of the *opa* genes present in a single genome contain variant structural sequences. Similar to the pilus gene arrangement these sequence variations are clustered in three major hypervariable domains: HV_a, HV_b, and HV_c (Stern & Meyer, 1987). According to our computer predictions the variable domains are exposed outside the gonococcal cell surface; one conserved domain however may be also surface oriented (Meyer *et al.*, 1987*b*).

Recently we described the detection of genes related to the gonococcal *opa* genes in *N. meningitidis* and *N. lactamica* (Stern & Meyer, 1987). DNA sequence analysis of such Opa-related gene sequences (*opr*) reveals a strong homology to the constant regions found in the gonococcal *opa*E genes. Like the *opa*E genes all *opr* genes seem to be complete structural entities. Together they form a large interspecies family with related structural characteristics.

Individual members of the *opa* gene family were observed to exchange variant sequences between each other. This was shown by Southern hybridisation using defined oligonucleotide probes of the HV_b domains of an individual *opa* locus, *opa*E1, found as the smallest band in a typical *Cla*I-restriction pattern. The HV_b probe to the parental gonococcal variant reveals a corresponding sequence, in the *opa*E1 locus, which apparently is lost in the progeny variants. Conversely, the progeny variants in this particular locus acquire another HV_b sequence originating from another *opa*E locus, as specified by the corresponding probe. There are no size changes associated with this recombination. Just as for the pilin genes, we can therefore explain this phenomenon best by internal or external gene conversion.

REPLICATION SLIPPAGE AT A CODING REPEATED
SEQUENCE IS THE PROBABLE CAUSE OF OPACITY
VARIATION

The observation that *opa* genes interact with each other by means of gene conversion does not explain why at any one time none, or only one or two, *opa* genes are expressed in a distinct gonococcal variant. Schwalbe & Cannon (1986) provided initial evidence that the expression phases of individual *opa* (P. II) genes were *cis*-domi-

nant. This was concluded from transformation experiments in which
O⁻ phase gonococci were transformed with chromosomal DNA prepared from defined O⁺ variants. After transformation the recipient
gonococci became O⁺, producing Opa (P. II) characteristic of gonococci from which the donor DNA was originally derived. The phase
of individual *opa* genes must therefore be determined by a *cis*-acting
genetic element.

Our molecular studies showed that all *opa*/*opr* genes were constitutively transcribed (Stern *et al.*, 1986; Stern & Meyer, 1987). However, not all of the *opa*/*opr* transcripts in a cell were successfully
translated into a functional protein. The clue to this puzzling expression regulation was a peculiar sequence found in all *opa*/*opr* genes,
namely, the so-called coding repeat (CR) that codes for the hydrophobic portion of the *opa*/*opr* signal sequence. This sequence was
found to have different numbers of repeat units in most of the loci,
ranging from 7 up to 28. It was appealing to speculate that the number
of CR pentamer units is the critical element by which the reading
frame, and therefore translation, of individual *opa*/*opr* loci is controlled (Fig. 6A).

To test this interesting hypothesis we designed a specific primer
extension experiment that should tell us the number of CR units
in each of the different *opa*/*opr* transcripts of a single clone (Fig. 6B).
The primer used was complementary to the region proximal to the
3′-end of the CR sequence in *opa*/*opr* transcripts and was extended
with reverse transcriptase in the presence of only two nucleoside
triphosphates, dATP and dGTP. These triphosphates were sufficient
to produce the complementary strand to the *opa*/*opr* RNAs, but
caused an abrupt stop since the primer could not be extended beyond
the 5′-end of the CR sequence. Accordingly, the extended primers
formed distinct bands in a polyacrylamide gel corresponding to the
number of CR units in the respective transcripts (Fig. 6).

This style of experiment was performed with a series of isogenic
Opa variants of *N. gonorrhoeae*, *N. meningitidis*, and of *N. lactamica*,
which were either positive (O⁺) or negative (O⁻) for the production
of Opa or Opa-related proteins. By this method, which reveals individual transcripts for each *opa*/*opr* locus, we could demonstrate that
N. gonorrhoeae harbours at least 8 and probably 12 *opa* loci (K.
Muralidharan & T. F. Meyer, unpublished), and that *N. meningitidis*
harbours 4 *opr* loci and *N. lactamica* two. More importantly, we
could show that changes in the expression of *opa*/*opr* are inherently
associated with changes in the number of CR units in distinct tran-

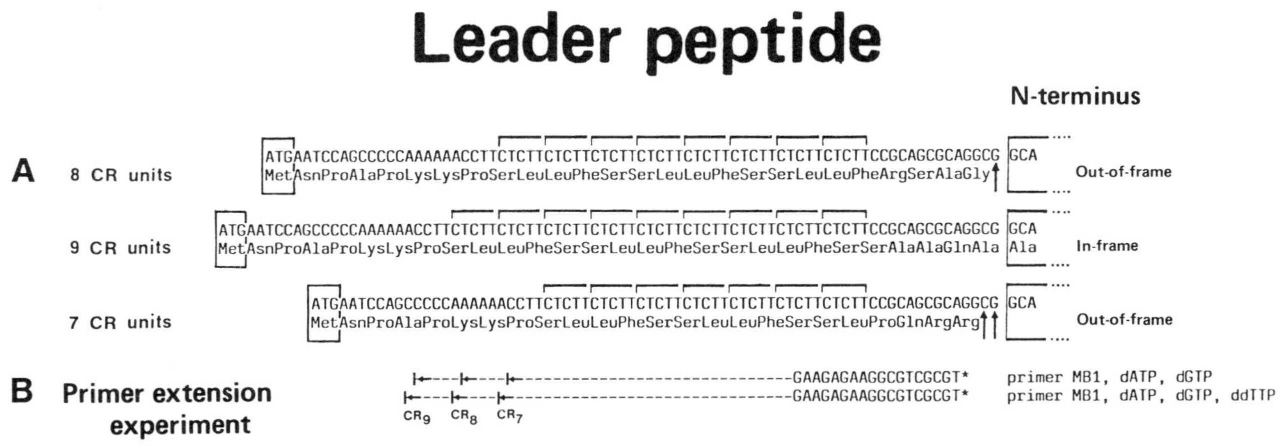
Leader peptide
N-terminus
A
8 CR units
ATGAATCCAGCCCCCAAAAAACCTTCTCTTCTCTTCTCTTCTCTTCTCTTCTCTTCTCTTCTCTTCCGCAGCGCAGGCG GCA
MetAsnProAlaProLysLysProSerLeuLeuPheSerSerLeuLeuPheSerSerLeuLeuPheArgSerAlaGly
Out-of-frame
9 CR units
ATGAATCCAGCCCCCAAAAAACCTTCTCTTCTCTTCTCTTCTCTTCTCTTCTCTTCTCTTCTCTTCTCTTCCGCAGCGCAGGCG GCA
MetAsnProAlaProLysLysProSerLeuLeuPheSerSerLeuLeuPheSerSerLeuLeuPheSerSerAlaAlaGlnAla Ala
In-frame
7 CR units
ATGAATCCAGCCCCCAAAAAACCTTCTCTTCTCTTCTCTTCTCTTCTCTTCTCTTCTCTTCCGCAGCGCAGGCG GCA
MetAsnProAlaProLysLysProSerLeuLeuPheSerSerLeuLeuPheSerSerLeuProGlnArgArg
Out-of-frame
B Primer extension experiment
GAAGAGAAGGCGTCGCGT* primer MB1, dATP, dGTP
GAAGAGAAGGCGTCGCGT* primer MB1, dATP, dGTP, ddTTP
CR9 CR8 CR7

Fig. 6. Model for the control of opa gene expression in pathogenic *Neisseriae*. Section A shows three possible sequence compositions in the 5′-coding region of a single *opa* transcript. Depending on the number of repeat units (which together code for the hydrophobic portion of the variable-length Opa leader peptide) the reading frame is shifted in or out of frame. Section B outlines the primer extension experiment used to determine the number of repeat units in *opa* transcripts. A labelled oligonucleotide primer was annealed to the conserved region downstream of the CR sequence and extended by reverse transcriptase in the presence of dATP and dGTP, with or without the chain terminator ddTTP. After resolution in sequencing gels the extended primers formed distinct bands, corresponding to the number of CR units present in individual transcripts.

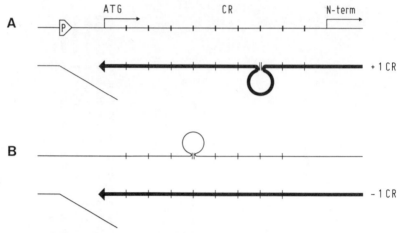

Fig. 7. A possible mechanism causing CR-variation. During the process of DNA replication of an *opa/opr* gene either the template (A) or the newly synthesised DNA strand (B) may slip back, generating an apparent insertion or deletion, respectively, of one or few coding repeat (CR) units. Depending on the number of CR units between the translational start codon (ATG) and the amino terminus of native Opa (N-term) a particular *opa/opr* gene will be in or out of frame. A constitutive promoter (P) causes transcription of *opa/opr* genes regardless of their reading frames. Frame slippage, by a mechanism similar to the one discussed here, may also occur at the transcriptional and/or translational levels and cause residual expression of out-of-frame *opa/opr* genes when cloned in *E. coli* as well as in *Neisseriae* (see Conclusions).

scripts. Precise evaluation at the nucleotide sequence level confirmed that changes in the repeat number of individual gene loci affect the reading frame of these genes and are therefore ultimately responsible for both phase and antigenic transitions in the pathogenic *Neisseriae* (Stern *et al.*, 1986; Stern & Meyer, 1987).

It is questionable if recombination is involved in CR-variation, although CR-variation formally causes deletions or insertions of repetitive elements in the *opa*E structural genes. In our view it is more likely that CR-variation is caused by a slippage process of DNA polymerases at the level of replication (Fig. 7). This assumption is confirmed by our observation that CR-variation is usually not associated with recombination events in the regions flanking the CR repeat. Initial findings by Koomey & Falkow (1985) with their gonococcal *rec*A mutant of strain MS11 led them to suggest that variation of opacity protein ('P. II') expression may be due to a mechanism different from the ones involved in pilin variation. More definitive results with this mutant MS11 strain were recently obtained by Swanson (pers. comm.) demonstrating that CR-variation is *rec*A-independent.

CR-variation

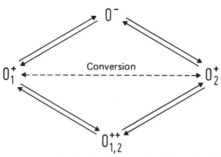

Fig. 8. Possible phase transitions for Opa in pathogenic *Neisseriae*. The principal mechanism causing phase transitions in Opa expression is CR variation. CR variation occurs independently in each *opa* locus, rendering individual loci in the on or off state. Therefore antigenic variation in the *opa* gene system, i.e. the switch from O_1 to O_2, is usually a two-step event requiring, for example, O_1 to be turned off and then O_2 to be turned on. The basic repertoire of antigenic variation is restricted to the number of variant *opa* in a cell. Gene conversions among the *opa* genes occurring at a lower rate may provide a means of extending the basic repertoire (Stern *et al.*, 1986).

Summarising our observations, we would conclude that CR-variation is the main mechanism for *opa/opr* phase transitions in pathogenic *Neisseriae*, while occasional gene conversion events may contribute to increase the basic repertoire of Opa expression, which would otherwise be limited to the (relatively small) number of variant *opa/opr* genes within these bacteria (Fig. 8).

ANTIGENIC VARIATION IN *BORRELIA*

Borrelia hermsii, the causative agent of relapsing fever, expresses an extensive antigenic repertoire during infection of a host (Stoenner *et al.*, 1982), due to the variable major protein (VMP or Protein I) which covers most of the spirochaetae. An isogenic *Borrelia* population spontaneously gives rise to cells with altered serotypes allowing a few organisms to escape the predetermined immune response and to proliferate into a new population. Concomitant with such a relapsing infection, the VMP undergoes dramatic size and serological changes.

The genetic analysis of this interesting phenomenon reveals striking similarities with the variation of the trypanosome surface glycoprotein (VSG) (for reviews see Borst & Greaves, 1987; Pays, this volume). Meier *et al.* (1985) have identified and cloned silent and expressed genes coding for different VMPs. From pulse field electrophoresis and subsequent Southern hybridisation experiments

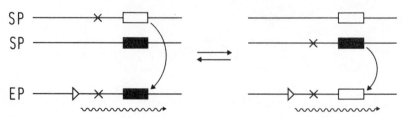

Fig. 9. Antigenic variation in *Borrelia hermsii* by recombination of linear plasmids. Linear plasmids harbour the genes coding for the variable major proteins (VMPs). Some plasmids (silent plasmids, SP) contain storage genes, one plasmid (expression plasmid, EP) harbours the expressed copy. By recombination at a site in front of the structural genes, different VMPs (black and white boxes) are set under control of the promoter (arrowhead) that is located on the expression plasmid.

Plasterk *et al.* (1985) concluded that the VMP genes are located on extra-chromosomal elements (linear plasmids) comparable with mini-chromosomes carrying the VSG genes in trypanosomes. Furthermore, the switch from one VMP serotype to another involves the linkage of a silent gene with an expression site present on a second linear plasmid, the expression plasmid (Plasterk *et al.*, 1985). The authors interpreted this process as a reciprocal recombination event that occurs between one silent linear plasmid and the expression plasmid (see Fig. 9). Subsequent segregation of the recombined plasmids into progeny cells should then lead to variant cell clones. According to the published data, this recombination might also be explained by duplicative transposition or gene conversion (see Borst & Greaves, 1987).

INVERTIBLE ELEMENTS

A genetic mechanism quite different from those utilised in *Neisseria* and *Borrelia* was originally described in *Salmonella* (Zieg *et al.* 1977; see van de Putte, this volume). This Gram-negative organism uses an invertible DNA segment to control the alternate expression of two flagella genes (H1 and H2) which encode different forms of flagellin, the flagellum protein subunit. In various modifications the inversion principle, which is *rec*A-independent, has been revealed in the bacteriophages Mu and Pl (Kamp *et al.*, 1978; Van de Putte *et al.*, 1980; Hiestand-Nauer & Iida, 1983), in *Staphylococcus aureus* (Murphy & Novick, 1979), and in *E. coli*. Indeed, *E. coli* can harbour two site-specific inversion systems, one that controls the expression of common pili (Abraham *et al.*, 1985), and a second of the defective phage e14 (Greener & Hill, 1980) which, in structural terms, closely

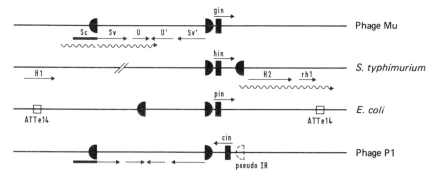

Fig. 10. Variation systems based on DNA inversion. Close structural and functional relationships are shared between the Gin, Hin, Pin and Cin systems that allow alternate switching between two phases. The invertible segments are flanked by indirect repeats (half moons) within which site-specific recombination occurs. Each system carries its own invertase gene which can be located outside (*gin*, *pin* and *cin*) or inside (*hin*) the invertible segment. Within the listed systems the DNA invertases are *trans*-complementable. At their 5'-ends the invertase genes carry the *sis* sequence (black-boxed) which allows binding of the host factor FIS thereby enhancing the DNA inversion reaction. As a consequence of DNA inversion in phage Mu protein splicing (S_c/S_v or $S_c/S_{v'}$) occurs. In *Salmonella typhimurium* the Hin segment directly controls expression of H2 and of the H1-repressor (*rh*1), while H1 expression is controlled indirectly via *rh*1. Note the evolutionary homology of the listed systems as suggested by the relative locations of an inverted repeat (*Pin* has a pseudo IR) upstream of the DNA invertase gene. (Figure modified according to Plasterk & van de Putte, 1984.)

resembles those of *Salmonella* and the phages Mu and Pl (Fig. 10). Furthermore, gene inversion was recently found to constitute the basis of pilin antigenic variation in Moraxella bovis (C. F. Marrs, pers. com.).

Owing to the fact that it has been possible to set up *in vitro* recombination assays the detailed molecular mechanism of the *Salmonella*/Mu-like inversion systems is emerging (Johnson *et al.*, 1986; Mertens *et al.*, 1984; Plasterk *et al.*, 1984; van de Putte, this volume). Each system possesses its own invertase (Hin, Gin, Pin and Cin). These enzymes are essential for the inversion reaction and, due to their high degree of homology, can complement each other. The invertases act at the invertible segment on two distal recombination sites (IR), each of which carries two invertase binding domains. *In vitro*, highly purified DNA invertase alone yields very poor inversion frequencies (Mertens *et al.*, 1986). Full recombination activity is restored by addition of a second factor recently purified from bacterial extracts, termed FIS or Factor II (Johnson *et al.*, 1986; Koch & Kahmann, 1986). For the *Salmonella* Hin system a third factor, the histone-like protein HU, has been implicated to assist in recombination of certain constructs (Johnson *et al.*, 1986).

FIS has been shown to specifically interact with a second sequence element associated with inversion systems, the so-called enhancer sequence (*sis*) (Huber *et al.*, 1985; Johnson & Simon, 1985; Kahmann *et al.*, 1985). The enhancer owes its name to its striking ability to stimulate *in vivo* inversion 20- to 200-fold, independently of its distance from and orientation relative to the inverted repeats. In the various systems the enhancer is located in the 5′-coding region of the respective DNA invertase genes and comprises between 60 and 100 bp. It has been shown that FIS induces bending within *sis* and that the degree of bending is critical for enhancer function (Kahmann, pers. comm.). Kahmann and coworkers further suggest that FIS and the DNA invertase interact when they are bound to their respective sites. This interaction may exert specific topological constraints on the invertible substrate in order to support an efficient site-specific recombination reaction.

It appears that invertible DNA systems, associated with site-specific invertases, have evolved rather specialised forms in order to conduct recombination independent of *rec*A. Interestingly, the host factor FIS may possibly perform further functions in the bacterium besides facilitating DNA inversion. This has been implied by the apparent DNA sequence homologies of the FIS binding domains within *sis* with other procaryotic cistrons (Kahmann, pers. comm.). Significant sequence homology with *sis* can also be found with the *pil*S1 and *pil*E loci of *N. gonorrhoeae* in a region including the RS1 element (unpublished, see Fig. 1). However, the mechanistic relevance of *sis*-related sequences as well as other possible homologies among the various site-specific procaryotic recombination systems, remains to be determined.

CONCLUSIONS

Inversion of a DNA segment seems to be a widely used mechanism among procaryotes which allows them to diverge into heterogeneous populations. The principle of DNA inversion is simple and economical, and in nature several structural versions are realised. The common version allows the alternative production of two different proteins, or, at most two different sets of proteins. The mechanism also permits protein splicing, giving rise to variant proteins with an identical amino terminus. But the range of phenotypes generated by any one inversion mechanism apparently has a narrow upper limit, as one invertible segment never gives rise to more than two different phases. Although in theory one could anticipate more complex systems of variability, simply by the introduction of additional recombination sites, such complex inversion systems have not been found.

Instead, other mechanisms, more suitable for the specific requirements of individual organisms, have evolved for generating greater variability. For example, the location of the VMP genes of *Borrelia hermsii* on extra-chromosomal elements might be important to guarantee mass-production of the surface component that comprises about half of the total protein of the organism. To achieve both mass-production and a high degree of surface antigen variability, intermolecular recombination between the linear plasmids may have proven to be the most convenient mechanism.

In this context it seems intriguing that the gonococcus has evolved two highly complex variation systems, which act in the same cell, but rely on two different molecular mechanisms. While the pilin variation might primarily depend on recombination, the opacity expression might be determined during the course of DNA replication. Common to both systems, however, is the large economic effort made by the gonococcus for their maintenance, using control mechanisms that only affect the translational or post-translational levels. One might therefore speculate that constitutive transcription and translation of the expression genes somehow is advantageous for the gonococcus, even when cells reside in a reversible P^- phase and/or the various O^- phases.

For the opacity system, where the contradiction between transcription and productive protein expression is most apparent, an interesting hypothesis can be raised. This relates to the observation that cloned *opa* genes in the out-of-frame mode (O^-) are usually expressed in *E. coli*, probably due to a transcriptional slipping or translational frame shifting (Stern *et al.*, 1986). The residual expression level of *opa*E (O^-) genes though is very low, so that in *Neisseria* it is not possible to discriminate O^+ heterogeneities in seemingly homogeneous O^- colonies. Still it may be possible that all *opa*E are expressed at a low level even though they are out-of-frame, while in the O^+ phase usually one or two proteins become over-expressed. The biological significance of such a situation is purely speculative, but it would explain why this unusual and seemingly wasteful control mechanism has been evolved. In *N. lactamica* we have found a two-gene system controlled by the same CR-variation mechanism. Why does this bacterium not simply use an invertible element to vary its protein expression? An answer to this question might also bring to light the specific physiological relevance of the principal CR-variation.

ACKNOWLEDGEMENTS

We would like to thank Th. Gretzinger and K. Lamberty for the photography and the art work, respectively. We gratefully appreciated critical comments on the manuscript by Dr P. Manning and the kind communication of unpublished information by Dr R. Kahmann, Dr C. F. Marrs, Dr P. F. Sparling and Dr J. Swanson, The authors' work reviewed in this article was supported by a grant from the Deutsche Forschungsgemeinschaft (Me705/2–3).

REFERENCES

ABRAHAM, J. M., FREITAG, C. S., CLEMENTS, J. R. & EISENSTEIN, B. I. (1985). An invertible element of DNA controls phase variation of type 1 fimbriae of *Escherichia coli*. *Proceedings of the National Academy of Sciences, USA*, **82**, 5724–7.

BERGSTRÖM, S., ROBBINS, K., KOOMEY, J. M. & SWANSON, J. (1986). Piliation control mechanisms in *Neisseria gonorrhoeae*. *Proceedings of the National Academy of Sciences, USA*, **83**, 3890–4.

BLACK, W. G., SCHWALBE, R. S., NACHAMRIN, I. & CANNON, J. G. (1984). Characterization of *Neisseria gonorrhoeae* protein II phase variation by use of monoclonal antibodies. *Infection and Immunity*, **45**, 453–7.

BLAKE, S. M. & GOTSCHLICH, E. C. (1983). Gonococcal membrane proteins: speculation on their role in pathogenesis. *Progress in Allergy*, **33**, 298–313.

BLAKE, M. S. & GOTSCHLICH, E. C. (1984). Purification and partial characterization of the opacity-associated proteins of *Neisseria gonorrhoeae*. *Journal of Experimental Medicine*, **159**, 452–62.

BORST, P. & GREAVES, D. R. (1987). Programmed gene rearrangements altering gene expression. *Science*, **235**, 658–67.

DIAZ, J.-L. & HECKELS, J. E. (1982). Antigenic variation of outer membrane protein II in colonial variants of *Neisseria gonorrhoeae* P9. *Journal of General Microbiology*, **128**, 585–91.

GREENER, A. & HILL, C. W. (1980). Identification of a novel genetic element in *Escherichia coli* K12. *Journal of Bacteriology*, **144**, 312–21.

HAAS, R. & MEYER, T. F. (1986). The repertoire of silent pilus genes in *Neisseria gonorrhoeae*: evidence for gene conversion. *Cell*, **44**, 107–15.

HAAS, R., SCHWARZ, H. & MEYER, T. F. (1987). Release of soluble pilin antigen coupled with gene conversion in *Neisseria gonorrhoeae*. *Proceedings of the National Academy of Sciences, USA*, **84**, in press.

HAGBLOM, P., SEGAL, E., BILLYARD, E. & SO, M. (1985). Intragenic recombination leads to pilus antigenic variation in *Neisseria gonorrhoeae*. *Nature*, **315**, 156–8.

HECKELS, J. E. (1981). Structural comparison of *Neisseria gonorrhoeae* outer membrane proteins. *Journal of Bacteriology*, **145**, 736–42.

HERMODSON, M. A., CHEN, K. C. S. & BUCHANAN, T. M. (1978). *Neisseria pili* proteins: amino-terminal aminoacid sequences and identification of an unusual amino acid. *Biochemistry*, **17**, 442–45.

HIESTAND-NAUER, R. & IIDA, S. (1983). Sequence of the site-specific recombinase gene cin and of its substrates serving in the inversion of the C segment of bacteriophage P1. *The EMBO Journal*, **2**, 1733–40.

HUBER, H. E., IIDA, S., ARBER, W. & BICKLE, T. A. (1985). Site-specific DNA inversion is enhanced by a DNA sequence element in *cis*. *Proceedings of the National Academy of Sciences, USA*, **82**, 3776–80.

IIDA, S. & HIESTAND-NAUER, R. (1986). Localized conversion at the crossover sequences in the site-specific DNA inversion system of bacteriophage P1. *Cell*, **45**, 71–9.

JAMES, J. F., LAMMEL, C. J., DRAPER, D. L. & BROOKS, G. F. (1980). Attachment of *Neisseria gonorrhoeae* phenotype variants to eukaryotic cells and tissues. In *Genetics and Immunobiology of Pathogenic* Neisseria, D. Danielsson and S. Normark, eds. (Sweden: University of Umea), pp. 213–16.

JOHNSON, R. C. & SIMON, M. I. (1985). Hin-mediated site-specific recombination requires two 26 bp recombination sites and a 60 bp recombinational enhancer. *Cell*, **41**, 781–91.

JOHNSON, R. C., BRUIST, M. F. & SIMON, M. I. (1986). Host protein requirements for *in vitro* site-specific DNA inversion. *Cell*, **46**, 531–9.

KAHMANN, R., RUDT, F., KOCH, C. & MERTENS, G. (1985). G inversion in bacteriophage Mu DNA is stimulated by a site within the invertase gene and a host factor. *Cell*, **41**, 771–80.

KAMP, D. R., KAHMANN, R., YIPSER, D., BROKER, T. R. & CHOW, L. T. (1978). Inversion of the G DNA segment of phage Mu controls phage infectivity. *Nature*, **271**, 577–80.

KELLOG, D. S., PEACOCK, W. L., DEACON, W. E., BROWN, L. & PIRKLE, C. I. (1963). *Neisseria gonorrhoeae* I: Virulence genetically linked to clonal variation. *Journal of Bacteriology*, **85**, 1274–9.

KOCH, C. & KAHMANN, R. (1986). Purification and properties of the *Escherichia coli* host factor required for inversion of the G segment in bacteriophage Mu. *Journal of Biological Chemistry*, **261**, 15673–8.

KOOMEY, J. M. & FALKOW, S. (1985). Pilus/Pilin expression and homologous recombination in *Neisseria gonorrhoeae*. In *The Pathogenic* Neisseriae, G. K. Schoolnik, ed. (Washington, D.C.: American Society for Microbiology), pp. 180–7.

KOOMEY, M., GOTSCHLICH, E. C., ROBBINS, K., BERGSTRÖM, S. & SWANSON, J. (1987). Effects of *recA* mutations on pilus antigenic variation and phase transitions in *Neisseria gonorrhoeae*. *Genetics*, in press.

LAMBDEN, P. R. & HECKELS, J. E. (1979). Outer membrane protein composition and colonial morphology of *Neisseria gonorrhoeae* strain P9. *FEMS Microbiology. Letters*, **5**, 263–5.

LAMBDEN, P. R., HECKELS, J. E., JAMES, L. T. & WATT, P. J. (1979). Variations in surface protein composition associated with virulence properties in opacity types of *Neisseria gonorrhoeae*. *Journal of General Microbiology*, **114**, 305–12.

LAMBDEN, P. R., HECKELS, J. E., McBRIDE, H. & WATT, P. J. (1981). The identification and isolation of novel pilus types produced by variants of *Neisseria gonorrhoeae* P9 following selection *in vivo*. *FEMS Microbiology Letters*, **10**, 339–41.

MEIER, J. T., SIMON, M. I. & BARBOUR, A. G. (1985). Antigenic variation is associated with DNA rearrangements in a relapsing fever *Borrelia*. *Cell*, **41**, 403–9.

MERTENS, G., HOFFMANN, A., BLOECKER, H., FRANK, R. & KAHMANN, R. (1984). Gin-mediated site-specific recombination in bacteriophage Mu DNA: overproduction of the protein and inversion *in vitro*. *The EMBO Journal*, **3**, 2415–21.

MERTENS, G., FUSS, H. & KAHMANN, R. (1986). Purification and properties of the DNA invertase Gin encoded by bacteriophage Mu. *Journal of Biological Chemistry*, **261**, 15668–72.

MEYER, T. F., MLAWER, N. & SO, M. (1982). Pilus expression in *Neisseria gonorrhoeae* involves chromosomal rearrangement. *Cell*, **30**, 45–52.

MEYER, T. F., BILLYARD, E., HAAS, R., STÖRZBACH, S. & SO, M. (1984). Pilus genes of *Neisseria gonorrhoeae*: chromosomal organization and DNA sequence. *Proceedings of the National Academy of Sciences, USA*, **81**, 6110–14.

MEYER, T. F., HAAS, R. & STERN, A. (1987*a*). Antigenic variation of proteins on the surface of pathogenic *Neisseriae*. *Biological Chemistry* Hoppe-Seyler, in press.

MEYER, T. F., HAAS, R., STERN, A., FIEDLER, H., FROSCH, M., JÄHNIG, F., MURALIDHARAN, K. & VEIT, S. (1987*b*). Variability of proteins on the surface of pathogenic *Neisseriae*. In *Bacterial Vaccines and Local Immunity*, R. Rappuoli, A. Tagliabue, eds., Siena.

MURPHY, E. & NOVICK, R. P. (1979). Physical mapping of *Staphylococcus aureus* penicillinase plasmid pI524: characterization of an invertible region. *Molecular and General Genetics*, **175**, 19–30.

NICOLSON, I. J., PERRY, A. C. F., HECKELS, J. E. & SAUNDERS, J. R. (1987*a*). Genetic analysis of variant pilin genes from *Neisseria gonorrhoeae* P9 cloned in *Escherichia coli*: physical and immunological properties of encoded pilins. *Journal of General Microbiology*, **133**, 553–61.

NICOLSON, I. J., PERRY, A. C. F., VIRJI, M., HECKELS, J. E. & SAUNDERS, J. R. (1987*b*). Localization of antibody-binding sites by sequence analysis of cloned pilin genes from *Neisseria gonorrhoeae*. *Journal of General Microbiology*, **133**, 825–33.

NORGREN, M., NORMARK, S., LARK, D., O'HANLEY, P., SCHOOLNIK, G., FALKOW, S., SVANBORG-EDEN, C., BAGA, M. & UHLIN, B. E. (1984). Mutations in *E. coli* cistrons affecting adhesion to human cells do not abolish Pap pili fiber formation. *The EMBO Journal*, **3**, 1159–65.

PEARCE, W. A. & BUCHANAN, T. M. (1978). Attachment role of gonococcal pili. *Journal of Clinical Investigation*, **61**, 931–43.

PERRY, A. C. F., HART, C. A., NICOLSON, I. J., HECKELS, J. E. & SAUNDERS, J. R. (1987). Interstrain homology of pilin gene sequences in *Neisseria meningitidis* isolates that express markedly different antigenic pilus types. *Journal of General Microbiology*, **133**, 1409–18.

PLASTERK, R. A., KANAAR, R. & VAN DE PUTTE, P. (1984). A genetic switch in vitro: DNA inversion by Gin protein of phage Mu. *Proceedings of the National Academy of Sciences, USA*, **81**, 2689–92.

PLASTERK, R. H. A. & VAN DE PUTTE, P. (1984). Genetic switches by DNA inversions in procaryotes. *Biochimica Biophysica Acta*, **782**, 111–19.

PLASTERK, R. H. A., SIMON, M. I. & BARBOUR, A. G. (1985). Transposition of structural genes to an expression sequence on a linear plasmid causes antigenic variation in the bacterium *Borrelia hermsii*. *Nature*, **318**, 257–63.

VAN DE PUTTE, P., CRAMER, S. & GIPHART-GASSLER, M. (1980). Invertible DNA determines host specificity of bacteriophage Mu. *Nature*, **286**, 218–22.

ROTHBARD, J. B., FERNANDEZ, R., WANG, L., TENG, N. N. H. & SCHOOLNIK, G. K. (1985). Antibodies to peptides corresponding to a conserved sequence of gonococcal pilins block bacterial adhesion. *Proceedings of the National Academy of Sciences, USA*, **82**, 915–19.

SCHOOLNIK, G. K., TAI, J. Y. & GOTSCHLICH, E. C. (1983). A pilus peptide vaccine for prevention of gonorrhea. *Progress in Allergy*, **33**, 314–31.

SCHWALBE, R. S. & CANNON, J. G. (1986). Genetic transformation of genes for protein II in *Neisseria gonorrhoeae*. *Journal of Bacteriology*, 167, 186–90.

SEGAL, E., BILLYARD, E., So, M., STÖRZBACH, S. & MEYER, T. F. (1985). Role of chromosomal rearrangement in *N. gonorrhoeae* pilus phase variation. *Cell*, 40, 293–300.

SEGAL, E., HAGBLOM, P., SEIFERT, H. S. & So, M. (1986). Antigenic variation of gonococcal pilus involves assembly of separated silent gene segments. *Proceedings of the National Academy of Sciences, USA*, 83, 2177–81.

SPARLING, P. F. (1966). Genetic transformation of *Neisseria gonorrhoeae* to streptomycin resistance. *Journal of Bacteriology*, 92, 1364–71.

STERN, A., NICKEL, P., MEYER, T. F. & So, M. (1984). Opacity determinants of *Neisseria gonorrhoeae*: gene expression and chromosomal linkage to the gonococcal pilus gene. *Cell*, 37, 447–56.

STERN, A., BROWN, M., NICKEL, P. & MEYER, T. F. (1986). Opacity genes in *Neisseria gonorrhoeae*: control of phase and antigenic variation. *Cell*, 47, 61–71.

STERN, A. & MEYER, T. F. (1987). Common mechanism controlling phase and antigenic variation in pathogenic *Neisseriae*. *Molecular Microbiology*, 1, 5–12.

STOENNER, H. G., DODD, T. & LARSEN, C. (1982). Antigenic variation of *Borrelia hermsii*. *Journal of Experimental Medicine*, 156, 1297–311.

SWANSON, J., STEPHEN, M. D., KRAUS, J. & GOTSCHLICH, E. C. (1971). Studies on gonococcus infection. Pili and zones of adhesion: their relation to gonococcal growth patterns. *Journal of Experimental Medicine*, 134, 886–906.

SWANSON, J. (1973). Studies on gonococcus infection. Pili: their role in attachment of gonococci to tissue culture cells. *Journal of Experimental Medicine*, 137, 571–89.

SWANSON, J. & BARRERA, O. (1983). Immunological characteristics of gonococcal outer membrane protein II assessed by immunoprecipitation, immunoblotting and coagglutination. *Journal of Experimental Medicine*, 157, 1405–20.

SWANSON, J., BERGSTRÖM, S., BARRERA, O., ROBBINS, K. & CORWIN, D. (1985). Pilus⁻ gonococcal variants: evidence for multiple forms of piliation control. *Journal of Experimental Medicine*, 162, 729–44.

SWANSON, J., BERGSTRÖM, S., ROBBINS, K., BARRERA, O., CORWIN, D. & KOOMEY, J. M. (1986). Gene conversion involving the pilin structural gene correlates with pilus⁺ − pilus⁻ changes in *Neisseria gonorrhoeae*. *Cell*, 47, 267–76.

SWANSON, J., ROBBINS, K., BARRERA, O. & KOOMEY, J. M. (1987a). Gene conversion variations generate structurally distinct pilin polypeptides in *Neisseria gonorrhoeae*. *Journal of Experimental Medicine*, 165, 1016–25.

SWANSON, J., ROBBINS, K., BARRERA, O., CORWIN, D., BOSLEGO, J., CIAK, J., BLAKE, M. & KOOMEY, J. M. (1987b). Gonococcal pilin variants in experimental gonorrhoea. *Journal of Experimental Medicine*, 165, 1344–57.

VIRJI, M. & HECKELS, J. E. (1983). Antigenic cross reactivity of *Neisseria* pili: investigations with type- and species specific monoclonal antibodies. *Journal of General Microbiology*, 129, 2761–68.

WATT, P. J. & WARD, M. E. (1981). Adherence of *Neisseria gonorrhoeae* and other *Neisseria* species to mammalian cells. In *Bacterial Adherence*, E. Beachey, ed. (New York: Chapman and Hall), pp. 255–88.

ZIEG, J., SILVERMAN, H., HILMEN, M. & SIMON, M. (1977). Recombinational switching for gene expression. *Science*, 196, 170–2.

EUKARYOTIC SYSTEMS

THE YEAST RETROTRANSPOSON TY AND RELATED ELEMENTS

ALAN J. KINGSMAN[1], SALLY E. ADAMS[2], SANDRA M. FULTON[1], MICHAEL H. MALIM[1], PETER D. RATHJEN[1], WILMA WILSON[1] AND SUSAN M. KINGSMAN[1].

[1] Department of Biochemistry, South Parks Road, Oxford, OX1 3QU, UK
[2] Department of Molecular Biology, British Biotechnology Ltd, Brook House, Watlington Road, Cowley, Oxford OX4 5LY, UK

INTRODUCTION

The term retrotransposon refers to a class of eukaryotic transposons that move to new genomic locations via an RNA intermediate and a reverse transcriptase reaction. They are structurally and functionally similar to higher eukaryotic retroviral proviruses (Varmus, 1983). Members of this class were isolated first by Finnegan *et al.* (1977) from the *Drosophila* genome although at that time nothing was known of their mode of transposition or their internal genetic organisation. During the past three years a detailed analysis of a yeast retrotransposon, the Ty element, has led to a full description of its genetic organisation and mode of transposition. Indeed, the term retrotransposon was coined following the demonstration that Ty transposes via an RNA intermediate (Boeke *et al.*, 1985). Subsequently, comparison of Ty genetic organisation with other elements, particularly from *Drosophila*, has shown that this class of element is probably widespread.

In this review the yeast Ty element will be concentrated on as this retrotransposon is the best analysed to date. Similar elements in *Drosophila* will be mentioned and comparisons drawn with retroviruses. A thorough treatment of retro-elements in the mammalian genome will be found in the chapter by Kuff *et al.*

THE YEAST TY ELEMENT

Ty was discovered by Cameron, Loh & Davis (1979) as a result of the polymorphism caused by the transposon in the *SUP4* region of chromosome X. There are about 30–35 copies of the element

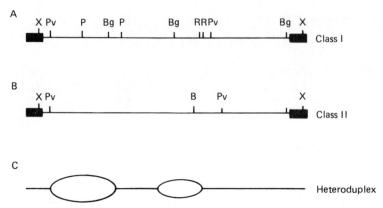

Fig. 1. The structure of class I and class II Ty elements. (a) Class I element Tyl-15. (b) Class II element Ty1-17 (c) Schematic representation of a heteroduplex between Ty1-15 and Tyl-17 (Kingsman *et al.*, 1981). In (a) and (b): thin line = internal epsilon region; closed box = delta sequences; X = XhoI; Pv = PvuII; P = PstI; Bg = BglII; R = EcoRI; B = BamHI.

in the haploid genomes of most laboratory strains of *Saccharomyces cerevisiae* although 'wild' strains vary substantially in their Ty complement. There have been no published reports, however, of a strain which lacks Ty sequences. Ty elements are dispersed, apparently randomly, throughout the genome (Kingsman *et al.*, 1981; Cameron *et al.*, 1979; Klein & Petes, 1984).

General structure

All of the Ty elements isolated to date are about 5.9 kb in length (Fig. 1). They comprise a 5.2 kb unique region, called the epsilon region, flanked by long direct terminal repeats (LTRs), called delta sequences (Cameron *et al.*, 1979). The genomic complement of Ty sequences constitutes a somewhat heterogeneous family (Kingsman *et al.*, 1981; Williamson, 1983) although they fall into two broad classes, Class I and Class II. The two classes differ by two large substitutions (Fig. 1) according to electron microscopic heteroduplex analyses (Kingsman *et al.*, 1981; Williamson *et al.*, 1983). Within each class there are minor variations recognised usually by differences in restriction sites.

The major Ty RNA species is a 5.7 kb 'full-length' transcript that starts in the 5′ or left delta and ends in the 3′ or right delta such that the RNA has a 50 nucleotide terminal redundancy (Elder, Loh & Davis, 1983) (Fig.1). This RNA is both the major message and the intermediate in Ty transposition via a reverse transcriptase reaction (Dobson *et al.*, 1984; Boeke *et al.*, 1985).

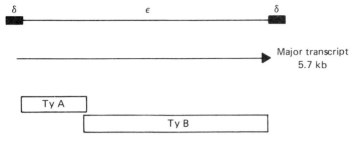

Fig. 2. The genetic organisation of the Ty element.

Several Ty elements have been sequenced recently (Hauber, Nelbock-Hochstetter & Feldmann, 1985; Clare & Farabaugh, 1985; Warmington *et al.*, 1985; Fulton *et al.*, unpublished data). In the case of class I elements this has shown that the 5.7 kb Ty transcriptional unit is divided into two open reading frames, *TYA* and *TYB* (Fig. 2). *TYA* starts at residue 291, within the 5′ delta, and ends at residue 1610. The first ATG within *TYA* is the first ATG of the 5.7 kb RNA. *TYB* overlaps *TYA* by 38 nucleotides starting with an ACA (threonine) codon at residue 1572 and ending at 5556 close to the end of the epsilon region. *TYB* is therefore in the +1 reading phase with respect to *TYA*. The general organisation of class II elements appears to be very similar with minor differences such as *TYA* and *TYB* overlap by 44 nucleotides and *TYB* starts with a GCG (alanine) codon (see later).

DNA/DNA interactions

Ty, and other retrotransposons, transpose through a DNA/RNA/DNA pathway (see later) unlike prokaryotic and other classes of eukaryotic transposons that move only through DNA/DNA interactions (Shapiro, 1983). Nevertheless, because Ty is a middle repetitive, dispersed sequence homologous recombination and conversion events can be mediated by Ty to give rise to a variety of genomic rearrangements. These include deletions, inversions and translocations (Fig. 3).

As well as these interactions that involve Ty:Ty recombination the elements also interconvert at detectable frequencies and delta–delta recombination within a single element can lead to excision of that element to leave a solo delta behind (for a review, see Roeder & Fink, 1982).

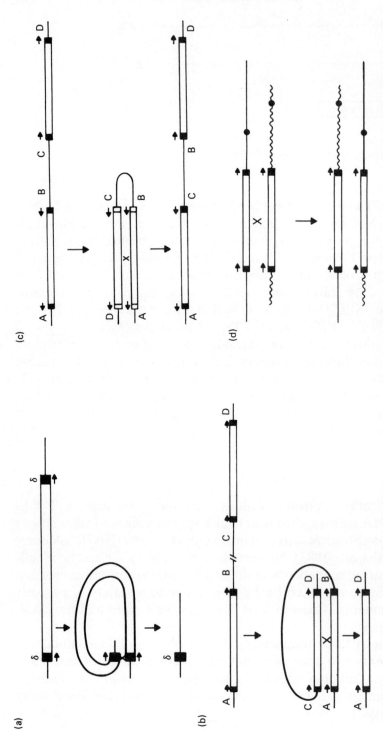

Fig. 3. Recombination interactions of Ty. (a) Deletion of Ty by delta–delta recombination to excise Ty leaving a solo delta. (b) Deletion of chromosomal segment (B–C) between two Ty elements in the same orientation on the same chromosome. (c) Inversion of chromosomal segment (B–C) between two Ty elements in opposite orientation on the same chromosome. (d) Reciprocal translocation by recombination between two Ty elements on different chromosomes. Ty elements are represented as boxes with closed ends (delta sequences). Arrows represent orientation of delta sequences. In each case A, B, C and D are genetic markers.

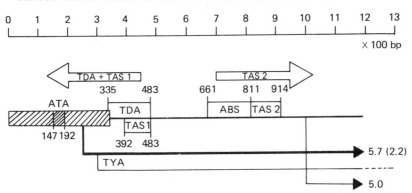

Fig. 4. Ty transcription signals. The first 1200 bp of Ty are shown. Hatched box = 5′ delta sequence. Start sites for the 5.7, 2.2 and 5.0 kb RNA species are marked together with the position of the *TYA* gene. ATA = 'TATA' box; TDA = Ty Downstream Activator; TAS1 = Ty Activator Sequence 1; TAS2 = Ty Activator Sequence 2; ABS = Activation Blocker Sequence. Open arrows indicate the direction of influence of the activator regions.

Transcription signals and regulation

In addition to the 5.7 kb RNA there are two other minor species of 2.2 kb and 5.0 kb (Mellor *et al.*, 1985c; Winston, Durbin & Fink, 1984). The 2.2 kb RNA appears to be 5′ coterminal with the 5.7 kb species (Wilson *et al.*, unpublished data) whereas the 5.0 kb species starts about 800 nucleotides downstream (Fig. 4). The signals that control these transcripts are complex. As might be expected, the 5′ delta sequence has some promoter activity and there are conserved sequences, including a TATA box, within all deltas that are good candidates for functional components (Williamson *et al.*, 1983; Elder *et al.*, 1983; Bowen *et al.*, 1984). In addition, Fulton *et al.* (manuscript submitted) have shown that a transcriptional activator is present between nucleotides 147 and 192. Interestingly, when this region is deleted and the levels of the 5.7 and 2.2 kb RNAs are reduced there is no reduction in the levels of the 5.0 kb RNA. The significance of this is unclear at present although it does suggest that the 5.0 kb RNA is controlled independently. Full transcription from the 5′ delta is not obtained, however, unless a sequence located between nucleotides 335 and 483 is also present. This sequence has been called TDA for Ty downstream activator (Fulton *et al.*, manuscript submitted). Data available to date suggest that the delta signals and the TDA drive the production of the 2.2 kb and the 5.7 kb species.

Clearly, the nature of the 5.7 kb RNA suggests that deltas possess both transcription promoters and terminators but that only the promoter is active in the 5′ delta and only the terminator is active in the 3′ delta. It is not clear how the termination signals are suppressed in the 5′ delta but it seems likely that the promoter activity of the 3′ delta is suppressed by a '5′ promoter-dominance' phenomenon similar to that decribed for avian leukosis virus (Cullen, Lomedico & Tu, 1984). Presumably the 2.2 kb RNA is produced by relatively infrequent premature termination of the 5.7 kb species although this low efficiency terminator has not yet been identified.

The promoter for the 5.0 kb RNA has not been identified. Given its 5′ terminus it seems likely that it will be located downstream from the 5′ delta. A good candidate for a component of the signals that drive this transcript is a sequence called TAS2 (Ty activator sequence) located between nucleotides 811 and 914 (Fig. 4). TAS2 was identified as a sequence that could reactivate the promoter of the yeast *PGK* gene that lacked its own UAS (upstream activator sequence). TAS2 is a bidirectional transcriptional activator when isolated alone but in its usual Ty environment it directs transcription only towards the 3′ delta because of the presence of an activation blocker sequence (ABS) located between nucleotides 661 and 811 (Fig. 4) which prevent activation upstream (Rathjen *et al.*, unpublished data).

Ty transcription is regulated by the mating status of the cell. In mating competent *MATa* or *MATα* haploids steady state RNA levels are about 20-fold higher that in non-mating *MATa/α* diploids (Elder *et al.*, 1980). Sequences responsible for this mating type sensitivity are located between nucleotides 335 and 483, close to the TDA, and it is possible that mating type regulation is, in fact, mediated through the TDA. There are some partial homologies to sequences that mediate mating type control in this region but their function has not been tested.

Transcription is also influenced by other 'host' genes. The most notable of these is the *SPT3* gene (Winston *et al.*, 1984). *spt3* mutants were originally identified as suppressors of certain *his4* mutations caused by Ty insertion into the 'promoter' region of *HIS4*. The mechanism of this suppression is not known but in *spt3* mutants the 5.7 kb RNA is no longer produced and the 5.0 kb species becomes the major transcript (Winston, Durbin & Fink, 1984; Winston & Minehart, 1986).

Gene disruption and gene activation

Like other transposons Ty can have a profound effect on 'host' gene expression as it moves around the genome. Clearly, integration of a 5.9 kb piece of DNA can disrupt gene activity by acting as a sort of 'molecular brick' (Roeder *et al.*, 1980; Eibel & Philipsen, 1984). However, more interesting than this insertional inactivation is the 'host' gene activation that can occur in some circumstances when Ty transposes into the 5' flanking region of a gene. Not only is the 'host' gene activated but it is also brought under mating type control. For example, when Ty integrates into the promoter region of the *ADR2* gene of a haploid cell, the gene is no longer regulated by its normal control circuits that respond to carbon source but is expressed at constitutive high levels (Williamson *et al.*, 1983). However these high levels are only maintained in mating competent haploids. In *MATa/α* diploids expression drops 5- to 20-fold. The mutants that express this complex phenotype are known as ROAM (Regulated Overproducing Alleles responding to Mating-type) mutants (Errede *et al.*, 1980). ROAM mutant derivatives of other genes such as *CYC7* have been isolated (Errede *et al.*, 1980) or, as in the case of the *PGK* gene, created *in vitro* (Rathjen *et al.*, unpublished data) (Fig. 5). ROAM mutations are not due to Ty providing a mating-type sensitive promoter because in every case Ty transcription is away from the activated gene (Williamson *et al.*, 1983; Elder *et al.*, 1983; Errede *et al.*, 1984). The relationship of Ty to the activated gene is, in fact, similar to the category III mode of activation of *c-myc* by ALV (Payne, Bishop & Varmus, 1982) and it has been assumed that an enhancer-like phenomenon is involved. This assumption has proven correct but unlike the analogous situation with ALV the enhancer-like sequences within Ty that are responsible for the ROAM effect are not located within the deltas (Errede *et al.*, 1985; Roeder, Rose & Pearlman, 1985; Rathjen *et al.*, unpublished data). Several preliminary analyses have localised 'ROAM' sequences to various portions of the first 1.0 kb of the element. However, recently Rathjen *et al.* (unpublished data) have created ROAM mutations of the yeast *PGK* gene *in vitro* by inserting various Ty fragments into the 5' promoter region. This has definitively localised the 'ROAM' activator, designated TAS1 (Ty activator sequence) and regulator to the region between nucleotides 392 and 483. Interestingly, this activator is unidirectional and only activates transcription upstream with respect to Ty transcription. It is conceivable therefore that TDA, TAS1, the ROAM mating type effector and the mating type responsive sequence controlling Ty expression are all the same. Certainly, at present, they have not been separated. The unidirectional TAS1 would activate anything upstream irrespec-

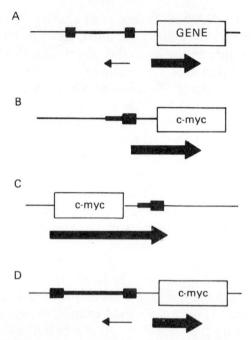

Fig. 5. The relationship of Ty gene activation to retroviral activation of oncogenes. (a) Ty activation, the ROAM phenomenon (see text). Ty is shown as a thick line flanked by closed boxes (delta sequences) integrated upstream from a yeast gene. The small arrow represents Ty transcription. The large arrow represents the activated transcript. (b) Activation of c-myc by ALV type I. ALV DNA is shown as a thick line with the LTR represented by a closed box. The ALV provirus is partially deleted and the 3' LTR is providing a powerful promoter for c-myc expression. Thick arrow represents activated transcript. (c) Activation of c-myc by ALV type II. Symbols as in (b). c-myc transcription is activated by the presence of the LTR enhancer 3' to the c-myc transcriptional unit. (d) Activation c-myc by ALV type III. Symbols as in (b). Thin line = ALV transcription. c-myc transcription is activated by ALV enhancer. (b), (c) and (d) follow the results of Payne *et al.* (1982).

tive of the orientation of the RNA start site that was being affected. This would be within the 5' delta or any flanking genes.

Ty proteins

Ty encoded proteins have been identified and characterised by massively overexpressing all or part of the 5.7 kb transcriptional unit (Dobson *et al.*, 1984; Mellor *et al.*, 1985a). In particular either the *TYA* region alone or both *TYA* and *TYB* from the class I element Ty1–15 have been overexpressed from a high efficiency yeast expression vector to the extent that Ty proteins become sufficiently abundant that they are easily visualised on simple SDS–PAGE gels stained

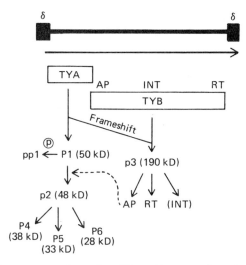

Fig. 6. Ty expression strategy and proteins. AP = acid protease homology in *TYB* or protein from p3; INT = retroviral integrase homology in *TYB* or protein from p3; RT = reverse transcriptase homology in *TYB* or protein from p3. INT protein is in parentheses because it has not yet been identified. (see text for explanation).

with coomassie blue or by autoradiography. Six Ty proteins have been identified to date (Fig.6) The primary translation product of *TYA* is protein p1. This protein is produced by the simple translation of *TYA* (Dobson *et al.*, 1984). p1 has a molecular weight of about 50 kD. It is slightly basic with a pI of 7.8. It is proline rich, can be phosphorylated and binds DNA and RNA in crude binding assays (Mellor *et al.*, 1985*a*). The sequence of *TYA* shows amino acid homology with Tn*3* resolvase and bacterial DNA binding proteins (Clare & Farabaugh, 1985; Warmington *et al.*, 1985).

TYB is translated by a complex mechanism. Translation starts at the beginning of *TYA* and proceeds to the *TYA*:*TYB* overlap region (Fig. 2). Within the overlap region a ribosomal frameshift occurs moving the reading phase to that of *TYB* (Mellor *et al.*, 1985*b*; Clare & Farabaugh, 1985; Wilson *et al.*, 1986). Translation then proceeds to the end of *TYB* to produce a *TYA*:*TYB* fusion protein of about 190 kD called p3 (Mellor *et al.*, 1985*a*). About 1/20 translation initiations at the start of *TYA* proceed through the frameshifting event (Wilson *et al.*, unpublished data). The nucleotide sequence of *TYB* reveals amino acid sequence homology with retroviral proteases, integrases and reverse transcriptases (Fig. 6) (Clare & Farabaugh, 1985; Mount & Rubin, 1985; Warmington *et al.*, 1985). Both p1 and p3 are precursor proteins that are subsequently cleaved to produce mature Ty proteins of which p2, p4, p5 and p6 are those identified to date. p2, p4, p5 and p6 are all processed products of p1 (Adams *et al.*, 1987). Although it is known that p3 is rapidly cleaved *in vitro* (Mellor *et al.*, 1985a) no *TYB* specific products have been found to date. This is probably because they are present at relatively low levels as their production is dependent on the relatively rare frameshifting event. As illustrated in Fig. 6, p2 is the primary cleavage product from p1 and Adams *et al.* (1987) have shown that this processing event is mediated by an enzyme encoded by the region of *TYB* that shows amino acid homology to retroviral proteases.

Ty virus-like particles

Ty employs an obviously viral strategy to produce its protein products in that genetic information is compact, differential product dosage is achieved through a genetically economical procedure and precursor proteins are used. This suggested that a search for virus-like structures composed of Ty proteins might be rewarding. Electron microscopy of thin sections of yeast cells that were massively overproducing the entire Ty transcriptional unit revealed very large numbers of virus-like particles (Ty-VLPs) (Mellor *et al.*, 1985c; Garfinkel, Boeke & Fink, 1985) (Fig. 7). These Ty-VLPs were easily purified on sucrose gradients and shown to contain mainly p2 but also p1, p4, p5 and p6 (Adams *et al.*, 1987). The particles are roughly spherical with a diameter of about 60 nm. They have a doughnut-like appearance with a dark centre and a light shell. They resemble to some extent the intracisternal A-type particles seen in mice (Kuff, Wixel & Leuders, 1968). They contain Ty RNA and reverse transcriptase and are capable of synthesising, *de novo*, Ty DNA in the absence of any exogenous primer:template (Mellor *et al.*, 1985c). The reverse transcriptase in the Ty-VLPs is encoded by the region of *TYB* that shows amino acid homology to retroviral reverse transcriptases (Adams *et al.*, 1987).

A model for Ty transposition

Boeke *et al.* (1985) inserted, *in vitro*, an intron into the 3' end of *TYB* and then showed that when the modified Ty element transposed the intron was removed. This demonstrated clearly the long suspected involvement of a RNA intermediate in Ty transposition. Implicit in the role of RNA is the involvement of reverse transcrip-

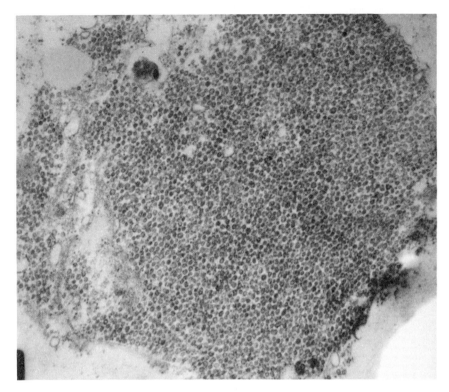

Fig. 7. Electron micrograph of Ty-VLPs. A cross section of a cell that is overexpressing Ty proteins is shown. It is full of 60 nm virus-like particles. Electron micrograph produced by Professor Keith Gull, University of Kent.

tase and a mechanism for priming the synthesis of a doubled stranded DNA copy of Ty from the 5.7 kb transcript. By analogy with retroviral systems (see later) priming of 'plus' strand synthesis is probably by a tRNAmet molecule as there is homology to the 3' end of this tRNA at the extreme 5' end of the epsilon region (Eibel *et al.*, 1984). The mechanism of negative strand priming is unclear but Warmington *et al.* (1985) have suggested that a 'reverse complement' sequence located at the 3' end of the epsilon region and in the delta sequences may be involved.

Clearly, the Ty-VLPs contain the key components for reverse transcription and on this basis we suggest that they function as a 'transposisome'. Transposition is then seen as a cycle of autonomous replication involving particle associated reverse transcriptase and Ty RNA (Fig. 8). A transposition event starts with transcription of Ty to produce the 5.7 kb transcript. This transcript is used to

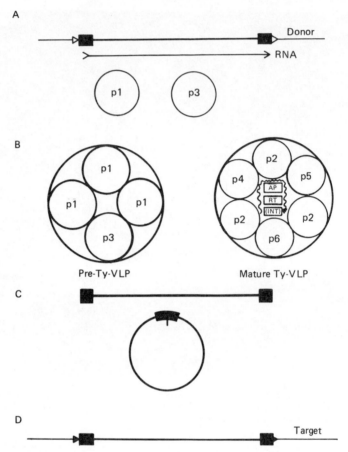

Fig. 8. A model for Ty transposition. (a) Ty is shown at a donor site prior to transposition to a target site. Open arrows flanking the element represent the short (5 bp) direct repeats created when Ty integrates. The 5.7 kb RNA is produced that acts both as message and as a transposition intermediate. p1 and p3 are produced from this message. (b) p1 and p3 assemble into a pre-Ty-VLP and then undergo maturation to produce a mature Ty-VLP containing predominantly p2 but also p4, p5 and p6, acid protease (AP), reverse transcriptase (RT), Ty RNA and presumably an integrase (INT) and a tRNAmet primer. (c) The Ty RNA is copied into either a linear or circular DNA molecule and that molecule then integrates into a target site (d) creating via the integration event site specific 5 bp direct repeats flanking the element. Transposition is, therefore, replicative.

produce proteins p1 and p3. Adams *et al.* (1987) have shown that p1 can produce precursor Ty-VLPs (pre-Ty-VLPs) without proteolytic processing. It is likely, therefore, that *in vivo* these pre-Ty-VLPs form and that proteolytic maturation takes place within the particle. The mature particle would then contain the various p1 related proteins, reverse transcriptase and the 5.7 kb Ty RNA as template. Reverse transcription would lead to the production of a double stranded, probably circular DNA that would then integrate into the target site. This integration event, like almost all similar non-homologous events, creates a short duplication of the target site (Farabaugh & Fink, 1980). Although homology with retroviral integrases has been observed within *TYB* (Mount & Rubin, 1985; Warmington *et al.*, 1985) the integrase has not been identified. To date this model has not been tested rigorously. It is possible, although we believe unlikely, that the Ty-VLPs are irrelevant to transposition and that all the reactions involved in retrotransposition go on free in solution unassociated with Ty-VLPs. These considerations will be resolved shortly.

Ty variability

An intriguing property of yeast Ty sequences is the variation between different members of the family. As mentioned previously these variations can be divided into the major differences between class I and class II elements (Fig. 1) (Kingsman *et al.*, 1981; Williamson *et al.*, 1983) and the minor differences within the classes. Superficial considerations of this variability have led to the suggestion that Ty elements are inactive and are therefore free to drift. In fact Boeke *et al.*, (1985) have suggested that the reason for the relatively low frequency of Ty transposition in wildtype strains is that most of the 30–35 copies of Ty in the genome are inactive. These arguments are based on the observation that transposition frequencies can be dramatically increased as a result of a relatively minor increase in expression of a single 'active' element. However, it seems that this is not correct. Many distinct elements have been shown to transpose suggesting that many elements are at least *cis* active (Williamson *et al.*, 1983; Boeke *et al.*, 1985). In addition, five Ty elements sequenced so far show the same genetic organisation and expression strategies. This is not compatible with most elements being free to drift but rather suggests that the integrity of the *TYA* and *TYB* genes has been selected and that these genes are, therefore, functional and active.

A more interesting feature of Ty variability is the maintenance of both class I and II elements within the same cell. The problem is best illustrated by consideration of the *TYA* genes of the two classes (Fulton *et al.*, 1985). The *TYA* regions roughly correspond to the first substitution loop seen in heteroduplex between class I and class II elements (Fig. 1) (Kingsman *et al.*, 1981; Williamson *et al.*, 1983). The expectation was that these two regions would be substantially different at the DNA level. Nucleotide sequence information confirms this (Fulton *et al.*, 1985; Warmington *et al.*, 1985) but rather than the differences being due to a large-scale substitution the two *TYA* genes differ by a large number of point mutations such that the total aligned homology is about 64% but this homology is totally dispersed. This means that even though Ty elements are known to interact by gene conversion (Roeder & Fink, 1982), a process that should lead to homogeneity, the two classes seem to have evolved by many small mutations rather that a single, block event. These considerations become even more complex when the *TYA* encoded protein sequences of class I and class II elements are compared. They share about 48% homology with 52% of the differences being due to conservative changes. This shows that even though there has been substantial variation at the DNA level that variation has been constrained by selection for the function of p1. Clearly then, both classes are functional as far as analyses to date can tell. Both tranpose and both form Ty-VLPs (Adams, Kingsman & Kingsman, 1987). However, it is possible that the two classes fulfil subtly different, complementing functions. Alternatively the two classes may have evolved separately and only recently come together within the same strain.

A COMPARISON OF TY AND RETROVIRAL PROVIRUSES

Ty obviously resembles retroviral proviruses in a number of ways (Figs 2, 6, 7, 9) (Varmus, 1983; Adams *et al.*, 1987; Fulton *et al.*, 1987). At the DNA level both are a few kilobases in length, both possess LTRs of a few hundred nucleotides and both create short direct repeats in target DNA. They both produce a terminally redundant genomic and messenger RNA that starts in the 5' LTR and ends in the 3' LTR. This RNA is a template for reverse transcription and has tRNA primer binding sites at the 5' end of the internal unique regions. Expression strategies are similar. Retroviral *gag* and *pol* genes have the same relationship as *TYA* and *TYB* respectively. In a virus such as Rous Sarcoma Virus *gag* is expressed as a primary translation product, $Pr76^{gag}$, of the full-length genomic RNA whereas *pol* is expressed as a *gag:pol* fusion protein, $Pr180^{gag-pol}$, via a frameshift event analogous but not identical to the frameshift event required to produce p3 (Jacks & Varmus, 1985). All retroviruses employ some translational 'misreading' event to fuse *gag* and *pol*.

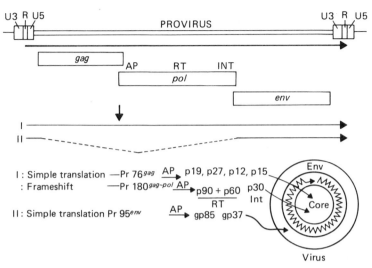

Fig. 9. Organisation and expression of a retrovirus such as ALV. The proviral form of the virus is shown with its LTRs divided into U3, R and U5 regions. The full length transcript running from the 5' LTR to the 3' LTR is also marked. The provirus is divided into three coding sequences, *gag, pol* and *env*. Within *pol* are regions encoding a protease (AP), reverse transcriptase (RT) and an integrase (INT). These coding sequences are translated from two messages, I and II, shown below the coding sequences. Message I is the full length primary transcript and is translated simply to produce a *gag* protein Pr76^{gag}. It is also translated via a frameshifting event that fuses *gag* and *pol* to produce a *gag–pol* fusion protein Pr180$^{gag-pol}$. Both translation products are precursor proteins that are subsequently cleaved to produce the mature viral proteins shown. *gag* derived proteins form the viral core and *pol* proteins provide the enzyme activities required for viral replication. Message II is a spliced RNA that removes the *gag* and *pol* coding sequences to leave only *env*. This is translated simply to produce another precursor protein Pr95^{env} which is subsequently cleaved to produce the mature viral envelope proteins (see Varmus, 1983 for review).

Also, like Pr76^{gag} and Pr180$^{gag-pol}$, p1 and p3 are precursor proteins with p1 and Pr76^{gag}-related products being involved in virus-like or virus particle core structure and p3 and Pr180$^{gag-pol}$ providing the enzyme activities required by the retrosystem (Adams *et al.*, 1987). The enzymes show amino acid homology in key functional regions although the order in which the enzyme activities are encoded by *TYB* and *pol* is different (Fig. 9).

The third gene, *env*, common to retroviruses is, however, absent from Ty (Clare & Farabaugh, 1985; Warmington *et al.*, 1985; Hauber *et al.*, 1985). *TYA* is a *gag* analogue and *TYB* is a *pol* analogue, but there is no *env* analogue. This may be the only significant difference between a retrotransposon and a retrovirus. If a retrovirus lost its *env* gene the particle would not assemble at the cell membrane, presumably would not bud from the cell and would certainly not infect another cell as it would have lost the components that

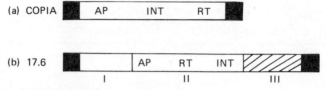

Fig. 10. The genetic organisation of the *Drosophila* retrotransposons copia and 17.6. LTRs are shown as closed boxes flanking the internal part of each element. Homologies with retroviral *pol* sequences and *TYB* are shown as AP for acid protease, INT for integrase and RT for reverse transcriptase. Copia has a single long open reading frame (open box). 17.6 has three open reading frames marked I, II (open boxes) and III (hatched box).

interact with cell surface receptors. It would be, therefore, trapped within the cell able only to transpose rather than infect. Infection may be regarded then as intergenomic transpositions. In this context it is noteworthy that a recently sequenced intracisternal A-type particle element from Hamster (IAP-H18) possesses an *env* region but the gene is riddled with stop codons (Ono *et al.*, 1985).

Despite the differences between Ty and retroviruses the close similarity between this lower eukaryotic transposon and mammalian and avian viruses is remarkable.

RETROTRANSPOSONS IN *DROSOPHILA*

The *Drosophila melanogaster* genome contains several families of dispersed, repetitive transposons the structure and functions of which suggest that they are retrotransposons. However, it has not been formally proven that they move via an RNA intermediate. These elements are generically termed copia-like elements after the first isolated family (Finnegan, 1985) and several have been shown to cause mutations by integrating into genes. The first demonstration of this was for the white-apricot mutation (Bingham & Judd, 1981; Bingham, Lewis & Rubin, 1981; Gehring & Paro, 1980; Goldberg, Paro & Gehring, 1982). Copia-like sequences include the elements copia, 412, 17.6, gypsy, 297, mdg1 and mdg3 (Finnegan, 1985). We will discuss copia and 17.6.

The copia element is 5146 bp in length with LTRs of 276 nucleotides (Mount & Rubin, 1985) (Fig. 10*a*). The most abundant transcript is a full-length species starting in the left LTR and terminating in the right LTR. Unlike Ty this transcriptional unit contains only a single open reading frame running the length of the transcript (Fig. 10*a*). However, this open reading frame would produce proteins with homologies to retroviral *gag* proteins, proteases, integrases and reverse transcriptases. Interestingly, the order of these homology domains in copia is the same as Ty, and unlike retroviruses, in that the positions of the integrase and reverse transcriptase regions are inverted (Figs. 6, 9 & 10*a*). Copia resembles, therefore, a Ty element in which *TYA* and *TYB* have become fused at the DNA level. Also like Ty, copia RNA and proteins are packaged into virus-like particles that contain reverse transcriptase activity (Shiba & Saigo, 1983) and retrovirus-like intermediates in the reverse transcription reaction have been identified (Arkhipova *et al.*, 1986).

17.6, although classified as a copia-like element, has a structure almost identical to that of a retroviral provirus (Fig. 10*b*) complete with an *env* region (Saigo *et al.*, 1984) and a *pol* gene with protease, reverse transcriptase and integrase homologies in the same order as a retrovirus. The element is 7439 bp in length with LTRs of 512 bp. There are three open reading frames (ORFs) designated ORF1, ORF2 and ORF3 which have the same relationship to each other as *gag, pol* and *env* in a retrovirus respectively. ORF1 and ORF2 overlap by 48 nucleotides and ORF2 is in the -1 reading phase with respect to ORF1. This is the most common relationship amongst retroviruses in which there is a frameshift (e.g. Schwartz, Tizard & Gilbert, 1983; Muesing *et al.*, 1985). In fact 17.6 so closely resembles a retroviral provirus that it might truly be either an infectious virus or an endogenous virus. What is clear is that these elements should not be grouped strictly with copia.

Copia and Ty obviously form a subclass of simple, basic units of 'retro-biology' capable only of intragenomic transposition. On the other hand 17.6 and retroviruses are a more elaborate version of Ty and copia that have acquired components, *env* or ORF3, necessary for inter-cellular transposition or infection.

EXPLOITATION OF RETROTRANSPOSONS

Multicopy integrative vectors

Two approaches have been taken to use the yeast Ty element as a vector to carry multiple copies of a DNA sequence into the yeast genome. The first (Wilson *et al.*, unpublished data) placed a defective selectable marker that was required at more than single copy between the delta sequences of a Ty element. This marker was a derivative of the yeast *LEU2* gene that could not supply, at single copy, sufficient isopropyl-malate dehydrogenase to a *leu2* auxotroph for growth. Alongside this marker, but still within the Ty element, was placed a cassette within which a human interferon gene was expressed from the yeast *PGK* promoter. When a *leu2* auxotroph was transformed with this construction and selection was imposed for leucine prototrophy, transformants were selected in which the vector had recombined with several of the endogenous Ty elements. These transformants contained, therefore, multiple integrated copies of the interferon expressing cassette dispersed throughout the genome. Interferon levels from these transformants were consistent with there being as many as 20 copies in some of the transformants. This provides, therefore, a multicopy alternative to the 2u-based vectors normally used.

The second approach is ongoing and arises from the recent detailed analysis of the transposition functions of Ty. Work in several laboratories is in progress to produce a Ty vector that would carry DNA sequences into the genome by transposition with the advantages that copy number is not limited to the number of endogenous elements as is the case for the system described above. The transposition vectors would be analogous to the retroviral vector systems in that the vector would require the presence of a helper element (Mann, Mulligan & Baltimore, 1983) to provide the components of the Ty-VLPs and the enzymes that mediate the retrotransposition pathway. No such system has been reported to date.

Polyvalent antigen carriers

The production of antigens for vaccines, diagnostics or research material by recombinant DNA technology has obvious advantages and applications. However, a substantial disadvantage of using recombinant DNA procedures for vaccines is that antigens are invariably produced as simple monomeric units (Brown, Schild & Ada, 1986). This may not be the ideal configuration for an immunising antigen as it is freely soluble and therefore easily dispersed and it does not permit the cross-linking of the components of the immune system that may be required for strong immune reactions. For this reason there has been considerable interest in finding polyvalent, particulate carrier systems for immunising antigens. The most notable of these being based on the 22 nm HBsAg particles produced in yeast (Valenzuela *et al.*, 1985). Unfortunately, when other

(a) Ty-VLPs

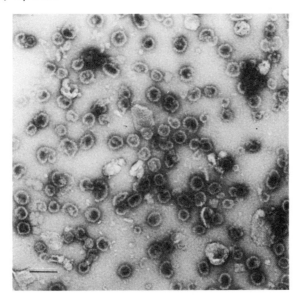

(b) HA:Ty-VLPs

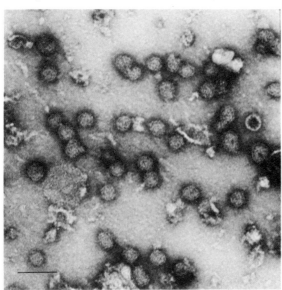

Fig. 11. Purified Ty-VLPs and hybrid Ty-VLPs. (a) Electron micrograph of purified 60 nm Ty-VLPs. (b) Electron micrograph of hybrid HA:Ty-VLPs carrying part of an influenza virus HA. Electron micrographs produced by Professor Keith Gull, University of Kent.

antigens are linked to these particles by expressing appropriate gene fusions in yeast the HBsAg component exerts immunodominance and rather poor responses to the added antigen are obtained.

The observation that the Ty protein p1 could, alone, form Ty-VLPs (Adams *et al.*, 1987) and that these particles contain about 30–100 p1 units suggested that Ty-VLPs might function as polyvalent, particulate antigen carriers. To test this idea fusion genes comprising the first 388 codons of *TYA* and various coding sequences have been expressed at high levels in yeast. These include a whole alpha-2 inteferon cDNA (Malim *et al.*, 1987), the Bovine Papilloma virus E1 and E2 open reading frames (Harrison *et al.*, unpublished data), parts of an Influenza virus HA cDNA and various regions of the HIV *env* gene (Adams *et al.*, 1987). All of these hybrid genes direct the synthesis of p1 fusion proteins and these proteins assemble into hybrid Ty-VLPs. Interestingly the hybrid particles have a distinct morphology (Fig. 11). The hybrid VLPs are able to induce a strong immune response, in rabbits, against the non-Ty components and the antiserea react with native antigen in solution (Adams *et al.*, unpublished data). There is, therefore, every reason to believe that any antigen or part of an antigen can be attached to Ty-VLPs. Hybrid Ty-VLPs are currently being tested as novel vaccines.

ACKNOWLEDGEMENTS

The authors would like to thank Frank Caddick for the excellent artwork and Keith Gull and Ray Newsam for producing the electronmicrographs. Ty research in Oxford is supported by the MRC, SERC and British Biotechnology Ltd.

REFERENCES

ADAMS, S. E., KINGSMAN, S. M. & KINGSMAN, A. J. (1987). The yeast Ty element: recent advances in the study of a model retroelement. *Bioessays* 7, 3–9.

ADAMS, S. E., DAWSON, D. M., GULL, K., KINGSMAN, S. M. & KINGSMAN, A. J. (1987). The expression of hybrid HIV:Ty virus-like particles in yeast. *Nature*, in press.

ADAMS, S. E., MELLOR, J., GULL, K., SIM, R. B., TUITE, M. F., KINGSMAN, S. M. & KINGSMAN, A. J. (1987). The functions and relationships of Ty-VLP proteins in yeast reflect those of mammalian retroviral proteins. *Cell*, 49, 111–19.

ARKHIPOVA, I. R., MAZO, A. M., CHERKASOVA, U. A., SORELOVA, T. V., SCHUPPE, N. G. & ILGIN, Y. V. (1986). The steps of reverse transcription of *Drosophila*. Mobile dispersed genetic elements and U3-R-U5 structure of their LTRs. *Cell*, 44, 555–63.

BENOIST, C., O'HARE, K., BEATHNACH, R. & CHAMBON, P. (1980). The ovalbumin gene – sequence of putative control regions. *Nucleic Acids Research*, **8**, 127–42.

BINGHAM, P. M. & JUDD, B. H. (1981). A copy of the *copia* transposable element is very tightly linked to the *W*ᵃ allele at the *white* locus of *D. melanogaster*. *Cell*, **25**, 705–11.

BINGHAM, P. M., LEWIS, R. & RUBIN, G. M. (1981). Cloning of DNA sequences from the *white* locus of *D. melanogaster* by a novel and general method. *Cell*, **25**, 693–704.

BOEKE, J. D., GARFINKEL, D. J., STYLES, C. A. & FINK, G. R. (1985). Ty elements transpose through an RNA intermediate. *Cell*, **40**, 491–500.

BOWEN, B. A., FULTON, A. M., TUITE, M. F., KINGSMAN, S. M. & KINGSMAN, A. J. (1984). Expression of Ty-lacZ fusions in *Saccharomyces cerevisiae*. *Nucleic Acids Research*, **12**, 1627–40.

BROWN, F., SCHILD, G. C. & ADA, G. L. (1986). Recombinant vaccinia viruses as vaccines. *Nature*, **319**, 549–550.

CAMERON, J. R., LOH, E. Y. & DAVIS, R. W. (1979). Evidence for transposition of dispersed repetitive DNA families in yeast. *Cell*, **16**, 739–751.

CLARE, J. & FARABAUGH, P. J. (1985). Nucleotide sequence of a yeast Ty element: evidence for an unusual mechanism of gene expression. *Proceedings of the National Academy of Sciences*, USA, **82**, 2829–38.

CULLEN, R., LOMEDICO, P. T. & TU, G. (1984). Transcriptional interference in avian retroviruses – implications for the promoter insertion model of leukaemogenesis. *Nature*, **307**, 241–5.

DOBSON, M. J., MELLOR, J., FULTON, A. M., ROBERTS, N. A., BOWN, B. A., KINGSMAN, S. M. & KINGSMAN, A. J. (1984). The identification and high level expression of a protein encoded by the yeast Ty element. *The EMBO Journal*, **3**, 1115–19.

EIBEL, H., GAFNER, J., STOTZ, A. & PHILIPPSEN, P. (1980). Characterisation of the yeast mobile element Ty. *Cold Spring Harbor Symposium on Quantitative Biology*, **45**, 609–17.

EIBEL, H. & PHILIPSEN, P. (1984). Preferential integration of yeast transposable element Ty into a promoter region. *Nature*, **307**, 386–8.

ELDER, R. T., ST JOHN, T. P., STINCHCOMB, D. T. & DAVIS, R. W. (1980). Studies on the transposable element Ty1 of yeast I. RNA homologous to Ty1. *Cold Spring Harbor Symposium on Quantitative Biology*, **45**, 581–91.

ELDER, R. T., LOH, E. Y. & DAVIS, R. W. (1983). RNA from the yeast transposable element Ty1 has both ends in the direct repeats, a structure similar to retrovirus RNA. *Proceedings of the National Academy of Sciences, USA*, **80**, 2432–6.

ERREDE, B., CARDILLO, T. S., TEAGUE, N. & SHERMAN, F. (1984). Identification of regulatory regions within the Ty1 transposable element that regulates iso-2-cytochrome C production in the *CYC7*-H2 yeast mutant. *Molecular and Cellular Biology*, **4**, 1393–1401.

ERREDE, B., CARDILLO, T. S., WEVER, G. & SHERMAN, F. (1980). Studies on the transposable elements in yeast I. ROAM mutations causing increased expression of yeast genes: their activation by signals directed towards conjugation functions and their formation by insertion of Ty1 repetitive elements. *Cold Spring Harbor Symposium on Quantitative Biology* **45**, 593–607.

ERREDE, B., COMPANY, M., FERCHAK, J. D., HUTCHINSON III, C. A. & YARNELL, W. S. (1985). Activation regions in the yeast transposon have homology to mating type control sequences and to mammalian enhancers. *Proceedings of the National Academy of Sciences, USA*, **82**, 5428–32.

FARABAUGH, P. J. & FINK, G. R. (1980). Insertion of the eukaryotic transposable element Ty1 creates a 5 bp duplication. *Nature*, **286**, 352–6.

FINNEGAN, D. J. (1985). Transposable elements in eukaryotes. *International Review of Cytology*, **93**, 281–326.

FINNEGAN, D. J., RUBIN, G. M., YOUNG, M. W. & HOGNESS, D. S. (1977). Repeated gene families in *Drosophila melanogaster*. *Cold Spring Harbor Symposium on Quantitative Biology*, **42**, 1053–63.

FULTON, A. M., MELLOR, J., DOBSON, M. J., CHESTER, J., WARMINGTON, J. R., INDGE, K. J., OLIVER, S. G., DE LA PAZ, P., WILSON, W., KINGSMAN, A. J. & KINGSMAN, S. M. (1985). Variants within the yeast Ty sequence family encode a class of structurally conserved proteins. *Nucleic Acids Research*, **13**, 4097–112.

FULTON, A. M., ADAMS, S. E., MELLOR, J., KINGSMAN, S. M. & KINGSMAN, A. J. (1987). The organisation and expression of the yeast retrotransposon, Ty. *Microbiological Sciences*, **4**, 180–5.

GARFINKEL, D. J., BOEKE, J. D. & FINK, G. R. (1985). Ty element transposition: reverse transcription and virus-like particles. *Cell*, **42**, 507–17.

GEHRING, W. J. & PARO, R. (1980). Isolation of a hybrid plasmid with homologous sequences to a transposing element of *Drosophila melanogaster*. *Cell*, **19**, 897–904.

GOLDBERG, M. L., PARO, R. & GEHRING, W. J. (1982). Molecular cloning of the *white* locus region of *Drosophila melanogaster* using a large transposable element. *The EMBO Journal*, **1**, 93–8.

GROSVELD, G. C., DE BOER, E., SHEWMAKER, C. K. & FLAVELL, R. P. (1982). DNA sequences necessary for transcription of the rabbit β-globin gene *in vivo*. *Nature*, **295**, 120–6.

HAUBER, J., NELBOCK-HOCHSTETTER, P. & FELDMANN, H. (1985). Nucleotide sequence and characteristics of a Ty element from yeast. *Nucleic Acids Research*, **13**, 2745–58.

JACKS, T. & VARMUS, H. E. (1985). Expression of Rous Sarcoma *pre* gene by ribosomal frameshifting. *Science*, **230** 1237–42.

KINGSMAN, A. J., GIMLICH, R. L., CLARKE, L., CHINAULT, A. C. & CARBON, J. (1981). Sequence variation in dispersed repetitive sequences in *Saccharomyces cerevisiae*. *Journal of Molecular Biology*, **145**, 619–32.

KLEIN, H. L. & PETES, T. D. (1984). Genetic mapping of Ty elements in *Saccharomyces cerevisiae*. *Molecular and Cellular Biology*, **4**, 329–39.

KUFF, E. L., WIXEL, N. A. & LEUDERS, K. K. (1968). The extraction of intracisternal A-type particles from a mouse plasma cell tumour. *Cancer Research*, **28**, 2137–48.

LINIAL, M., MEDEIROS, E. & HAYWARD, W. S. (1978). An avian oncovirus mutant (SE21Q1b) deficient in genomic RNA: Biological and biochemical characterisation. *Cell*, **15**, 1371–81.

MALIM, M. H., ADAMS, S. E., GULL, K., KINGSMAN, A. J. & KINGSMAN, S. M. (1987). The production of hybrid Ty:IFN virus-like particles in yeast. *Nucleic Acids Research*, in press.

MANN, R., MULLIGAN, R. C. & BALTIMORE, D. (1983), Construction of a retrovirus packaging mutant and its use to produce helper-free defective retrovirus. *Cell*, **33**, 153–9.

MELLOR, J., FULTON, A. M., DOBSON, M. J., ROBERTS, N. A., WILSON, W., KINGSMAN, A. J. & KINGSMAN, S. M. (1985a). The Ty transposon of *Saccharomyces cerevisiae* determines the synthesis of at least three proteins. *Nucleic Acids Research*, **13**, 6249–63.

MELLOR, J., FULTON, S. M., DOBSON, M. J., WILSON, W., KINGSMAN, S. M. & KINGSMAN, A. J. (1985b). A retrovirus-like strategy for expression of a fusion protein encoded by the yeast transposon, Tyl. *Nature*, **313**, 243–6.

MELLOR, J., MALIM, M., GULL, K., TUITE, M. F., McCREADY, S., DIBBAYAWAN, T., KINGSMAN, S. M. & KINGSMAN, A. J. (1985c). Reverse transcriptase activity and Ty RNA are associated with virus-like particles in yeast. *Nature*, **318**, 583–6.

MOUNT, S. M. & RUBIN, G. M. (1985). Complete nucleotide sequence of the *Drosophila* transposable element copia: homology between copia and retroviral proteins. *Molecular and Cellular Biology*, 5, 1630–8.

MUESING, M. A., SMITH, D. H., CABRADILLA, C. D., BECTON, C. V., LASKY, M. A. & CAPON, D. J. (1985). Nucleic acid structure and expression of the human AIDS/lymphadenopathy virus. *Nature*, 313, 450–7.

ONO, M., TOH, H., MIYATA, T. & AWAYA, T. (1985). Nucleotide sequence of the Syrian hamster intracisternal A-particle gene: close evolutionary relationship of type A particle gene to types B and D oncovirus genes. *Journal of Virology*, 55, 387–94.

PAYNE, G. S., BISHOP, J. M. & VARMUS, H. E. (1982). Multiple arrangements of viral DNA and an activated host oncogene in bursal lymphomas. *Nature*, 295, 209–14.

ROEDER, G. S., FARABAUGH, P. J., CHALEFF, D. T. & FINK, G. R. (1980). The origins of gene instability in yeast. *Science*, 209, 1375–80.

ROEDER, G. S. & FINK, G. R. (1982). Movement of yeast transposable elements by gene conversion. *Proceedings of the National Academy of Sciences, USA*, 79, 5621–5.

ROEDER, G. S., ROSE, A. B. & PEARLMAN, R. E. (1985). Transposable element sequences involved in the enhancement of yeast gene expression. *Proceedings of the National Academy of Sciences USA*, 82, 5428–32.

SCHWARZ, D. E., TIZARD, R. & GILBERT, W. (1983). Nucleotide sequence of Rous Sarcoma virus. *Cell*, 32, 853–69.

SHAPIRO, J. A. (1983). ed. Mobile Genetic Elements. New York, Academic Press.

SHIBA, T. & SAIGO, K. (1983). Retrovirus-like particles containing RNA homologous to the transposable element *copia* in *Drosophila melanogaster*. *Nature*, 302, 119–24.

VALENZUELA, P., COIT, D. & KUO, C. H. (1985). Synthesis and assembly in yeast of hepatitis B surface antigen particles containing the polyalbumin receptor. *Biotechnology*, 3, 317–20.

VARMUS, H. E. (1983). Retroviruses. In Mobile Genetic Elements ed. J. A. Shapiro, pp. 1599–1607. New York, Academic Press.

WARMINGTON, J. R., WARING, R. B., NEWLONG, C. S., INDGE, K. & OLIVER, S. G. (1985). Nucleotide sequence characterisation of Tyl-17, a class II transposon from yeast. *Nucleic Acids Research*, 13, 6679–93.

WILLIAMSON, V. M. (1983). Transposable elements in yeast. *International Review of Cytology*, 83, 1–25.

WILLIAMSON, V. M., COX, D., YOUNG, E. T., RUSSELL, D. W. & SMITH, M. (1983). Characterisation of transposable element-associated mutations that alter yeast alcohol dehydrogenase II expression. *Molecular and Cellular Biology*, 3, 20–31.

WILSON, W., MALIM, M. H., MELLOR, J., KINGSMAN, A. J. & KINGSMAN, S. M. (1986). Expression strategies of the yeast retrotransposon Ty: a short sequence directs ribosomal frameshifting. *Nucleic Acids Research*, 14, 7001–16.

WINSTON, F., DURBIN, K. & FINK, G. R. (1984). The *SPT 3* gene is required for normal transcription of Ty elements in *Saccharomyces cerevisiae*. *Cell*, 39, 675–82.

WINSTON, F., CHALEFF, D. T., VALENT, B. & FINK, G. R. (1984). Mutations affecting Ty-mediated expression of the *HIS4* gene of *Saccharomyces cerevisiae*. *Genetics*, 107, 179–97.

WINSTON, F. & MINEHART, P. L. (1986). Analysis of the yeast *SPT3* gene and identification of its product, a positive regulator of Ty transcription. *Nucleic Acid Research*, 14, 6885–900.

YOSHINAKA, Y., KATOH, I., COPELAND, T. D. & OROSZLAN, S. (1985). Murine leukaemia virus protease is encoded by the *gag-pol* gene and is synthesised through suppression of an amber termination codon. *Proceedings of the National Academy of Sciences, USA*, **82**, 1618–22.

STRUCTURE AND EVOLUTION OF RETROVIRUSES WITH HOMOLOGY TO THE TRANSPOSABLE INTRACISTERNAL A-PARTICLE ELEMENTS OF MICE

EDWARD L. KUFF and KIRA K. LUEDERS

Laboratory of Biochemistry, National Cancer Institute, National Institutes of Health, Bethesda, Maryland 20892, USA

INTRODUCTION

Retrotransposons are sources of genetic variation in many, perhaps most, organisms. Therefore, it is important to understand the factors that govern their persistence, expression and mobility in the genome. Class I retrotransposons (Fanning & Singer, 1987) are regulated by nucleotide sequences contained in (long) terminal repeat units (LTRs), and integrated by a mechanism which produces for each type a fixed-sized duplication of host sequences at the insertion site. Typical class I retrotransposons include the transmissible avian and mammalian retroviruses, and non-infectious intracellular retroviral particles such as the murine intracisternal type-A particle, copia-like elements in *Drosophila* and Ty in yeast. Class II retrotransposons lack LTRs and produce host sequence duplications of variable length at sites of integration. The related LINE-1 elements of rodents and primates (Singer & Skowronski, 1985; Fanning & Singer, 1987), and the I and F elements in *Drosophila* (Fawcett *et al.*, 1986; DiNocera *et al.*, 1983) are examples of Class II retrotransposons. This paper will deal primarily with Class I retrotransposons, with occasional reference to the mammalian LINE-1 elements.

The components encoded by Class I retrotransposons typically assemble in virus or virus-like particulate forms. The particles contain specific RNA transcripts and the structural and enzymatic elements required for particle assembly and maturation, reverse transcription of the RNA and integration of the new DNA copy. Horizontally transmissible class I retrotransposons also encode envelope proteins needed for recognition and penetration of the target cells.

A salient feature of retrotransposons is their tendency to amplify in the host genome. Proviral forms related to type C (murine leukaemia virus) and type B (mouse mammary tumour virus) retroviruses accumulate to copy numbers of 10–50 per haploid genome, with new copies generally thought to result from horizontal infection of oocytes or embryonic cells of the germ line lineage. Vertically transmitted retrotransposons that do not produce transmissible particles can amplify to a much greater extent, reaching copy numbers of 10^3 (IAPs) to 10^5 (LINE 1 elements) per haploid genome in mammalian cells. Generation of new elements by reverse transcription and integration of these into the germ line is believed to be a major source of amplification. However, since transposable elements in several species have been shown to accumulate in regions containing other repetitive sequences (see Dowsett & Young, 1982, and Tchurikov et al., 1980, for copia; Kuff et al., 1986, and Lueders, 1987, for Syrian hamster and mouse IAP elements), other mechanisms such as unequal crossing-over and translocation could significantly affect the copy numbers and chromosomal position. Steele et al. (1986) have suggested that a process other than typical reverse transcription has been involved in the amplification and chromosomal dispersion of some human endogenous MuLV-related retroviral elements, since the amplification events have apparently involved both viral and flanking sequences.

Families of endogenous retrotransposons generally contain a high proportion of defective forms. This is true whether the integrated elements are moderately reiterated, as with the MuLV- and MMTV-related proviruses in mice, or much more abundant as in the case of the IAPs or LINE-1 elements. Defectiveness may result from gross alterations (deletions, rearrangements) or single base mutations that terminate translation in one or more of the open reading frames. Of the 1000 or more IAP elements in the BALB/c mouse genome, about 40% contain major deletions involving the gag and pol genes (Shen-Ong & Cole, 1982; Lueders & Mietz, 1986; see also Fig. 4 and related text). In addition, apparently intact elements can contain multiple small mutations that abolish function (e.g. Toh et al., 1985; Mietz et al., 1987). Boeke et al. (1985) found that retrotransposition of an introduced Ty element in yeast was highly mutagenic in terms of single base changes. Their data suggested that most of the 35 endogenous Ty elements were defective in terms of protein production, even though they could be actively transcribed. A majority of the endogenous MuLV proviral copies are defective for virus production, although some defective elements may be able to express limited portions of their genomes (e.g. gp70). Almost without exception, the new IAP transpositions detected in mouse somatic and germ line DNA have involved grossly deleted and otherwise mutated proviral elements (references in Mietz *et al.*, 1987; Kuff & Lueders, in press 1988). Of the some 10^5 LINE-1 elements in mice and primates, more than 90% are truncated and/or rearranged (Singer & Skowronski, 1985; Fanning & Singer, 1987), and it is not yet clear if, or how many, fully functional copies are retained.

Physical defectiveness is one factor that can restrict expression of endogenous retrotransposons. Sequestration in transcriptionally inactive regions of the genome (e.g. blocks of constitutive heterochromatin or small intercalated clusters of repetitive sequence elements) is another. The activity of integrated mammalian retrotransposons, both newly acquired and endogenous, can be suppressed by methylation of CpG sites within their 5' LTRs (Jahner & Jaenisch, 1985). These and other factors that limit retrotransposon activity are considered at greater length elsewhere (Kuff, 1988). Taking IAPs as an example, the net result is that transcription from this abundant gene family is ordinarily low in normal mouse tissues (Kuff & Fewell, 1985), and the probability of mutagenic retrotranspositions correspondingly reduced. Extensive genomic demethylation, whether associated with the neoplastic state or induced by treatment with 5'-azacytidine, is accompanied by large increases in IAP expression (reviewed in Kuff & Lueders, 1988).

In the following discussion, we will focus on the IAP-related transposable elements and consider (1) the evolutionary relationships between the rodent IAP-related elements and other Class I mammalian and avian retrotransposons, and (2) data suggesting that multiple independent expansions of IAP-related elements have occurred in various rodent species over evolutionary time.

RESULTS AND DISCUSSION

Properties of IAPs

IAPs are intracellular retrovirus-like particles found typically in cells of various rodents (mice, rats, gerbils), and known to be encoded by members of a large family of endogenous proviral elements (Kuff & Lueders, in press 1988). The intracisternal R-particles often found in cultured Syrian hamster cells are almost certainly the expression of an abundant IAP-homologous family of elements in that species

(Lesser *et al.*, 1986). Both the mouse and hamster particles are expressed in preimplantation embryos of the respective species (reviewed by Yotsuyanagi & Szollosi, 1984). IAPs and R-particles both assemble on the endoplasmic reticulum membranes (ER) and bud into the cisternal spaces where they remain without leaving the cells. The IAP gene families in both mice and Syrian hamsters have been intensively studied. Mouse IAPs contain a major 73 kDa *gag*-related structural protein which is equivalent to the uncleaved *gag*-precursor proteins of conventional retroviruses. The genome also contains coding domains for protease, reverse transcriptase and inte-grase (endonuclease) functions. DNA in the usual *env* coding region is closed in all three reading frames in several sequenced examples of mouse and hamster IAP elements. Isolated IAPs contain a Mg-dependent reverse transcriptase activity (Wilson *et al.*, 1974). An ATP-stimulated endonuclease activity was found in IAP fractions prepared from a mouse myeloma (Nissen-Meyer & Eikhom, 1981), but has not yet been shown to be encoded by the provirus. No viral envelope has been detected in IAPs. Fragments of mouse IAP ele-ments and elements with major deletions have been cloned and sequenced (e.g. Burt *et al.*, 1984; Ymer *et al.*, 1986; Moore *et al.*, 1986; Grossman *et al.*, 1987). In addition, a full-sized IAP proviral copy has been cloned and sequenced from the genomes of both *M. domesticus* (BALB/c) and the Syrian hamster, and the gene order and reading frames of both elements have been established by comparison with homologous regions in RSV and the type D retroviruses (Ono *et al.*, 1985; Mietz *et al.*, 1987).

Evolutionary relationships between IAPs and infectious retroviruses

The diagram in Fig. 1 is freely adapted from those prepared by Ono *et al.* (1986) and Sonigo *et al.* (1986) on the basis of sequence relatedness within the *pol* genes. In Fig. 1, distances between the nodal points on the upper horizontal line reflect the relative diver-gence between the indicated taxa. The morphology of the expressed particle(s) and the intracellular site of viral core assembly are shown for each type of retroviral element. Thus, Mason-Pfizer monkey virus (MPMV), of which SRV-1 is a specific subtype (Power *et al.*, 1986), and squirrel monkey retroviruses (SMRV) have type D extra-cellular virions whose cores first appear as dispersed type A particles in the cytoplasmic matrix (cytoplasmic A-particles, CAPs) and subse-quently acquire an outer envelope on budding through the plasma

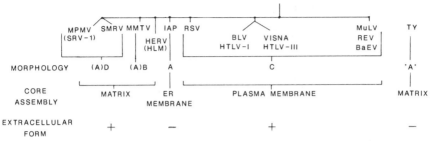

Fig. 1. Evolutionary relationships, morphology, and site of assembly of mammalian and avian retroviruses and retrovirus-like elements. The Ty retrotransposon of yeast is included for comparison. Distances between the nodal points along the upper horizontal line are roughly proportional to the evolutionary divergence between the indicated taxa as calculated by Sonigo *et al.* (1986) and Ono *et al.* (1986) from sequence homologies in the *pol* gene region. Sonigo *et al.* provided additional evolutionary details within the various mammalian type C groups which have been omitted here, and not all members of these groups have been included. Abbreviations are: MPMV, Mason-Pfizer monkey virus; SRV-1, a type of MPMV isolated from Macaque monkeys with an AIDS-like syndrome (Power *et al.*, 1986): MMTV, mouse mammary tumour virus; HERV-K, a human endogenous retrovirus-like element with homology to IAP, recently isolated by Ono *et al.* (1986); HLM, HLM-2, isolated by Callahan *et al.* (1985), a member of the HERV-K family; IAP, intracisternal type-A particle; RSV, Rous sarcoma virus; BLV, bovine leukaemia virus; HTLV-III, now referred to as HIV; MuLV, murine leukaemia virus; REV, avian reticuloendotheliosis virus; BaEV, baboon endogenous virus.

membrane. The inner cores of the type B mouse mammary tumour virus (MMTV) generally appear as clustered CAPs which subsequently move to the plasma membrane and leave the cell. MMTV virions can also assemble entirely at the plasma membrane, as is universally the case with type C retrovirions. The Ty-encoded particles in yeast appear in clusters in the cytoplasmic matrix without obvious relationship to either the ER or plasma membranes (Garfinkel *et al.*, 1985; Adams *et al.*, 1987). Ty particles are not known to leave the cells.

IAPs (and R-particles) are thus far unique among known retrotransposon-encoded particles in their intracellular site of assembly and their final sequestration within the ER cisternae. (Intracisternal type-A particles of unknown genetic origin appear in cells of some non-rodent species; see Yotsuyanagi & Szollosi (1984).) The fact that the IAP genome contains a non-functional segment of DNA in the putative envelope region suggests that these intracellular retrotransposons may have evolved from an originally transmissible precursor. Ty elements, on the other hand, lack an *env* domain entirely (Clare & Farabaugh, 1985; Adams *et al.*, 1987) consistent with an intracellular evolutionary origin.

HLM-2 and HERV-K (human endogenous retrovirus-K) are representatives of a family of proviral elements with about 50 homologous but non-identical copies (Callahan *et al.*, 1985; Ono *et al.*, 1986). A 9.18 kbp provirus, HERV-K10, has been entirely sequenced (Ono *et al.*, 1986) and shown to have homology with both MMTV and Syrian hamster IAP (see also Callahan *et al.*, 1985). It is not known whether a fully competent HERV-K (HLM) proviral copy exists in the human genome; even if not, specific genes could still be expressed from partially functional elements.

The genetic organisation of a mouse IAP proviral element, MIA14, is compared in Fig. 2 with those of the closely homologous Syrian hamster element H18 (Ono *et al.*, 1986) and several more distantly related retroviral genomes. The IAP elements are unique among the mammalian class I retrotransposons in incorporating the protease coding domain within the *gag* open reading frame; however, they share this arrangement with the avian RSV. The mouse and Syrian hamster IAP elements are also unique in having completely closed *env* genes.

The solid bars in Fig. 2 indicate regions of at least 50% amino acid homology between the IAP genomes and the other elements, based on comparisons of their translated nucleotide sequences (Callahan, *et al.*, 1985; Ono, *et al.*, 1985, 1986; Deen & Sweet, 1986; Power *et al.*, 1986; Sonigo *et al.*, 1986; Mietz *et al.*, 1987; K. K. Lueders & J. A. Mietz, unpublished analyses). In general, homologies are much more extensive in the region beginning with p27[gag] and extending through the *pol* gene than in the adjacent N-terminal *gag* and *env* domains. For example, the N-terminal portion of the mouse IAP element MIA14 has no detectable amino acid homology with the equivalent region of the hamster IAP-H18 or any other proviral element. H18 appears to have some homology with the entire *gag* region of MMTV (Ono *et al.*, 1985) but this is much less in the N-terminal region than in the p27 domain. Thus for the N-terminal *gag* domain, 54% of the amino acids in MMTV could be matched with those in H18, but only 22% were identical and the matching required many large and small gaps in both sequences. In p27, however, overall homology was 62%, 38% were identical residues, and only a few small gaps were needed for alignment. HERV-K10 shows good homology with SRV-1 over the entire C-terminal half of the *gag* domain (p27), but only over part of the N-terminal half (Ono *et al.*, 1986). Callahan *et al.* (1985) showed

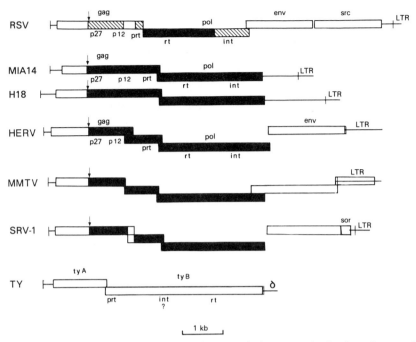

Fig. 2. Genetic organisation of representative retroviral genomes having homology to the mouse IAP element (the yeast Ty retrotransposon is included for comparison). The bars indicate open reading frames defined by nucleotide sequencing. Solid bars represent amino acid homologies of 50% or greater with the mouse IAP element, MIA14. Hatched regions indicate more distant but still detectable homology. Open bars are regions of no apparent homology with MIA14. Thin lines represent non-coding regions of the genomes. The diagrams begin at the 3' boundary of the 5' LTR and extend through the 3' LTR. The boundaries of the 3' LTRs are indicated by short vertical lines. MIA14 and H18 are cloned mouse and Syrian hamster genomic IAP elements that have both been fully sequenced (Mietz et al., 1987; Ono et al., 1985). Other abbreviations are explained in the legend of Fig. 1. A vertical arrow shows the position of the peptide sequence phe–pro–val, which marks the 5' boundary of the p27^{gag} coding domain. p27 is the major capsid protein; p12 a nucleic acid binding protein; prt, protease; rt, reverse transcriptase; int, an integrase (endonuclease) activity. MIA14 and H18 have the same organisation of coding domains; and the arrangement of the *gag, prt* and *pol* open reading frames is the same in HERV, MMTV and SRV-1. δ is the terminal repeat unit of the Ty element; the coding domains of Ty are ordered according to Adams et al. (1987). The position of the integrase coding domain is uncertain at present writing. Nucleotide sequence of RSV (Prague C strain) was reported by Schwartz et al. (1983), that of HERV-K by Ono et al. (1986); MMTV (milk factor) by Moore et al. (1987); SRV-1 by Power et al. (1986); and Ty by Clare & Farabaugh (1985).

by hybridisation that the major regions of homology between HLM-2 and MMTV, SMRV, and the M432 retrovirus spanned the 3' half of the *gag* gene through the entire *pol* region. M432 is an extracellular retrovirus of *M. cervicolor* in which IAP-related sequences have contributed the region p27 through *pol*, and otherwise unknown retroviral sequences make up the rest (Kuff *et al.*, 1978; Callahan *et al.*, 1981).

In terms of the *env* region, SRV-1 (MPMV) is homologous to none of the other elements shown in Fig. 2, but instead to the type C avian reticulo-endotheliosis virus, REV-A (Sonigo *et al.*, 1986). MMTV and HERV-K (HLM-2) share homology over the 3′ third of their *env* genes (Callahan *et al.*, 1985; Ono *et al.*, 1986) and over a short 5′ *env* region, but little or none over the gp52 domain.

A cassette of information for the major capsid protein p27, the nucleic acid binding protein p12, and the three enzymes required for protein processing, reverse transcription and integration has evidently been conserved during evolution of the elements shown in Fig. 2, while the N-terminal *gag* proteins and the envelope components have undergone much more extensive changes. This cassette can be regarded as the basic evolutionary unit of the IAP-related retrotransposons, subject to recombinational events with cellular or other retroviral elements that introduce novel *env* and N-terminal *gag* sequences. The M432 retrovirus of *M. cervicolor* mentioned above is a good case in point. The central IAP-related cassettes of *Mus domesticus* IAP and M432 have the expected species-related nucleotide divergence (ΔT_m of $-6°C$), but the other portions of the two genomes are not homologous (Kuff *et al.*, 1978; Callahan *et al.*, 1981). Similarly, MPMV (SRV-1) has no homology with the *M. musculus* IAP in the N-terminal *gag* region and has an envelope sequence acquired from an otherwise remotely related avian type C virus. Recombination within type C retroviruses is well known. For example, the baboon endogenous virus (BaEV) is a chimeric element with homology to MuLV throughout the *gag* and *pol* regions but related to the Simian type D retroviruses in *env* (Kato *et al.*, 1987).

The proteins encoded by the N-terminal portion of the *gag* gene of extracellular retroviruses are thought to participate in targeting of the *gag* polyprotein precursor to the plasma membrane and in the association between the core and envelope components during assembly of the particles. Myristylation of the N-terminal glycine of the *gag* precursor protein (Henderson *et al.*, 1983) may be essential for plasma membrane localisation in a majority of the known mammalian retroviruses. For example, *in vitro* mutagenesis of the myristylation site in MPMV still permits assembly of the type A cytoplasmic particles but prevents their subsequent association with the plasma membrane (Rhee & Hunter, 1987).

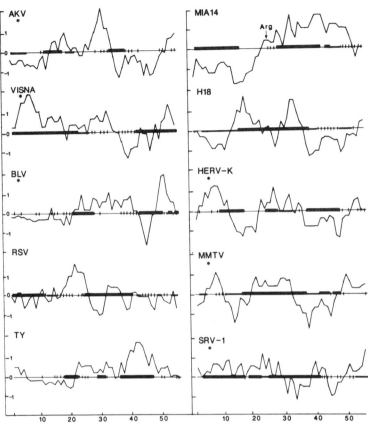

Fig. 3. Hydrophilicity profiles and predicted secondary structures for the N-terminal 55 amino acids encoded by the *gag* regions of the indicated retroviral genomes and the yeast Ty element. The nucleotide sequence for AKV (AKR endogenous ecotropic leukaemia virus) was reported by Etzerodt *et al.* (1984); for visna, by Sonigo *et al.* (1985); and for BLV (bovine leukaemia virus), by Rice *et al.* (1985). Other sequences were reported as described in the legend of Fig. 2. Hydrophilicity profiles and secondary conformation were analysed using the algorithms of Hopp & Woods (1981) and Garnier *et al.* (1978), respectively, as incorporated in the programme of Queen & Korn (1984). Amino acid residue number is shown on the abscissa and the relative hydrophilicity indices on the ordinate. Secondary conformations are indicated along the horizontal lines as follows: α-helix, heavy line; β-sheet, intermediate thickness line; turn, short vertical line; random, thin line. Asterisk indicates that the sequence contains a functional myristylation site (Shultz & Oroszlan, 1984).

Fig. 3 shows, for a number of infectious and intracellular elements, the secondary structures and hydrophilicity profiles predicted for the first 55 amino acids of the *gag* translation product. The corresponding configuration of the yeast TyA gene product is also shown. Virion assembly is clearly consistent with a variety of N-terminal conformations. However, a common motif among mammalian extracellular retroviral proteins is a hydrophobic region in the vicinity of residues 30 to 45. HERV-K and MMTV have similar profiles throughout the entire 55 amino acid segment, and other resemblances can be seen on inspection of the patterns.

The extensive N-terminal hydrophobic domain predicted for the mouse IAP *gag* protein is unique among known retrotransposons. We postulate that this segment can function as a signal peptide by which the nascent IAP protein is brought into association with the ER membrane (Mietz *et al.*, 1987). Arginine was found as the N-terminal residue in p73 from mouse myeloma IAPs (Marciani & Kuff, 1974). Assuming that this protein in myeloma is encoded by a gene(s) with the same or similar N-terminal sequence as MIA14, the position of the first arginine residue (see arrow, MIA14 panel in Fig. 3) is consistent with proteolytic removal of the hydrophobic leader. Since the sequence near the cleavage point is somewhat atypical for a signal peptide cleavage site, we suggested (Mietz *et al.*, 1987) that it might be cleaved slowly if at all by the usual signal peptidase, thus allowing the hydrophobic leader to persist during the few minutes required to assemble the IAPs (Lueders & Kuff, 1975). The hydrophobic leader may ultimately be removed by signal peptidase or other proteases during the day or two required for turnover of the particle proteins.

The sequence of the Syrian hamster IAP-equivalent genomic element, H18, predicts a short moderately hydrophobic N-terminal segment and a more extensive internal hydrophobic region (Fig. 3), with some resemblance to a transmembrane domain. Either or both of these regions might be involved in association of the nascent R-type particle (the presumed expression of Syrian hamster IAP elements) with the ER membrane. Although there is a distant homology between the N-terminal amino acid sequences of the predicted H18 protein and MMTV *gag* (Ono *et al.*, 1985; see remarks above), and a similarity in the internal hydrophilicity profiles (Fig. 3), a 5-residue gap found in the H18 sequence when it is aligned with that of MMTV markedly alters the character of the predicted extreme N-terminal region. Introduction of a short N-terminal hydrophobic segment need not in itself retarget a nascent *gag* protein from the plasma membrane to the ER, if the myristylation site is intact (see, for example, the AKV profile, Fig. 3). However, the myristylation site is also absent in the H18 sequence. Thus, a relatively small change in the N-terminal coding region could have major consequences in terms of the intracellular site of retrovirus assembly, and by extension, the transmissibility of the viral genome. Similar types of small protein sequence changes in the envelope region may not have major consequences of the same sort, since assembly and release of particles at the cell surface can persist in mutants of infectious retroviruses that do not form envelope proteins at all (Linial & Blair, 1985). The complete loss of *env* gene function found in the mouse and hamster IAP elements may have been secondary to changes in the N-terminus of the *gag* gene which retargeted the particle proteins to the endoplasmic reticulum.

Aspects of the recent evolution of IAP-related elements in rodent species

Figure 4 summarises data on the evolution of IAP-related elements in rat, mouse and hamster. Numerals at some branch points of the diagram show the approximate times of divergence in units of 10^6 years (Myr) from the present. An approximate copy number of the IAP-related sequence elements per haploid genome is given for each species. These were taken from the literature (Lueders & Kuff, 1977, 1981, 1983; Suzuki *et al.*, 1982; Ono *et al.*, 1984) or, for the mouse species, determined by dot blot analysis (Kuff, unpublished experiments) using a BALB/c (*M. domesticus*) IAP probe for hybridisation under relaxed conditions (hydridisation at ΔT_m of $-35°C$). Ono *et al.* (1984) reported that *M. cervicolor* (and *M. caroli*) contains several hundred copies of IAP-related sequences rather than the approximately 25 copies previously detected by us (Kuff *et al.*, 1978). Our recent dot blot estimates are in general agreement with their data on these Southeast Asia *Mus* species (see further discussion below).

All of the six species in the group that contains *M. musculus* ('*M. musculus* group') have some 10^3 copies per haploid genome (Fig. 4). The Syrian hamster, but not the Chinese hamster (a different genus) contains about the same number. Approximately 500 copies have been ascribed to the rat (Lueders & Kuff, 1983).

Divergence among members of the IAP-related sequence families within a given species was estimated from the melting point depression ($-\Delta T_m$) of hybrids formed between cloned IAPs and their respective genomic DNAs, relative to the T_m of the self-annealed clones; these data are shown in Fig. 4. Finally, Fig. 4 includes genomic restriction patterns obtained with the DNA of the various mouse species, using a BALB/c genomic probe under relaxed stringency (see legend, Fig. 4), and with hamster DNAs using a cloned Syrian hamster element (Lueders & Kuff, 1983).

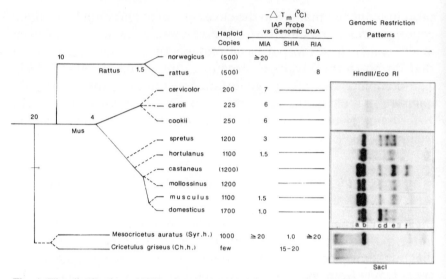

Fig. 4. The distribution of IAP-related sequence elements among various rodent species. Numbers at the bifurcation points of the evolutionary tree indicate the approximate times of divergence in 10^6 years (Myr). Estimates of divergence time for the two *Mus* groups range as high as 6 Myr. The two hamster genera are believed to have diverged 14–15 Myrs ago. Copy numbers for the *Mus* species were determined by dot blot analysis using a cloned *M. domesticus* (BALB/c) IAP element as a hybridisation probe. Dot hybridisations were carried out for 48 hours at a ΔT_m of $-35°C$, and washes at a ΔT_m of $-25°C$. Values were calibrated by reference to a dilution series containing salmon sperm DNA mixed with varying amounts of the cloned IAP sequence. Ono *et al.* (1984) have provided additional data on copy numbers in *Mus domesticus*, *M. caroli* and *M. cervicolor*. Copy numbers for rat and Syrian hamster were from Lueders & Kuff (1983) and Suzuki *et al.* (1982). Values for ΔT_m were taken from Lueders & Kuff (1983) or subsequently determined. Restriction enzyme digests of genomic DNAs (5 μg) were electrophoresed in agarose gels, transferred by blotting to nitrocellulose membranes and hybridised and washed under the conditions described above, except that hybridisation was for only 24 hours, 18 hours at a ΔT_m of $-35°C$ and 6 hours at a ΔT_m of $-25°C$. These conditions may be somewhat more stringent than those used for the dot blot hybridisation. Electrophoretic migration was from left to right. Bands a to e in the patterns presented by the *Mus musculus*-related species are formed from known structural variants of the IAP gene family, as discussed in the text. MIA, SHIA and RIA refer to cloned IAP-related elements from *M. domesticus*, Syrian hamster and rat (*R. norvegicus*), respectively.

The divergence between the mouse and rat IAP sequences (Fig. 4) indicates an independent evolution of IAP-related genes present in the common ancestral line; in fact, the very large ΔT_m of $-20°C$ or more observed for hybrids between the mouse probe and rat genomic DNA suggests that already divergent sequence elements were selected for amplification in the two genera. (The divergence of IAP-related sequences within the two rat species was similar when tested with a cloned *R. norvegicus* probe (ΔT_ms of $-6°C$ and $-8°C$ for *R. norvegicus* and *R. rattus*, respectively.) Heteroduplex analysis of cloned IAP elements from *R. norwegicus* has confirmed the internal heterogeneity of this family (Lueders & Kuff, 1983). Thus, the last major amplification of IAP-related elements in *Rattus* seems to have occurred well before the separation of the two rat species.

The hamster species have markedly different families of IAP sequences (Lueders & Kuff, 1981, 1983). The approximately 10^3 copies in *Mesocricetus auratus* (Syrian hamster) form an internally homogeneous family as judged by the low ΔT_m of $-1°C$ of hybrids formed with a cloned hamster probe (Fig. 4), and by analysis of multiple individual cloned samples (Suzuki *et al.*, 1982; Lueders & Kuff, 1983; Ono *et al.*, 1984). In *Cricetulus griseus* (Chinese hamster), the IAP sequence population is grossly heterogeneous as evidenced by the absence of discrete fragment sizes when genomic DNA is cut with a variety of restriction enzymes (Lueders & Kuff, 1981, 1983; see also Fig. 4). These findings have been interpreted to indicate that IAP-related elements have undergone a recent amplification in the Syrian hamster genome, while in the Chinese hamster they consist primarily of heterogeneous remnants of a more ancient event (Suzuki *et al.*, 1982; Lueders & Kuff, 1983).

The IAP gene family in *M. domesticus* (laboratory mice) is also relatively homogeneous at the nucleotide sequence level, with a ΔT_m of $-1°C$ for hybrids formed between cloned IAP probes and genomic DNA (Fig. 4). Sequencing of various cDNA and genomic clones has revealed a somewhat greater actual divergence, between 2 and 5% at the individual nucleotide level. The ΔT_m could underestimate average divergence within a multigene family if it is biased by the melting behaviour of the more closely homologous members.

About 40% of the IAP elements in inbred laboratory strains contain major deletions which form discrete size-subsets. These contribute to, and can be identified in, genomic restriction patterns (Lueders & Kuff, 1980; Kuff *et al.*, 1981; Shen-Ong & Cole, 1982). Characteristic restriction site polymorphisms also affect these patterns. Thus components *a* and *b* in the *Hind*III/*Eco*RI patterns in Fig. 4 are 5.5 and 5.2 kb in length, respectively, and represent fragments derived from full-sized (7 kb) IAP elements polymorphic for *Eco*RI restriction sites at the 5′ end (Kuff *et al.*, 1981; Mietz *et al.*, 1987). Component *c* is derived from a large class of elements with 1.9 kb deletions spanning the *gag-pol* junction region (Shen-Ong & Cole, 1982; Kuff *et al.*, 1986b; Grossman *et al.*, 1987), and components *d* and *e* from so-called type II IAP elements (Sheng-Ong & Cole, 1982; Lueders & Mietz, 1986), which have a specific small sequence insertion as well as major deletions of several size classes.

The amplified IAP families in the group of mouse species more closely related to *M. musculus* can be distinguished by the relative intensities of the common restriction bands and/or by the presence of certain additional fragments not detected in others (Fig. 4). The pattern shown for *M. musculus* was found in two feral representatives (Czech II and Skive), and the *M. domesticus* pattern was common to inbred laboratory strains and members of three feral populations in Europe and the United States.

The restriction patterns show that a common set of IAP-related structural variants was already defined when *M. spretus* diverged from the lineage of *M. musculus*. On the other hand, the relative homogeneity of the IAP gene family at the nucleotide level indicates that amplification has continued in *M. domesticus*, and by inference in the other related species. Differences in the proportions of IAP structural variants among the *M. musculus* related species could reflect the favoured retrotransposition of particular elements in germ cells or early embryos. An example of retrotranspositional amplification in somatic cells was provided by Shen-Ong & Cole (1982), who found a nearly two-fold increase in the copy number of a specific set of type II IAP elements in a BALB/c myeloma as compared to the germ line DNA and showed that this represented insertion of the elements at novel positions in the genome. Relative copy numbers can also be affected by concerted gene conversions, or by unequal crossing over in the case of IAP elements located in regions of predominantly repetitive sequences (Kuff *et al.*, 1986; Lueders, 1987). Without precise information on the sequence divergence within the IAP families in other *M. musculus*-related species, it is difficult to decide whether the major amplification of these elements occurred in the common ancestral line or at a similar rate in the individual species as they subsequently evolved.

What is the relationship between the IAP elements amplified in the group of six species of the *M. musculus* group and those found in the three Southeast Asian species of the *M. cervicolor* group? Even though the hybridisations shown in Fig. 4 were carried out at greatly reduced stringency (see legend, Fig. 4), the reactive bands were much lighter in the patterns of the *M. cervicolor* group than would have been expected had the entire complement of 200–250 IAP-related sequence equivalents comprised a relatively homogeneous population. Comparable results were obtained on hybridisation of patterns obtained from *Pst*I-digested genomic DNAs (not shown). The relative intensity of the discrete bands in the *M. cervicolor* pattern was consistent with the roughly 30 IAP-sequences per haploid genome estimated previously for this species by liquid hybridisation kinetics, using probes derived from the RNA of both BALB/c IAPs and the M432 IAP-related endogenous retrovirus of *M. cervicolor* (Kuff *et al.*, 1978). Altogether, our findings suggest that the great majority of the IAP-related elements in the *M. cervicolor* group are structurally heterogeneous and/or widely divergent at the nucleotide sequence level.

On the other hand, among seven IAP-related clones selected from a *M. caroli* gene library using a BALB/c IAP probe, Ono & co-workers (1984) found three complete proviral copies and four incomplete elements apparently truncated during the cloning procedure. The complete elements were all 6.5 kb in length, 0.5 kb less than those found in *M. musculus* related species. Their LTRs had 80% sequence homology with the LTRs of several cloned BALB/c elements, and their internal regions (not sequenced) had a number of restriction sites, but not all, in common with the BALB/c genes. Using cloned probes from the same internal region of both BALB/c and *M. caroli* elements, Ono *et al.* (1984) found by hybridisation kinetics 80 and 180 IAP gene copies, respectively, in *M. caroli* DNA, and 180 and 300 copies in *M. cervicolor*. As noted earlier, we have obtained comparable copy numbers for these species by dot blot hybridisation with a BALB/c IAP probe (Fig. 4).

It is difficult to reconcile all of the above results. If a great majority of the IAP sequences in the species of the *M. cervicolor* group are quite heterogeneous at the gross structural and/or nucleotide levels, they might be fully measured by dot and liquid hybridisations of whole DNA, but more difficult to detect in restriction digest patterns. Although the kinetic data of Ono *et al.* (1984) show little evidence of heterogeneity within the reannealing sequence population in *M. caroli* DNA, the presence of a 10–15% subgroup with greater sequence homology to BALB/c IAP elements could be difficult to detect. Our working hypothesis is that the genomes of the *M. cervicolor* group do contain some 30 relatively homogeneous IAP elements, more closely related to those in *M. musculus* and detected under our particular hybridisation conditions (see legend to Fig. 4), and roughly 200 sequence equivalents of a more heterogeneous nature. Although many aspects of the basic structure of the IAP genome may thus have been determined before divergence of the two groups of *Mus* species, the IAP elements in the *M. cervicolor* group have not undergone the massive amplification characteristic of the *M. musculus*-related species.

The data summarised in Fig. 4 show that IAP-related elements have undergone independent amplifications in various rodent species. We speculate that the central cassette of IAP-related genetic information has been carried in the evolutionary lineage of the rodents in combination with a variety of other sequence components (LTRs, envelope genes, N-terminal *gag* domains) required to complete a proviral element. Periodically, particular variants have appeared with properties favouring amplification in the germ line and the large expansions have occurred. Amplification stops when the burden of deleterious insertional mutations reaches an unacceptable level. Divergence of IAP-related sequence elements from one another by random mutations events may subsequently lead to a situation in which traces of an evolutionarily distant amplification cycle are difficult to detect (Chinese hamster?).

Lack of horizontal transmissibility may be a prerequisite for very large amplification of retrotransposons, as evidenced by the relatively great abundance of copia-like elements in *Drosophila*, Ty elements in yeast, and IAPs in mice and Syrian hamsters. Fully functional copies of transmissible retrotransposons (retroviruses) may accumulate to a lesser extent in mammalian or avian genomes because they are much more effective insertional mutagens.

What is the mechanism for germ-line amplification of endogenous retrotransposons? Infection of oocytes or germ line precursors probably occurs *in vivo* with MuLV shed from surrounding cells (see discussions in Jenkins & Copeland, 1985; Yotsuyanagi & Szollosi, 1984). Genetic factors can influence this process, as evidenced by the remarkable frequency with which new copies of an endogenous ecotropic MuLV provirus are inserted into the germ line of SWR/J-RF/J hybrid mice (Jenkins & Copeland, 1985). The expression of IAPs in mouse oocytes and preimplantation embryos, and of R-particles in early Syrian hamster development (references in Yotsuyanagi & Szollosi, 1984) provides a setting for retrotransposition of IAP elements into the germ line linage. The fixation of one new copy per 10^2 to 10^3 years in a population undergoing a 'burst' of amplification could lead to an accumulation of 1000 copies in an evolutionarily brief period. During this period, the overall rate of insertional mutations among individual animals might be significantly increased, since the total number of germ line retrotranspositions will be greater than the number of those which are ultimately selected to become fixed in the population as a whole.

Gerasimova *et al.* (1985) described successive 'transposition explosions' of copia-like Class I retrotransposons in individuals of a destabilised strain (Ct^{MR2}) of *Drosophila* carrying a mdg4 insertion in the *cut* locus. In these occurrences, multiple transpositions, involving a number of different types of elements, can be found in the same germ line cell. A related situation is found in yeast, where the introduction of an active Ty element can facilitate multiple retrotranspositions of the endogenous Ty population (Boeke *et al.*, 1985; Garfinkel *et al.*, 1985).

Recombination of retrotransposons with cellular genes

Acquisition of cellular protooncogenes by infectious retroviruses is well known. In mammalian and some avian systems, such transforming recombinants are helper-dependent for replication, packaging and infectivity (Coffin, 1982). Similar types of recombinants have not been detected among the IAP elements, although these elements have no requirement for infectivity and in fact commonly express and co-package transcripts of incomplete (deleted) gene copies (Paterson *et al.*, 1978; Sheng-Ong & Cole, 1984). Since IAPs themselves are not infectious, transforming IAP variants might best be sought by transfection of indicator cells with the DNA of long-established IAP-expressing mouse tumours. Although incorporation of cellular oncogene sequences into the body of IAP element is not known, IAP transpositions have been shown to activate the c-mos gene in a transplantable IAP-rich BALB/c plasmocytoma (Canaani *et al.*, 1983) and to induce the constitutive production of interleukin-3 in a myelomonocytic leukaemia (Ymer *et al.*, 1985). The effect of the interleukin-3 mutation was to render the leukaemia cells independent of an outside source of this factor and to increase their growth autonomy. It has been often suggested that IAP transpositions could have a role in neoplastic transformation or progression by providing a source of genetic variability within the tumor cell population.

A recent surprising observation suggests that the mouse-specific N-terminal half of the IAP *gag* gene may have been contributed by coding sequences for a cellular protein. Martins *et al.* (1985a), in studies aimed at understanding the factors which regulate immunoglobulin E (IgE) synthesis, prepared cDNA clones from a rat-mouse T-cell hybridoma that produced a set of soluble IgE-binding factors (IgE-BF) resembling those formed in the whole animal in response to specific immunogens (Ishizaka, 1984). The authors selected several cDNA clones that directed IgE-BF production when transfected into heterologous monkey cells. Some of these factors had the additional property of enhancing IgE production in cultures of B-lymphocytes taken from appropriately immunised mice. Analysis of these clones by hybridisation, heteroduplex formation, nucleotide sequence and immunological characterisation of their protein products demonstrated that they each represented a different structural variant of the mouse IAP gene family (Moore *et al.*, 1986; Kuff *et al.*, 1986). The only coding region common to all of them was a 300–330 bp section which spanned the junction region between the conserved p27 domain and the upstream mouse-specific *gag* segment (Fig. 3). An N-glycosylation coding site essential for biological activity but not for IgE-binding was located in the p27 domain (Mietz *et al.*, 1987). As shown in Fig. 4, the mouse *gag* gene encodes a strongly hydrophobic N-terminal segment that is unique among the retroviral *gag* proteins. This could represent either the peptide signal sequence or truncated transmembrane domain of a precursor cellular gene, conceivably one that specifies a cell membrane protein with immunoglobulin-binding properties. The fortuitous location of an N-glycosylation site in the conserved p27 domain of some IAP elements (the site is mutated in others) could have contributed to the observed biological activity of the recombinant *gag* protein. If this concept is correct, *M. domesticus* and closely related species may be the only animals in which IgE-BF are encoded by retrovirus-like elements.

Martin *et al.* (1985), Singer & Skowronski (1985), d'Ambrosio *et al.* (1986) and Fanning & Singer (1987) have reviewed the evolution of LINE-1 repeat families, which have accumulated to very high copy numbers, about 10^5, in the genome of many species. In mouse and human, the great majority of these elements are truncated, scrambled or otherwise mutated to an obviously non-functional state. Some copies may be functionally intact, since transcripts have been found in a few cell types and transpositions of LINE-1 elements have been observed. LINE-1 elements are apparently evolving in concert in the genomes of *M. domesticus, M. caroli* and *M. platythrix* (which last shared a common evolutionary ancestor approximately 10–12 Myr ago). Using sequence data obtained for multiple clones from all three species, Martin *et al.* (1985) calculated that half of the elements turn over through accumulation of random mutations every 3–4 Myr. Since the total copy number remains roughly the same, the internal divergence among family members within each species changes at approximately the same rate. The rat LINE-1 family, in contrast to those in man and mouse, is relatively homogeneous and more widely transcribed (d'Ambrosio *et al.*, 1986). For reasons that are not yet understood, a majority of the LINE-1 copies in rat are full-sized rather than truncated. d'Ambrosio *et al.* (1986) have suggested that transposition of these elements may occur by a replicative mechanism other than by reverse transcription. Recombination with other transposons or novel cellular sequences is not a recognised aspect of LINE-1 evolution. This situation is different from the model of independent cycles of amplification proposed for the evolution of IAP-related elements.

SUMMARY

We have considered the evolutionary history of the IAP-related family of Class I retrotransposons, in which a conserved central cassette of genetic information essential for particle core formation, reverse transcription and proviral integration has been recombined with a variety of upstream (N-terminal *gag* domain) and downstream (envelope) sequences to form genomes that encode transmissible retroviruses or 'defective' intracellular particles. The number of proviral elements related to infectious retroviruses is usually no more than about 50 per haploid genome; but elements encoding intracellular particles such as IAPs have accumulated to much higher levels, about 10^3. Many of the endogenous class I retrotransposons in mammalian species appear to be mutated elements which lack full functional activity.

The available data on the relationships between IAP-related elements in rats, mice and hamsters are consistent with an evolutionary process in which periodic amplification of novel has occurred independently in different species. Amplification could involve gene conversion and unequal crossing over as well as duplicative retrotransposition. The observed expression of IAPs and the related R-particles of Syrian hamsters in oocytes and preimplantation embryos could facilitate the insertion of new proviral copies in germ line DNA. On a vastly expanded time scale, the episodes of IAP amplification are reminiscent of the bursts of endogenous retrotransposon activity in *Drosophila* and yeast, and like them may be associated with a significant increase in the incidence of insertional mutations.

ACKNOWLEDGEMENTS.

We thank Beverly Miller for her skilled help in preparing this manuscript.

REFERENCES

ADAMS, S. E., MELLOR, J., GULL, K., SIM, R. B., TUITE, M. F., KINGSMAN, S. M. & KINGSMAN, A. J. (1987). The functions and relationships of Ty-VLP proteins in yeast reflect those of mammalian retroviral proteins. *Cell*, **49**, 111–19.

BOEKE, J. D., GARFINKEL, D. J., STYLES, C. A. & FINK, G. R. (1985). Ty elements transpose through an RNA intermediate. *Cell*, **40**, 491–500.

BURT, D. W., REUTH, A. D. & BRAMMER, W. J. (1984). A retroviral provirus closely associated with the Ren-2 gene of DBA/2 mice. *Nucleic Acids Research*, **12**, 8579–93.

CALLAHAN, R., CHIU, I.-M., WONG, J. F. H., TRONICK, S. R., ROE, B. A., AARONSON, S. A. & SCHLOM, J. (1985). A new class of endogenous human retroviral genomes. *Science*, **228**, 1208–11.

CALLAHAN, R., KUFF, E. L., LUEDERS, K. K. & BIRKENMEIER, E. (1981). Genetic relationship between the *Mus cervicolor* M432 retrovirus and the *Mus musculus* intracisternal type A particle. *Journal of Virology*, **40**, 901–11.

CALLAHAN, R. & TODARO, G. D. (1978). Four major endogenous retrovirus classes each transmitted in various species of *Mus*. In *Origins of Inbred Mice*, ed. H. C. Morse, III, pp. 689–713. New York: Academic Press.

CANAANI, E., DREAZEN, O., KLAR, A., RECHAVI, G., RAM, D., COHEN, J. B. & GIVOL, D.. (1983). Activation of the c-*mos* oncogene in a mouse plasmacytoma by insertion of an endogenous intracisternal A particle genome. *Proceedings of the National Academy of Sciences, USA*, **80**, 7118–22.

CLARE, J. & FARABAUGH, P. (1985). Nucleotide sequence of a yeast Ty element: Evidence for an unusual mechanism of gene expression. *Proceedings of the National Academy of Sciences, USA*, **82**, 2829–33.

COFFIN, J. (1982). Structure of the retroviral genome. In *RNA Tumor Viruses*, eds. R. Weiss, N. Teich, H. Varmus & J. Coffin, pp. 261–368. Cold Spring Harbor Laboratory, New York.

D'AMBROSIO, E., WAITZKIN, S. D., WITNEY, F. R., SALEMME, A. & FURANO, A. V. (1986). Structure of the highly repeated, long interspersed DNA family (LINE or LIRn) of the rat. *Molecular and Cellular Biology*, **6**, 411–24.

DEEN, K. & SWEET, R. W. (1986). Murine mammary tumor virus *pol*-related sequences in human DNA: Characterization and sequence comparison with the complete murine mammary tumor virus *pol* gene. *Journal of Virology*, **57**, 422–32.

DINOCERA, P. O., DIGAN, M. E. & DAWID, I. (1983). A family of oligoadenylated transposable sequences in *Drosophila*. *Journal of Molecular Biology*, **168**, 715–27.

DOWSETT, A. P. & YOUNG, M. W (1982). Differing levels of dispersed repetitive DNA among closely related species of *Drosophila*. *Proceedings of the National Academy of Sciences, USA*, **79**, 4570–4.

ETZERODT, M., MIKKELSEN, T., PEDERSEN, F. S., KJELDGAARD, N. O. & JORGENSEN, P. (1984). The nucleotide sequence of the AKv murine leukemia virus genome. *Virology*, **134**, 196–207.

FANNING, T. G. & SINGER, M. F. (1987). LINE-1: A mammalian transposable element. *Biochimica et Biophysica Acta*, in press.

FAWCETT, D. H., LISTER, C. K., KELLETT, E. & FINNEGAN, D. J. (1986). Transposable elements controlling I-R hybrid dysgenesis in *D. melanogaster* and similar to mammalian LINEs. *Cell*, **47**, 1007–15.

GARFINKEL, D. J., BOEKE, J. D. & FINK, G. R. (1985). Ty element transposition: Reverse transcriptase and virus-like particles. *Cell*, **42**, 507–17.

GARNIER, J., OSGUTHORPE, D. J. & ROBSON, B. (1978). Analysis of the accuracy and implications of simple methods for predicting the secondary structure of globular proteins. *Journal of Molecular Biology*, **120**, 97–120.

GROSSMAN, Z., MIETZ, J. & KUFF, E. L. (1987). Nearly identical members of the heterogenous IAP gene family are expressed in thymus of different mouse strains. *Nucleic Acids Research*, **15**, 3823–34.

HENDERSON, L. E., KRUTZSCH, H. K. & OROSZLAN, S. (1983). Myristyl amino terminal acylation of murine retroviral proteins, an unusual post-translational protein modification. *Proceedings of the National Academy of Sciences, USA*, **80**, 339–43.

HOPP, T. P. & WOODS, K. R. (1981). Prediction of protein antigenic determinants from amino acid sequences. *Proceedings of the National Academy of Sciences, USA*, **78**, 3824–8.

JAHNER, D. & JAENISCH, R. (1985). Chromosomal position and specific demethylation in enhancer sequences of germ line-transmitted retroviral genomes during mouse development. *Molecular and Cellular Biology*, **5**, 2212–20.

JENKINS, N. A. & COPELAND, N. G. (1985). High frequency of germline acquisition of ecotropic MuLV proviruses in SWR/J-RF/J hybrid mice. *Cell*, **43**, 811–19.

JOHNSON, M. S., MCCLURE, M. A., FENG, D.-F., GRAY, J. & DOOLITTLE, R. F. (1986). Computer analysis of retroviral *pol* genes: Assignment of enzymatic functions to specific sequences and homologies with non-viral enzymes. *Proceedings of the National Academy of Sciences, USA*, **83**, 7648–52.

KATO, S., MATSUO, K., NISHIMURA, N., TAKAHASHI, N. & TAKANO, T. (1987). The entire nucleotide sequence of the baboon endogenous virus DNA: A chimeric genome structure of murine type C and Simian type D retrovirus. *Japanese Journal of Genetics*, **62**, 129–37.

KUFF, E. L. (1988). Factors affecting retrotransposition of intracisternal A-particle proviral elements. *Banbury Reports*, **30**, in press.

KUFF, E. L. & FEWELL, J. W. (1985). Intracisternal A-particle gene expression in normal mouse thymus tissue: Gene products and strain-related variability. *Molecular and Cellular Biology*, **5**, 474–83.

KUFF, E. L., FEWELL, J. W., LUEDERS, K. K., DIPAOLO, J. A., AMSBAUGH, S. C. & POPESCU, N. C. (1986a). Chromosome distribution of intracisternal A-particle sequences in the Syrian hamster and mouse. *Chromosoma*, **98**, 213–19.

KUFF, E. L. & LUEDERS, K. K. (1988). The intracisternal A-particle gene family: Structure and functional aspects. *Advances in Cancer Research*, **51**, in press.

KUFF, E. L., LUEDERS, K. K. & SCOLNICK, E. M. (1978). Nucleotide sequence relationship between intracisternal type-A particles of *Mus musculus* and an endogenous retrovirus (M432) of *Mus cervicolor*. *Journal of Virology*, **28**, 66–74.

KUFF, E. L., MIETZ, J. A., TROUNSTINE, M. L., MOORE, K. W. & MARTINS, C. L. (1986b). cDNA clones encoding murine IgE-binding factors represent multiple structural variants of intracisternal A-particle genes. *Proceedings of the National Academy of Sciences, USA*, **83**, 6583–7.

KUFF, E. L., SMITH, L. & LUEDERS, K. K. (1981). Intracisternal A-particle genes in *Mus musculus*: A conserved family of retrovirus-like elements. *Molecular and Cellular Biology*, 1, 216–27.

LESSER, J., LASNERET, J., CANIVET, M., EMANOIL-RAVIER, R. & PERIES, J. (1986). Simultaneous activation by 5-azacytidine of intracisternal-R particles and murine intracisternal-A particle related sequences in Syrian hamster cells. *Virology*, 155, 249–56.

LINIAL, M. & BLAIR, D. (1985). Genetics of retroviruses. In *RNA Tumor Viruses*, 2, eds. R. Weiss, N. Teich, H. Varmus & J. Coffin, pp. 147–85. Cold Spring Harbor Laboratory.

LUEDERS, K. K. (1987). Specific association between type-II intracisternal A-particle elements and other repetitive sequences in the mouse genome *Gene*, 52, 139–46.

LUEDERS, K. K. & KUFF, E. L. (1975). Synthesis and turnover of intracisternal A-particle structural proteins in cultured neuroblastoma cells. *Journal of Biological Chemistry*, 250, 5192–9.

LUEDERS, K. K. & KUFF, E. L. (1977). Sequences associated with intracisternal A-particles are reiterated in the mouse genome. *Cell*, 12, 963–72.

LUEDERS, K. K. & KUFF, E. L. (1981). Sequences homologous to retrovirus-like genes of the mouse are present in multiple copies in the Syrian hamster genome. *Nucleic Acids Research*, 9, 5917–30.

LUEDERS, K. K. & KUFF, E. L. (1983). Comparison of the sequence organization of related retrovirus-like multigene families in three evolutionarily distant rodent genomes. *Nucleic Acids Research*, 11, 4391–408.

LUEDERS, K. K. & MIETZ, J. A. (1986). Structural analysis of type II variants within the mouse intracisternal A-particle family. *Nucleic Acids Research*, 14, 1495–510.

MARCIANI, D. J. & KUFF, E. L. (1974). Structural proteins of intracisternal A-particles: Possible repetitive sequences. *Journal of Virology*, 14, 1597–9.

MARTIN, C. L., HUFF, T. F., JARDIEU, P., TROUNSTINE, M. L., COFFMAN, R. L., ISHIZAKA, K. & MOORE, K. W. (1985*a*). cDNA clones encoding IgE-binding factors from a rat-mouse T-cell hybridoma. *Proceedings of the National Academy of Sciences, USA*, 82, 2460–4.

MARTIN, C. L., VOLIVA, C. F., HARDIES, S. C., EDGELL, M. H. & HUTCHISON, C. A., III. (1985*b*). Tempo and mode of concerted evolution in the L1 repeat family of mice. *Molecular Biology Evolution*, 2, 127–40.

MIETZ, J. A., GROSSMAN, Z., LUEDERS, K. K. & KUFF, E. L. (1987). Nucleotide sequence of a complete mouse intracisternal A-particle genome: Relationships to known aspects of particle assembly and function. *Journal of Virology*, 61, 3020–9.

MOORE, K. W., JARDIEU, P., MIETZ, J. A., TROUNSTINE, M. L., KUFF, E. L., ISHIZAKA, K. & MARTINS, C. (1986). Rodent IgE-binding factor genes are members of an endogenous retrovirus-like gene family. *Journal of Immunology*, 136, 4283–90.

MOORE, R., DIXON, M., SMITH, R., PETERS, G. & DICKSON, C. (1987). Complete nucleotide sequence of a milk-transmitted mouse mammary tumor virus: Two frame-shift suppression events are required for translation of *gag* and *pol*. *Journal of Virology*, 61, 480–90.

NISSEN-MEYER, J. & EIKHOM, T. S. (1981). Properties of an intracisternal A-particle associated endonuclease activity which is stimulated by ATP. *Journal of Virology*, 40, 927–31.

ONO, M., KITASATO, H., OHISHI, H. & MOTOBAYASHI-NAKAJIMA, Y. (1984). Molecular cloning and long terminal repeat sequences of intracisternal A-particle genes in *Mus caroli*. *Journal of Virology*, 50, 352–8.

ONO, M., TOH, H., MIYATA, T. & AWAYA, T. (1985). Nucleotide sequence of the Syrian hamster intracisternal A-particle gene: Evolutionary relationship of type A particle gene to types B and D oncovirus genes. *Journal of Virology*, **55**, 387–94.

ONO, M., YASUNAGA, T., MIYATA, T. & USHIKUBO, H. (1986). Nucleotide sequence of human endogenous retrovirus genome related to the mouse mammary tumor virus genome. *Journal of Virology*, **60**, 589–98.

PATERSON, B., SEGAL, S., LUEDERS, K. K. & KUFF, E. L. (1978). RNA associated with murine intracisternal type-A particles codes for the main particle protein. *Journal of Virology*, **27**, 118–26.

POWER, M. D., MARX, P. A., BRYANT, M. L., GARDNER, M. B., BARR, P. J. & LUCIER, P. A. (1986). Nucleotide sequence of SRV-1, a type-D Simian acquired immune deficiency syndrome retrovirus. *Science*, **231**, 1567–72.

QUEEN, C. & KORN, L. J. (1984). A comprehensive sequence analysis program for the IBM personal computer. *Nucleic Acids Research*, **12**, 581–99.

RHEE, S. S. & HUNTER, E. (1987). Myristylation is required for intracellular transport but not for assembly of D-type retrovirus capsids. *Journal of Virology*, **61**, 1045–53.

RICE, N. R., STEPHENS, R. M., BURNY, A. & GILDEN, R. V. (1985). The *gag* and *pol* genes of bovine leukemia virus: Nucleotide sequence and analysis. *Virology*, **142**, 357–77.

SCHWARTZ, D. E., TIZARD, R. & GILBERT, W. (1983). Nucleotide sequence of Rous sarcoma virus. *Cell*, **32**, 853–69.

SHEN-ONG, G. L. & COLE, M. D. (1982). Differing populations of intracisternal A-particle genes in myeloma tumors and mouse subspecies. *Journal of Virology*, **42**, 411–21.

SHENG-ONG, G. L. & COLE, M. D. (1988). Amplification of a specific of intracisternal A-particle genes in a mouse plasmacytoma. *Journal of Virology*, **49**, 171–7.

SINGER, M. F. & SKOWRONSKI, J. (1985). Making sense out of LINES: Long interspersed repeat sequences in mammalian genomes. *Trends in Biochemical Sciences*, **10**, 119–22.

SONIGO, P., ALIZON, M., STASKUS, K., KLATZMANN, D., COLES, S., DANOS, O., RETZEL, E., TIOLLAIS, P., HAASE, A. & WAIN-HOBSON, S. (1985). Nucleotide sequence of the visna lentivirus: Relationship to the AIDS virus. *Cell*, **42**, 369–82.

SONIGO, P., BARKER, C., HUNTER, E. & WAIN-HOBSON, S. (1986). Nucleotide sequence of Mason-Pfizer monkey virus, an immunosuppressive D-type retrovirus. *Cell*, **45**, 375–85.

STEELE, P. E., MARTIN, M. A., RABSON, A. B., BRYAN, T. & O'BRIEN, S. J. (1986). Amplification and chromosomal dispersion of human endogenous retroviral sequences. *Journal of Virology*, **59**, 545–50.

STROMBERG, K., BENVENISTA, R. E., ARTHUR, L. O., RABIN, H., GIDDENS, W. E., JR., OCHS, H. D., MORTON, W. R. & TSAI, C.-C. (1984). Characterization of exogenous type-D retrovirus from a fibroma of a Macaque with Simian AIDS and fibromatosis. *Science*, **224**, 289–92.

SUZUKI, A., KITISATO, H., KAWAKAMI, M. & ONO, M. (1982). Molecular cloning of retrovirus-like genes present in multiple copies in the Syrian hamster genome. *Nucleic Acids Research*, **10**, 5733–46.

TCHURIKOV, N. A., ZELENTSOVA, E. S. & GEORGIOEV, G. P. (1980). Clusters containing different mobile dispersed genes in the genome of *Drosophila melanogaster*. *Nucleic Acids Research*, **8**, 1243–58.

TOH, H., ONO, M., SAIGO, K. & MIYATA, T. (1985). Retroviral protease-like sequence in the yeast transposon Ty1. *Nature*, **315**, 691.

WILSON, S. H., BOHN, E. W., MATSUKAGE, A., LUEDERS, K. K., & KUFF, E. L. (1974). Studies on the relationship between DNA polymerase activity and intracisternal A-type particles in mouse myeloma. *Biochemistry*, **13**, 1087–94.

YMER, S., TUCKER, W. Q. J., CAMPBELL, H. D. & YANG, I. G. (1986). Nucleotide sequence of the intracisternal A-particle genome inserted 5′ to the interleukin-3 gene of the leukaemia cell line WEHI-3B. *Nucleic Acids Research*, **14**, 5901–18.

YMER, S., TUCKER, W. Q. J., SANDERSON, C. J., HAPEL, A. J., CAMPBELL, H. D. & YANG, I. G. (1985). Constitutive synthesis of interleukin-3 by leukemia cell line WEHI-3B is due to retroviral insertion near the gene. *Nature*, **317**, 255–8.

YOTSUYANAGI, Y. & SZOLLOSI, D. (1984). Virus-like particles and related expressions in mammalian oocytes and preimplantation stage embryos. In *Ultrastructure of Reproduction*, eds. J. Van Blerkom and P. M. Motta, pp. 218–34. Boston: Martinus Nijhoff.

I-FACTORS IN *DROSOPHILA MELANOGASTER* AND SIMILAR ELEMENTS IN OTHER EUKARYOTES

DAVID J. FINNEGAN

Department of Molecular Biology, University of Edinburgh, King's Buildings, Mayfield Road, Edinburgh EH9 3JR, UK

I–R HYBRID DYSGENESIS AND THE PROPERTIES OF I FACTORS

Hybrid dysgenesis is the appearance of a set of abnormal characteristics in the progeny of crosses between particular strains of *Drosophila melanogaster* (Bregliano & Kidwell, 1983). These traits include reduced fertility and increased frequencies of apparent point mutations and cytologically visible chromosome rearrangements. Most strains of *D. melanogaster* can be classified as being one or other of two types with respect to I–R dysgenesis, I, or inducer, and R, or reactive. Dysgenesis is seen in the female progeny, known as SF females, of crosses between reactive strain females and inducer strain males. The male progeny of this cross, and the progeny of the reciprocal cross, are apparently normal, as are the progenies of crosses between any two reactive, or any two inducer strains.

The inducer state is controlled by chromosomal determinants called I-factors. Inducer strains contain a small number of active I-factors, and these may be found on all chromosomes (Pélisson & Picard, 1979; Pélisson, 1981). They are stable in inducer strains but are affected by hybrid dysgenesis, and in SF females transpose from inducer to reactive chromosomes at high frequency (Picard, 1976, 1978). The reactive state is controlled in a more complex way that seems to involve both cytoplasmic and chromosomal components (Bucheton & Bregliano, 1982).

The increased mutation rates associated with I–R hybrid dysgenesis, suggest that dysgenesis is due to activation of transposable elements, including I-factors themselves. These appear to be held in check by regulatory molecules in inducer strains, and to be activated in a dysgenic cross because they are introduced into an environment lacking the appropriate regulatory factors.

The DNA of the I-factor has been identified by analysing molecular lesions associated with eight *white* gene mutations, w^{IR1-8}, that

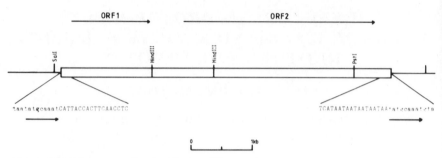

Fig. 1. The I-factor associated with the w^{IR1} mutation. The thin line indicates *white* gene sequences, and the box indicates the I-factor. The horizontal lines above the map indicate the two longest open reading frames of the I-factor, ORF1 and ORF2. These are oriented from left to right. The *white* gene is transcribed from right to left. The sequences at the ends of the I-factor are shown below the map. The lower case letters indicate *white* sequences, while the upper case letters are I-factor sequences. The target site duplication is underlined. This could be either 9 or 12 bp long as the sequence TAA occurs at both the target site and the end of the I-factor.

arose in SF females (Bucheton *et al.*, 1984; Sang *et al.*, 1984). Two of these mutations, w^{IR7} and w^{IR8}, determine a bleached white phenotype and are due to deletions of part of the *white* gene. The other six are associated with insertions of indistinguishable 5.4 kb elements. It is believed that these insertions are of I-factor DNA since each is linked to I-factor activity (Pélisson, 1981 and personal communication), and we have shown that the element associated with the w^{IR3} mutation can confer the inducer phenotype on a reactive strain into which it has been introduced by P-factor-mediated transformation (M. Prichard, J-M. Dura, A. Bucheton, A. Pélisson & DJF, unpublished data).

The complete base sequence of the I-factor associated with the w^{IR1} mutation has been determined (Fawcett *et al.*, 1986). It is 5371 bp long and is flanked by direct repeats of a 12 bp target sequence. It has no terminal repeats, but has four copies of the sequence TAA at the 3' end of one strand (Fig. 1). We have confirmed that this sequence organisation is a general property of I-factors by sequencing the ends of the five other insertions within the *white* gene, and one associated with a mutation of the Bithorax complex, bx^{F31} (Peifer & Bender, 1986). These elements are flanked by target site duplications, and these vary slightly in length from one element to another (Fig. 2). Three of these elements, those associated with mutations w^{IR1-3}, have inserted at exactly the same site. This suggests that I-elements may have preferred target sites, although it has not been possible to derive a convincing consensus

MUTATION	LEFT-HAND END	RIGHT-HAND END	TARGET-SITE DUPLICATION
w^{IR1}	CATTACC	$TCA(TAA)_4$	TAATATGCAAAT
w^{IR3}	CAGTACC	$TCA(TAA)_5$	TTAATATGCAAAAT
w^{IR4}	CAGTACC	$TCA(TAA)_5$	TTAATATGCAAAT
w^{IR2}	CATTACC	$TCA(TAA)_7$	TTTACTGCAGAG
w^{IR5}	CAGTACC	$TCA(TAA)_6$	TCCGAAATAACT
w^{IR6}	CAGTACC	$TCA(TAA)_6$	TAACAACCAG
Bx^{F31}	CAGTACC	$TCA(TAA)_6$	TAAAAGGCCGAAA

Fig. 2. The ends of I-factors associated with different mutations. The first column indicates the mutations concerned. The second and third columns show sequences at the left- and right-hand ends of the I-factors concerned. The last column shows the target site duplications associated with each insertion. We cannot be sure of the precise length of the duplications associated with w^{IR1}, w^{IR6} and bx^{F31}. In each case one copy of the TAA repeat could be either part of the target site duplication, or of the I-factor.

sequence from the duplications that have been sequenced so far. The ends of the elements themselves are highly conserved, and the only differences between them are in the third base, which is either G or T, and in the number of TAA repeats at the right-hand end. All the coding capacity of the I-factor appears to be in one strand that contains two long open reading frames, ORF1 of 1278 bp and ORF2 of 3258 bp, that are transcribed from left to right in Fig. 1. Both have methionine codons near their 5' ends that could be used to initiate translation. There is no indication from the sequence of the I-factor that ORF1 and ORF2 might be joined by RNA splicing, but this possibility can not be excluded. There is a 228 bp open reading frame in the 471 bp separating ORF1 and ORF2, and this could also be important for I-factor activity.

The genetic properties of I-factors suggests that they encode at least two functions, one required for transposition, and the other to regulate expression in inducer strains. This sequence organisation is similar to that of processed pseudogenes and other elements that are thought to move around the genome by reverse transcription of RNA intermediates (Rogers, 1985). If I-elements were to move in a similar way, then a reverse transcriptase would be a likely transposase. With this in mind the amino acid sequences encoded by ORF1 and ORF2 have been compared with those of known reverse transcriptases (Fig. 3; Fawcett *et al.*, 1986). Toh *et al.* (1985) have compared the amino acid sequences of the putative RNA-dependent DNA polymerases of three retroviruses, three hepatitis viruses and 17.6, a *copia*-like transposable element of *D. melanogaster*, and have

A

```
I Factor   LKGCAPGLNRISYQMIKNSSHTTKNRITKLFNEI-FNSHIPQAYKTSLIIP
F          spkksPGydlltpeMIiqlpHsavryITKLFNaItklgyfPQrvKmmkIIm
Ingi       psGsAaGpdclyneaIqhlgiTaINvvIrLFNesIrtgvvPpAwKTgvIIP
L1Md       ptkksPGpdgfSaefyqtfkedIipiIhKLFhkIevegtIPnsfyeatItI
Bm-R2      dwrtsPGpdgIrsgqwravpvhIKaem---FNawmargeIPeiIrqcrtvf
Sc-al      rikskPG---nitpgttIetIdgiNiI-yLnkIsneIgtgkfkFKpmrIvn
17.6       tKhnIPlyskySYpqayeqevesqiqdmInqgiI-rtSnsPynspiwv---
RSV        kwkpdhtpvwIdqwpIpegkIvaItqIvekeIqI-ghiepsIscwntpvfv
HBV        LqfrnskpcsdycIsIivnIIedwgpcaehgehh-iriprtpsrvTggvfI
```

B

```
             ▼ ▼▼
I Factor   ILKP-NTDKTKTSSYRPISLNCCIAKILDK     48aa
F          IpkP-gknhTvaSSYRPISLIsCIsKIfeK     51aa
Ingi       ILKa-gkkaedIdSYRPvtLtsCIcKvmer     38aa
L1Md       IpkP-qkDpTKienfRPISLmnidAKILnK     50aa
Bm-R2      vpKv-erpggpge-YRPISiasipIrhfhs     47aa
Sc-al      IpKPkgg-------iRPISvgnprdKIvqe     43aa
17.6       vpKkqdasgkqk--fRiv----idyrkLne     24aa
RSV        IrKasgs-------YRII----hdIravna     23aa
HBV        vdKnphntae----sRIv----vdfsqfsr     27aa
```

C   D

```
             ▼ ▼▼▼                    ▼ ▼ ▼▼                    ▽
I Factor   LVTLDFSRAFDRVGV   45aa   GIPQGSPISVILFLIAF-NKLSNIISLH   4aa
F          aVfLDvSqAFDkVwI   44aa   GvPQGSvIgptLyLIyt-adiptnsrLt
Ingi       vf-vDyekAFDtVdh   45aa   GvPQGtvpgsImFiIvm-NsLSq--rLa   8aa
L1Md       iisLDaekAFDkiqh   45aa   GtrQGcPISpyLFnIvI-evIaraIrqq  14aa
Bm-R2      vavLDFakAFDtVsh   45aa   GvrQGdPISpILFnvvmdIiLasIperv  10aa
Sc-al      fievDIkkcFDtish   40aa   GIPQGSIISpILcnIvitIvdnwIedyi  53aa
17.6       ftTiDIakgFhqiem   20aa   rmPfGIknapatFqrcmmd----IIrpI   4aa
RSV        LmvLDIkdcFfsipI   27aa   vIPQGmtcSpticqIvvgqvLe-pIrLk   5aa
HBV        wIsLDvSaAFyhIpI   44aa   kIPmGvgISpfLIaqftsaicSvvrraf   3aa
```

E   F   G

```
            ▼▼▼▼ ▼▼                ●                 ▼▼▼ ▼
I Factor   FNAYADDFFL   28aa   GASLSLSKCQ   24aa   TSLKILGITL
F          vstfADDtai   28aa   rikvneqKCk   24aa   devtyLGvhL
Ingi       --ffADDItL   16aa   GInvvLqwsk   10aa   TkctIfGcTe
L1Md       isIIADDmiv   23aa   GykinsnKsm   24aa   nniKyLGvTL
Bm-R2      aIAYADDIvL   22aa   GIrLncrKsa   10aa   kkhhyLtert
Sc-al      yvrYADDiIi   23aa   GItineeKtI    6aa   TparfLGyni
17.6       cIvYIDDiiv   22aa   nIkLqLdKCe    3aa   qettfLGhvL
RSV        mIhYmDDIIL   22aa   GftiSpdKvQ    2aa   pgvqyLGykL
HBV        afsYmDDvvL   22aa   GihLnpnKtk    3aa   ySLnfmGyvi
```

Fig. 3. Amino acid sequences of open reading frames known, or believed, to encode reverse transcriptases. Each line gives the amino acid sequence encoded by a different element. These are, I-Factor, ORF2 of the I-factor associated with the w^{IR3} mutation (Fawcett et al., 1986); F, the F element associated with the w^{i+A} mutation (Di Nocera & Casari, 1987); Ingi, the ingi-3 element from *T. brucei* (Kimmel et al., 1987); L1Md, the L1 element A2 from *Mus domesticus* (Loeb et al., 1986); Bm-R2, the R2 insertion from clone B131 of *B. mori* (Burke et al., 1987); Sc-al, the first intron of the mitochondrial gene for subunit 1 of cytochrome oxidase of *S. cerevisiae* (Bonitz et al., 1980); 17.6, the *copia*-like element 17.6 of *D. melanogaster* (Saigo et al., 1984); RSV, Rous sarcoma virus (Schwartz et al., 1983) and HBV, human hepatatitis B virus (Galibert et al., 1979).

Amino acid residues are designated by the one letter code. Regions B–G are known to be conserved in reverse transcriptases (Toh et al., 1985; Hattori et al., 1986). The triangles indicate positions at which Toh et al. (1985) found identical or similar amino acid residues when comparing the sequences of eight known or putative reverse transcriptases. A filled triangle indicates that the sequence of ORF2 of the I-factor codes for an identical or similar residue; an open triangle indicates that it does not. The filled circle indicates a position at which five viral reverse transcriptases and the I-factor code for the same residue. The sequence of the I-factor, and any other sequence that matches it exactly, are written in capital letters. Sequences that differ from that of the I-factor are in lower case.

```
I Factor    NFIFTDGSKI    98aa    GNELADQAAK
RNAase H    veIFTDGScl   118aa    cdELAraAAm
RSV         ptvFTDaSss   108aa    adsqAtfqAy
```

Fig. 4. The region of ORF2 of the I-factor that may encode a ribonuclease H. The second line gives the amino acid sequence from two regions of *E. coli* ribonuclease H (RNAase H; Kanaga & Crouch, 1983). The first and third lines show regions of ORF2 of the I-factor (Fawcett *et al.*, 1986) and the *pol* gene of Rous sarcoma virus (RSV; Schwartz *et al.*, 1983) that match the *E. coli* sequences. The I-factor sequence, and sequences that match it exactly are shown in upper case. Sequences that differ from that of the I-factor are shown in lower case.

found 27 positions at which there are identical or chemically similar amino acids. The sequence of ORF2 contains amino acids corresponding to 21 of these, 18 of which are shown in Fig. 3.

Retroviral reverse transcriptases are associated with at least three activities, an RNA-dependent DNA polymerase (Temin & Mizutani, 1970), a ribonuclase H (Verma, 1975), and an endonuclease required for proviral integration (Schwartzberg *et al.*, 1984). Sequences similar to the conserved amino acids in region E in Fig. 3 can be found in several non-retroviral polymerases including the α subunit of *E. coli* RNA polymerase (Kamer & Argos, 1984; Johnson *et al.*, 1986) strongly suggesting that this region is associated with polymerase activity. RNAase H activity is probably encoded at the C termini of these polypeptides where there is sequence homology between reverse transcriptases and *E. coli* RNAase H. (Johnson *et al.*, 1986). A similar sequence can be found at the 3' end of ORF2 (Fig. 4). These homologies suggest that the product of ORF2 also has both RNA-dependent DNA polymerase and RNAase H activities, and we think that these may be required to synthesise a double stranded DNA from an RNA intermediate during I-factor transposition. We have not been able to detect any homology between ORF2 and the region of reverse transcriptases believed to be responsible for their endonuclease activities.

There is no extensive homology between the amino acid sequence of ORF1 and any polypeptide in the data bases with which it has been compared, but it does contain a region that matches a conserved motif $CX_2CX_4HX_4C$, found in viral nucleic acid binding proteins including *gag* polypeptides of retroviruses and the coat protein of cauliflower mosaic virus (Covey, 1986). This is consistent with the ORF1 product regulating I-factor activity by binding to DNA or RNA. Alternatively, it could be the major structural protein of a ribonucleoprotein particle containing both I-factor-encoded reverse

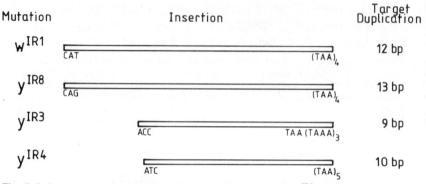

Fig. 5. I-elements associated with *yellow* gene mutations. The y^{IR8} mutation appears to be associated with a complete I-factor very similar to that of w^{IR1}. Mutations y^{IR3} and y^{IR4} are associated with I-elements truncated at their left-hand (5') ends. The y^{IR3} element differs from most I-elements in having tandem repeats of the sequence TAAA at its right-hand (3') end. Each element is associated with a target site duplication (Busseau & Pelisson, personal communication).

transcriptase and I-factor RNA. Reverse transcription of retroviral RNA takes place in intracellular particles, and, at least in the case of murine leukaemia virus, these may be required for proviral integration (Brown *et al.*, 1987). RNAs complementary to Ty elements in *S. cerevisiae* and *copia* elements in *D. melanogaster* can be found in virus-like particles associated with reverse transcriptase activity (Garfinkel *et al.*, 1985; Mellor *et al.*, 1985; Shiba & Saigo, 1983), and the same may be true of I-factor RNA.

Not all mutations induced by I–R dysgenesis and associated with insertions of I-factor sequences contain complete 5.4 kb elements. Busseau & Pélisson (personal communication, Finnegan *et al.*, 1987) have found that, of three I–R induced *yellow* mutations associated with I-factor sequences, only one, y^{IR8}, contains a 5.4 kb element. The others, y^{IR3} and y^{IR4}, have insertions of incomplete I-elements that are truncated at their 5' ends (Fig. 5). The 3' ends of these elements are like those of complete I-factors, except that one has the terminal sequence TAA(TAAA)$_3$ rather than simply a run of TAA repeats. Both of these elements are flanked by target site duplications suggesting that they inserted as truncated elements rather than having been deleted after insertion.

I-factors appear to play a direct role in the production of gross chromosome rearrangements that result from I–R dysgenesis. Busseau & Pelisson (personal communication; Finnegan *et al.*, 1987) have found that two I–R induced *yellow* mutations that they have studied are due to inversions that have I-elements at both ends.

In each case the two elements are in opposite orientations, one lying within the *yellow* gene and the other outside it. These mutations could have been produced by insertion of an I-element within the *yellow* gene and then recombination between it and an adjacent element in the opposite orientation. This recombination could have been stimulated by an I-factor-encoded product and/or by repair enzymes induced by chromosome damage resulting from transposition. A deletion would be produced by recombination between a newly inserted I-element and an adjacent element with the same polarity.

MECHANISM OF I-FACTOR TRANSPOSITION

The I-factor differs from most eukaryotic and prokaryotic transposable elements in that it has no terminal inverted repeats. In other transposable elements these repeats are believed to be recognition sites for enzymes involved in transposition, and may be required for cleavage at the ends of donor elements and/or for integration of transposition intermediates. The I-factor presumably transposes by a different mechanism since it does not have terminal repeats that could be used in this way, and our observation that it encodes a reverse transcriptase-like polypeptide suggests that it transposes via an RNA intermediate. This RNA must include all the sequences of a donor element since we have found that seven I-factors have precisely the same 5′ ends and differ at their 3′ ends only in the number of TAA repeats. The structure of the incomplete I-elements associated with y^{IR3} and y^{IR4} is consistent with this RNA being synthesised from left-to-right in Fig. 1, and then reverse transcribed in the opposite direction. The truncated elements associated with y^{IR3} and y^{IR4} could be the result of premature termination of reverse transcription. The promoter for this RNA must lie within the I-factor itself since we know that the element associated with the w^{IR1} mutation can transpose to new sites (Pélisson, 1981). It does not use the *white* gene promoter for transposition as this would direct transcription of the I-factor in the wrong direction (right-to-left in Fig. 1), and we have found that the w^{IR3} I-factor can transpose from a 6.2 kb restriction fragment that does not include any known promoters (Pritchard, Dura, Bucheton, Pélisson & DJF, unpublished data). It is unlikely that this internal promoter is recognised by pol III as the first 200 bp of the I-factor has no match with the box

A and box B sequences characteristic of pol III promoters (Galli *et al.*, 1981), and both strands contain oligo T regions that are known to serve as pol III terminators (Bogenhagen & Brown, 1981). It is concluded that if the I-factor does transpose *via* an RNA intermediate, then this must be synthesised from an internal pol II promoter.

It is not known whether the I-elements associated with y^{IR3} and y^{IR4} are exact copies of deleted donor elements, or whether they were truncated during transposition. The latter seems more likely as elements which are deleted at their 5' ends should have lost the promoter required for synthesis of the transposition intermediate. There is no information regarding the mechanism by which the products of reverse transcription integrate at new sites, but these 5' truncated elements indicate that the left-hand end of the I-factor is not essential for this. Since neither ORF1 nor ORF2 have sequence homology with the endonuclease domain of retroviral *pol* genes, I-elements may integrate at staggered single strand breaks that occur fortuitously in chromosomal DNA. This would explain the variable lengths of the target site duplications associated with different I-elements. On the other hand, the three independent insertions at the same site in the *white* gene suggest that integration is non-random, and may be at sites selected by an I-factor-encoded polypeptide.

SIMILARITY BETWEEN THE I-FACTOR AND OTHER TRANSPOSABLE ELEMENTS

The organisation of repeated sequences in eukaryotic genomes was first investigated by Britten & Davidson and their colleagues (Britten & Kohne, 1968; Davidson *et al.*, 1973). Their renaturation experiments indicated that all genomes contain a substantial fraction of moderately repeated sequences and that these are distributed throughout the genome interspersed with non-repeated DNA (Graham *et al.*, 1974; Davidson *et al.*, 1975). These interspersed repeats fall into two size classes, long sequences of a few kilobases, and short sequences of a few hundred base pairs (Davidson *et al.*, 1973). Singer (1982) has called these two classes LINEs and SINEs, respectively. There is a single major LINE family in all mammalian species that have been studied so far, and in most cases these were first detected as strong bands on ethidium bromide-stained agarose gels containing restriction digests of genomic DNA. They were originally named after the restriction enzymes that generated these bands, but are now called L1 repeats. There are about 10^4 members of primate L1 families (Grimaldi *et al.*, 1984), and these make up about 0.1% of the genome.

Several authors have suggested that L1 sequences may be transposable elements since they are distributed throughout the genome, and are often flanked by short direct repeats. Unlike most transposable elements, but like the I-factor, L1 elements have no terminal repeats themselves but have a very A rich region at the 3' end of one strand. Not all members of an L1 family are identical and many have lost varying amounts of DNA from the 5' end of this strand. L1 elements from several species have been found to contain open reading frames that encode amino acid sequences with homology to the reverse transcriptase domains of retroviral *pol* genes (Fig. 3; Hattori *et al.*, 1986). These observations suggest that L1 elements may also transpose by reverse transcription of an RNA intermediate.

Five families of transposable elements, D (Pittler & Davis, 1987), F (Di Nocera *et al.*, 1983), G (Di Nocera & Dawid, 1983), Doc (Schneuwly *et al.*, 1987) and Jockey (Mizrokhi *et al.*, 1985; Ilyin personal communication), that are structurally similar to L1 elements have been found in *D. melanogaster* (Finnegan & Fawcett *et al.*, 1986 and 1987). F elements are known to transpose because they are located at different sites on the chromosomes of individuals from different *Drosophila* strains (Dawid *et al.*, 1981; Pardue & Dawid, 1981), and one has inserted at the *white* locus in chromosomes carrying the w^{i+A} mutation. Di Nocera & Casari (1987) have determined the base sequence of this F element and have found that it contains an open reading frame with homology to viral reverse transcriptases (Fig. 3). There is no direct evidence that G elements transpose, but they also encode reverse transcriptase-like sequences (Di Nocera, personal communication).

Another family of elements of this type, called ingi, has been found in the genome of *Trypanosoma brucei* (Kimmel *et al.*, 1987). These elements are about 5.2 kb long, have an A rich region at the 3' end of one strand and are present in about 200 copies per haploid genome. There is no direct evidence that they transpose, but they are flanked by potential target site duplications (Kimmel *et al.*, 1987), and an element composed of a tandem duplication of an internally deleted ingi element has been found inserted into a 28S rRNA gene (Hasan *et al.*, 1984). This deleted element, called RIME, is also present in many copies elsewhere in the genome as a dispersed repeat (Hasan *et al.*, 1984; Kimmel *et al.*, 1987). The complete base sequence of one ingi element has been determined (Kimmel *et al.*, 1987) and has been found to encode an amino acid sequence that resembles viral reverse transcriptases in a similar way to those of F, G and L1 elements and the I-factor (Fig. 3).

Perhaps the most intriguing elements of this type are found within 28S rRNA genes of *D. melanogaster* and *Bombyx mori*. The 28S rRNA genes of several invertebrates and protozoa are interrupted by intervening sequences. In *D. melanogaster* and *B. mori*, these are of two types, type I and type II in *D. melanogaster* (Roiha *et al.*, 1981), and R1 and R2 in *B. mori* (Burke *et al.*, 1987). Type II and R2 insertions are present at equivalent positions within rRNA genes and have oligo A sequences at the 3' end of one strand. A complete R2 element has been sequenced and, like all the elements discussed so far, it encodes an open reading frame with homology to reverse transcriptases (Fig. 3). The sequence of a complete type II insertion from *D. melanogaster* has not yet been published, but it will be surprising if it does not have a similar coding capacity. There is no direct evidence that either R2 or type II insertions are transposable. They are not flanked by target site duplications, and most R2, and all type II, sequences are present at precisely the same point within rRNA genes. RIME elements are found in 28S rRNA genes of *T. brucei* but are not equivalent to R2 and type II elements as they are inserted in the opposite orientation (Hasan *et al.*, 1984; Roiha *et al.*, 1981; Burke *et al.*, 1987).

Sequences similar to type II and R2 insertions can be found in the mitochondria of *S. cerevisiae*, and some other fungi. These include the first intron of the gene for subunit 1 of cytochrome oxidase (Bonitz *et al.*, 1980; Michel & Lang, 1985; Hattori *et al.*, 1986). This intron encodes a reverse transcriptase-like sequence (Fig. 3) although there is no evidence that it can transpose, and it does not have an A rich sequence at one end.

All the elements discussed in this section have properties that suggest that they can move from place to place in the genome via an RNA intermediate, and thus fall into the group of sequences that Rogers (1983) has called 'retroposons'. The relationships between these elements is indicated by Table 1. This shows the number of exact matches between ORF2 of the I-factor, and open reading frames from five retroposons, the *copia*-like transposable element 17.6, Rous sarcoma virus and human hepatitis B virus. This is given separately for regions A–G of Fig. 3, as well as the total number of matches summed over all seven regions. Each element

Table 1. *The first line shows the number of amino acids in regions A–G of Fig. 3, individually and in total. The remaining lines show the number of exact matches between corresponding amino acids of the I-factor and each of the elements listed in Fig. 3. The numbers in brackets are the percentages of matches overall.*

| | REGION | | | | | | | |
	A	B	C	D	E	F	G	Total
I Factor	50	29	15	27	10	10	10	151
F	18	15	8	9	3	2	3	58 (38%)
Ingi	15	10	5	11	4	2	3	50 (33%)
L1Md	9	16	5	9	3	2	5	49 (32%)
Bm-R2	8	6	7	9	6	3	1	40 (27%)
Sc-al	7	8	3	8	4	2	3	35 (23%)
17.6	7	3	3	4	3	4	3	35 (23%)
RSV	1	4	4	6	4	4	3	26 (17%)
HBV	2	2	5	6	3	3	3	24 (16%)

has about the same number of matches with ORF2 as any other in regions thought to be associated with RNA dependent DNA polymerase activity (Fig. 3, regions C–G; Johnson *et al.*, 1986). The greatest variation in the number of exact matches is found in regions A and B where the retroposons are more like the I-factor than are 17.6 and the two viruses. The strongest homology, 33/79 matches, is between the I and F elements. This is not just because they are both from *D. melanogaster* as ingi elements from *T. brucei* and the L1 element from the mouse have only slightly fewer matches, 25/79 in each case.

The similarity between the I-factor and other retroposons in regions A and B indicates either that they have descended from a common ancestor, or that they have undergone convergent evolution, in either case, this degree of similarity must reflect functional constraints on these sequences. This function is presumably required specifically by retroposons, and might be involved in the priming of reverse transcription or integration at new sites. The region of ORF2 that matches retroposons so closely is shown in Fig. 6, together with the putative polymerase and ribonuclease H domains.

CONCLUSIONS

Transposable elements have been found in a wide variety of eukaryotes, and it is unlikely that any genome is free of them. There

ORF 2

?   RTase   RNAase H

Fig. 6. Possible functional domains of the ORF2 encoded polypeptide. The box represents the 1086 amino acids encoded by ORF2. The three shaded regions indicate sequences that may be associated with particular functions as described in the text. These are, ?, a function required for transposition of retroposons; RTase, an RNA dependent DNA polymerase; RNAase H, a ribonuclease H.

is no evidence that any transposable element controls a function that is important for individual cells, or whole organisms, but they can have long-term consequences. The effects of I–R dysgenic crosses are believed to be due to increased I-factor activity. At present there is no indiction why this results in reduced fertility, but a direct role has been found for I-elements in producing both point mutations and gross chromosomal rearrangements. The genomes of many species contain transposable elements that are structurally similar to the I-factor. Nothing is known about the regulation of these elements, but expression of a functional element might lead to increased mutation frequencies if it were stimulated by genetic or environmental factors. This might result directly from transposition of these elements, or indirectly because they encode reverse transcriptases that occasionally copy RNAs from conventional genes producing processed pseudogenes. This seems to happen rarely, perhaps because both enzyme and RNA have to be packaged into a particle to allow reverse transcriptase to proceed. In any event, it is clear that most, if not all, eukaryotic cells have the capacity to encode reverse trancriptases even though they may be expressed rarely.

ACKNOWLEDGEMENTS

Some of the experiments reported here were supported by Project Grants from the Medical Research Council. The author is grateful to Annie Wilson and Graham Brown for help in producing illustrations.

REFERENCES

BOGENHAGEN, D. F. & BROWN, D. D. (1981). Nucleotide sequences in Xenopus 5S DNA required for transcription termination. *Cell*, **24**, 261–70.

BONITZ, S. G., CORUZZI, G., THALENFELD, B. E., TZAGOLOFF, A. & MACINO, G. (1980). Assembly of the mitochondrial membrane system, structure and nucleotide sequence of the gene coding for subunit I of cytochrome oxidase. *Journal of Biological Chemistry*, **255**, 11927–41.

BREGLIANO, J. C. & KIDWELL, M. G. (1983). Hybrid dysgenesis determinants. in *Mobile Genetic Elements*, ed. J. Shapiro, pp. 363–410. New York, Academic Press.

BRITTEN, R. J. & KOHNE, D. E. (1968). Repeated sequences in DNA. *Science*, **161**, 529–40.

BROWN, P. O., BOWERMAN, B., VARMUS, H. E. & BISHOP, M. J. (1987). Correct integration of retroviral DNA *in vitro*. *Cell*, **49**, 347–56.

BUCHETON, A. & BREGLIANO, J. C. (1982). The I–R system of hybrid dysgenesis in *Drosophila melanogaster*: heredity of the reactive condition. *Cellular Biology*, **46**, 123–32.

BUCHETON, A., PARO, R., SANG, H. M., PELISSON, A. & FINNEGAN, D. J. (1984). The molecular basis of I–R hybrid dysgenesis: identification, cloning and properties of the I-factor. *Cell*, **38**, 153–63.

BURKE, W. D., CALALANG, C. C. & EICKBUSH, T. H. (1987). The site-specific ribosomal insertion element type II of *Bombyx mori* (R2Bm) contains the coding sequence for a reverse transcriptase-like enzyme. *Molecular and Cellular Biology*, **7**, 221–30.

COVEY, S. N. (1986). Amino acid sequence homology in *gag* region of reverse transcribing elements and the coat protein gene of cauliflower mosaic virus. *Nucleic Acids Research*, **14**, 623–33.

DAVIDSON, E. H., HOUGH, B. R., ARNESSON, C. S. & BRITTEN, R. J. (1973). General interspersion of repetitive with non-repetitive sequence elements in the DNA of *Xenopus*. *Journal of Molecular Biology*, **77**, 1–23.

DAVIDSON, E. H., GALAU, G. A., ANGERER, R. C. & BRITTEN, R. J. (1975). Comparative aspects of DNA organization in metazoa. *Chromosoma*, **51**, 253–9.

DAWID, I. B., OLONG, E. O., DI NOCERA, P. P. & PARDUE, M. L. (1981). Ribosomal insertion-like elements in *Drosophila melanogaster* are interpersed with mobile sequences. *Cell*, **25**, 399–408.

DI NOCERA, P. P. & DAWID, I. B. (1983). Interdigitated arrangement of two oligo(A)-terminated DNA sequences in *Drosophila*. *Nucleic Acids Research*, **11**, 5475–82.

DI NOCERA, P. P. & CASARI, G. (1987). Related polypeptides are encoded by Drosophila F elements, I factors and mammalian L1 sequences. *Proceedings of the National Academy of Sciences USA*, **84**, 5843–7.

DI NOCERA, P. P., DIGAN, M. E. & DAWID, I. (1983). A family of oligo-adenylated transposable sequences in *Drosophila melanogaster*. *Journal of Molecular Biology*, **168**, 715–27.

FAWCETT, D. H., LISTER, C. K., KELLETT, E. & FINNEGAN, D. J. (1986) Transposable elements controlling I–R hybrid dysgenesis in *D. melanogaster* are similar to mammalian LINEs. *Cell* **47**, 1007–15.

FINNEGAN, D. J., BUSSEAU, I., LYNCH, M., FAWCETT, D. H., LISTER, C. K., PELISSON, A., SANG, H. M. & BUCHETON, A. (1987).The structure of mutations induced by I–R hybrid dysgenesis in *Drosophila melanogaster*. in *Transposable Elements as Mutagenic Agents* ed. Cold Spring Harbor Laboratory, Cold Spring Harbor, in press.

GALIBERT, F., MANDART, E., FITOUSSI, F., TIOLLAIS, P. & CHARNAY, P. (1979). Nucleotide sequence of the hepatitis B virus genome (subtype ayw) cloned in *E. coli*. *Nature*, **281**, 646–50.

GALLI, G., HOFSTETTER, H. & BIRNSTEIL, M. L. (1981). Two conserved sequence blocks within eukaryotic tRNA genes are major promoter elements. *Nature*, **294**, 626–31.

GARFINKEL, D. J., BOEKE, J. D. & FINK, G. R. (1985). Ty element transposition: reverse transcriptase and virus-like particles. *Cell*, **42**, 507–17.

GRAHAM, D. E., NEUFELD, B. R., DAVIDSON, E. H. & BRITTEN, R. J. (1974). Interspersion of repetitive and non-repetitive DNA sequences in the sea urchin genome. *Cell*, **1**, 127–37.

GRIMALDI, G., SKOWRONSKI, J. & SINGER, M. (1984). Defining the beginning and end of *Kpn*I family segments. *The EMBO Journal*, **3**, 1753–9.

HASAN, G., TURNER, M. J. & CORDINGLEY, J. S. (1984). Complete nucleotide sequence of an unusual mobile element from *Trypanosoma brucei*. *Cell*, **37**, 333–41.

HATTORI, M., KUHARA, S., TAKENAKA, O. & SAKAKI, Y. (1986). L1 family of repetitive DNA sequences in primates may be derived from a sequence encoding a reverse transcriptase-related protein. *Nature*, **321**, 625–8.

JOHNSON, M. S., MCCLURE, M. A., FENG, D. F., GRAY, T. & DOOLITTLE, R. F. (1986). Computer analysis of retroviral *pol* genes: assignment of enzymatic functions to specific sequence and homologies with nonviral enzymes. *Proceedings of the National Academy of Sciences, USA*, **83**, 7648–52.

KAMER, G. & ARGOS, P. (1984). Primary structure comparison of RNA-dependent polymerases from plant, animal and bacterial viruses. *Nucleic Acids Research*, **12**, 7269–82.

KANAGA, S. & CROUCH, R. J. (1983). DNA sequence of the gene coding for *Escherichia coli* ribonuclease H. *The Journal of Biological Chemistry*, **258**, 1276–81.

KIMMEL, B. E., MOIYOI, O. K. & YOUNG, J. R. (1987). Ingi, a 5.2 kb dispersed sequence element from *Trypanosoma brucei* that carries half of a smaller mobile element at either end and has homology with mammalian LINEs. *Molecular and Cellular Biology*, **7**, 1465–75.

LOEB, D. D., PADGETT, R. W., HARDIES, S. C., SHEHEE, W. R., COMER, M. B., EDGELL, M. H. & HUTCHISON, C. A. (1986). The sequence of a large L1Md element reveals a tandemly repeated 5′ end and several features found in retrotransposons. *Molecular and Cellular Biology*, **6**, 168–82.

MELLOR, J., MALIM, M. H., GULL, K., TUITE, M. F., MCCREADY, S., DIBBAYAWAN, T., KINGSMAN, S. & KINGSMAN, A. J. (1985). Reverse transcriptase activity and Ty RNA are associated with virus-like particles in yeast. *Nature*, **318**, 583–6.

MICHEL, F. & LANG, B. F. (1985). Mitochondrial class two introns encode proteins related to the reverse transcriptases of retroviruses. *Nature*, **316**, 641–3.

MIZROKHI, L. J., OBOLENKOVA, L. A. PRIIMAGI, A. F., ILYIN, Y. V., GERASIMOVA, T. I. & GEORGIEV, G. P. (1985). The nature of unstable insertion mutations and reversions in the locus *cut* of *Drosophila melanogaster*: molecular mechanism of transposition memory. *The EMBO Journal*, **4**, 3781–7.

PARDUE, M. L. & DAWID, I. B. (1981). Chromosomal locations of two DNA segments that flank ribosomal insertion-like sequences in *Drosophila*: flanking sequences are mobile elements. *Chromosoma*, **83**, 29–43.

PEIFER, M. & BENDER, W. (1986). The anterobithorax and bithorax mutations of the bithorax complex. *The EMBO Journal*, **5**, 2293–303.

PÉLISSON, A. (1981). The I–R system of hybrid dysgenesis in *Drosophila melanogaster*: are I factor insertions responsible for the mutator effect of the I–R interaction? *Molecular and General Genetics*, **183**, 123–9.

PICARD, G. (1976). Non-Mendelian female sterility in *Drosophila melanogaster*: hereditary transmission of I-factor. *Genetics*, **83**, 107–23.

PICARD, G. (1978). Non-Mendelian female sterility in *Drosophila melanogaster*: further data on chromosomal contamination. *Molecular and General Genetics*, **164**, 235–47.

PITTLER, S. J. & DAVIS, R. L. (1987). A new family of the poly-deoxyadenylated class of *Drosophila* transposable elements identified by a representative member at the *dunce* locus. *Molecular and General Genetics*, **208**, 325–8.

ROGERS, J. E. (1983). Retroposons defined. *Nature*, **301**, 460.

ROGERS, J. E. (1985). The origin and evolution of retroposons. *International Review of Cytology*, **93**, 188–280.

ROIHA, H., MILLER, J. R., WOODS, L. C. & GLOVER, D. M. (1981). Arrangements and rearrangements of sequences flanking the two types of rDNA insertions in *D. melanogaster*. *Nature*, **290**, 749–53.

SAIGO, K., KUGIMYA, W., MATSUO, Y., INOUYE, S., YOSHIOKA, K. & YUKI, S. (1984). Identification of the coding sequence for a reverse transcriptase-like enzyme in a transposable genetic elementary in *Drosophila melanogaster*. *Nature*, **312**, 659–61.

SANG, H. M., PÉLISSON, A., BUCHETON, A. & FINNEGAN, D. J. (1984). Molecular lesions associated with *white* gene mutations induced by I-R hybrid dysgenesis in *Drosophila melanogaster*. *The EMBO Journal*, **3**, 3079–85.

SCHNEUWLY, S., KUROIWA, A. & GEHRING, W. J. (1987). Molecular analysis of the dominant homoeotic *Antennapedia* phenotype. *The EMBO Journal*, **6**, 201–6.

SCHWARTZ, D. E., TIZZARD, R. & GILBERT, W. (1983). Nucleotide sequence of Rous sarcoma virus. *Cell*, **32**, 853–69.

SCHWARTZBERG, P., COLICELLI, J. & GOFF, S. P. (1984). Construction and analysis of deletion mutants in the *pol* gene of Moloney murine leukemia virus: a new viral function required for productive infection. *Cell*, **37**, 1043–52.

SHIBA, T. & SAIGO, K. (1983). Retrovirus-like particles containing RNA homologous to the transposable element *copia* of *Drosophila melanogaster*. *Nature*, **302**, 119–24.

SINGER, M. F. (1982). SINEs and LINEs: highly repeated short and long interspersed sequences in mammalian genomes. *Cell*, **28**, 433–4.

TEMIN, H. M. & MIZUTANI, S. (1970). RNA-dependent DNA polymerase in virions of Rous sarcoma virus. *Nature*, **226**, 1211–13.

TOH, H., KIKUNO, R., HAYASHIDA, T., KUGIMIYA, W., INOUYE, S., YUKI, S. & SAIGO, K. (1985). Close structural resemblance between putative polymerase of a *Drosophila* transposable genetic element *17.6* and *pol* gene product of Moloney murine leukemia virus. *The EMBO Journal*, **4**, 1267–72.

VERMA, I. M. (1975). Studies on reverse transcription of RNA tumor viruses. I. Localisation of thermolabile DNA polymerase and RNase H activities on one polypetide. *Journal of Virology*, **15**, 121–6.

P ELEMENTS IN *DROSOPHILA MELANOGASTER*: THEIR MOLECULAR BIOLOGY AND USE AS VECTOR SYSTEMS

DONALD C. RIO

Whitehead Institute for Biomedical Research, Nine Cambridge Center, Cambridge, MA 02142, USA and Dept of Biology, Massachusetts Institute of Technology, Cambridge, MA 02139, USA

INTRODUCTION

P elements are one of four structurally distinct classes of transposable elements found in the fruitfly, *Drosophila melanogaster* (for review, see Rubin, 1983). The P element family of mobile genetic elements exhibits a number of interesting regulatory properties, including the fact that the high rates of P element transposition are controlled genetically and in a tissue-specific fashion. This review will summarise recent molecular genetic and biochemical studies directed toward understanding the regulation of P element transposition. It will also review the use of P elements as genetic tools in *Drosophila*. This article will emphasise recent results not covered in two reviews of P elements written a few years ago (Engels, 1983; O'Hare, 1985).

P ELEMENTS AND HYBRID DYSGENESIS

Hybrid dysgenesis is a syndrome of correlated genetic traits that occurs during certain interstrain matings of *Drosophila* (Sved, 1979; Bregliano *et al.*, 1980; Kidwell, *et al.*, 1977). The phenomenon was discovered by population geneticists who found that when certain *Drosophila* strains were interbred a series of abnormalities ensued in the germlines of the progeny from these matings. These defects included high rates of mutation, male recombination, chromosomal rearrangements, sterility, and abnormal germline development. These dysgenic traits were only observed when males isolated from wild populations (termed P or paternally-contributing strains) were mated to females derived from laboratory populations (termed M or maternally-contributing strains). No dysgenesis occurred when the reciprocal cross of mating laboratory males to wild females was performed or when wild males and females were mated to each other. Furthermore, the ability of P strains to induce or repress dysgenic traits (termed cytotype control, see below) exhibited an interesting pattern of inheritance that was dependent not only on genotype but also upon whether the female egg cytoplasm was derived from a P or M strain mother (Engels, 1979a, 1981). Also, a number of dygenesis-induced mutations were unstable and exhibited high reversion frequencies. These and other genetic observations led to the hypothesis that P strains carried a family of transposable elements, called P factors, that were the causative agents of hybrid dysgenesis and that M strains did not contain P factors. It is now thought that P factors have been introduced into *Drosophila* in the wild within the last 30 years. An excellent review of hybrid dysgenesis and the pattern of inheritance of dysgenic traits has appeared recently (Bregliano & Kidwell, 1983). In addition to the P–M system, there is a second independent system of hybrid dysgenesis in *Drosophila* known as I–R (Bregliano & Kidwell, 1982), which is the subject of another article in this volume (Finnegan). In fact, a third system of hybrid dysgenesis in *Drosophila* involving the hobo family of transposable elements has recently been described (Blackman *et al.*, 1987).

MOLECULAR ANALYSIS OF P ELEMENTS

The hypothesis that P–M hybrid dysgenesis was due to a P strain-specific family of transposable elements led Rubin, Kidwell & Bingham (1982) to test this idea at a molecular level. A number of dysgenesis-induced mutations at the white (w) locus were isolated in a dysgenic screen. The white locus had previously been cloned thus allowing a direct molecular analysis of the mutants that were isolated in this screen. Analysis of DNA from the dysgenesis-induced w^- alleles indicated that they carried insertions of foreign DNA segments at the white locus. Furthermore, reversion of these alleles occurred at high frequency (i.e. they were unstable) and was accompanied by the loss of the foreign DNA insertion. Molecular cloning of DNA from the w^- alleles identified these foreign DNA segments as being middle repetitive DNA sequences which were present at about 50 copies per genome in P strains but absent from M strains. These studies clearly demonstrated that P–M hybrid dysgenesis was caused by a P strain-specific transposon family that were termed P elements.

The isolation of P element insertions in the white locus facilitated a detailed molecular analysis of several independent P elements (O'Hare & Rubin, 1983). These studies showed that there were two classes of P elements: a complete element, 2907 bp in length, and a class of elements that is heterogeneous in size ranging from about 0.5–2.5 kb. The smaller elements appear to derive from the 2.9 kb element by heterogeneous internal deletions. It is known from genetic analysis (Engels, 1984) and embryo microinjection experiments (Spradling & Rubin, 1982), that complete 2.9 kb P elements encode a function required to catalyse transposition and excision (referred to as transposase). The smaller elements are incapable of encoding transposase themselves, but carry all of the *cis*-acting DNA sequences required for transposition and this can be activated to move by providing transposase activity in *trans*. There are about 10–15 complete 2.9 kb elements and 30–40 smaller, non-autonomous elements in a typical P strain (Rubin *et al.*, 1982). DNA sequence analysis indicated that all P elements carry perfect 31 bp inverted repeats at their termini. In addition, there is an 11 bp inverted repeat located 120–140 bp from each end (K. Moses, unpublished). It was shown by analysis of many different sites of insertion that an 8 bp duplication of target site DNA is created upon insertion (O'Hare & Rubin, 1983). A very weak 'consensus' sequence of the target sites was compiled and one common feature was their general G–C rich nature. In addition, analysis of several w^+ revertants of the dysgenesis-induced w^- alleles indicated that P elements can also undergo precise excision, leaving only a single copy of the target site duplication. P elements are also known to excise imprecisely, leaving a portion of the element behind or taking adjacent non-P element DNA sequences with it (Daniels *et al.*, 1985). In fact, imprecise excision appears to occur at a higher frequency than precise excision. DNA sequence analysis of the complete 2.9 kb P element indicated the presence of four extended open reading frames that were presumed to encode P element functions responsible for transposition of the element and also perhaps for a regulator of transposition.

The molecular characterisation of P elements facilitated their development as vectors for the transformation of *Drosophila*. Briefly, this technique (described in detail on p. 296) involves the microinjection of cloned DNA into preblastoderm *Drosophila* embryos and subsequent P element transposition from the injected plasmid onto the germline chromosomes. The ability to reintroduce altered P elements into the *Drosophila* genome allowed the analysis of P transposable element functions to be extended. In an elegant series of experiments (Karess & Rubin, 1984), frameshift mutations were made in each of the four P element open reading frames and these mutations were introduced separately into a complete 2.9 kb P element. The resulting modifed P elements were then transformed into the *Drosophila* germline using a helper plasmid which could provide transposase functions in trans but was itself unable to transpose. This helper plasmid was important since the frameshift mutations could inactivate transposase functions. The mutant P elements were then analysed for their ability to encode transposase using a very sensitive genetic assay for transposase activity developed by W. Engels (1984). This assay relies on singed-weak (sn^w), a hypermutable allele of the singed bristle (sn) locus (Engels, 1979b). sn^w is a dygenesis-induced allele that carries two tandem non-autonomous P elements in a head to head orientation (Roiha, Rubin & O'Hare (1988)). In the presence of transposase, sn^w mutates at high frequency (up to 50%) to either a wild type allele (sn$^+$) or an allele with a more extreme phenotype (sn^e) depending upon which one of the two tandem P elements is excised. This simple genetic test provides a sensitive assay for transposase function *in vivo*. The wild-type P$_c$[ry] element gave high levels of sn^w mutability. However, all of the frameshift mutations in the four P element open reading frames (ORF) failed to give sn^w mutational activity, indicating that each ORF was needed to encode transposase. Furthermore, each pairwise combination of the frameshift mutants failed to complement one another which suggested that the four ORF were spliced together in the P element transcript to encode one transposase polypeptide. Karess & Rubin (1984) also determined that a complete 2.9 kb P element encodes a 2.6 kb mRNA that starts at nucleotide 87 on the P element sequence (O'Hare & Rubin, 1983). In addition, RNA from dysgenic and non-dysgenic embryos was analysed by RNA blotting methods. These results indicated no change in the P element transcriptional pattern or level in P male × M female (dysgenic) versus M male × P female (non-dysgenic) embryos. However, transcriptional regulation could occur in the germ cells, yet any change in RNA from these cells would be masked by the large proportion (95%) of somatic cells in the embryonic cell population from which the RNA samples were derived.

GENETIC CONTROL OF P ELEMENT TRANSPOSITION

Genetic studies of hybrid dysgenesis, the unique inheritance patterns of P factors, and the nonreciprocity of P element transposition has led to the definition of two states (Engels, 1983). P strain eggs are said to possess the P 'cytotype' or 'the cytoplasmic environment capable of preventing P element transposition', e.g. in the case in which a P element carrying sperm enters an egg cytoplasm derived from a P strain mother and no transposition events occur. Alternatively, M strain eggs possess the M 'cytotype' or 'the cytoplasmic environment capable of allowing P element transposition', e.g. when a P element carrying sperm enters an egg whose cytoplasm was derived from an M strain mother and transposition occurs. A variety of genetic data following the inheritance and activity of P factors and cytotype suggest that something about the P elements themselves, either the physical DNA itself or a product that they encode is responsible for P cytotype regulation (Engels, 1979*a*, 1979*b*; Kidwell, 1981).

Several models have been proposed to account for cytotype regulation. One, called the transposase titration model, states that in P strain eggs the presence of endogenous P elements (perhaps as episomal circles generated by the action of transposase) titrate the transposase made by the elements introduced by a P strain sperm. Thus transposase becomes distributed at a wide variety of sites but never reaches an active concentration at any individual P element (Simmons & Bucholz, 1985). An alternative model has hypothesised the existence of two P element-encoded functions, one being the transposase that is required for P element transposition and excision and a second product that would function as a negative regulator of transposition in P strains (O'Hare & Rubin, 1983). This model predicts that the regulator must not only prevent transposase synthesis or activity in P strains but must also activate its own synthesis in some way. It should be pointed out that neither of these models is exclusive and cytotype regulation might be brought about by a combination of both mechanisms. There are other scenarios involving either transcriptional or post-transcriptional negative control of transposition in the germline that might account for cytotype control but none have been tested experimentally in a definitive way. For example, it is possible that a negative regulator could function in the germline to reduce transcription or proper splicing of transposase mRNA. In addition, it might be advantageous for the regulator to

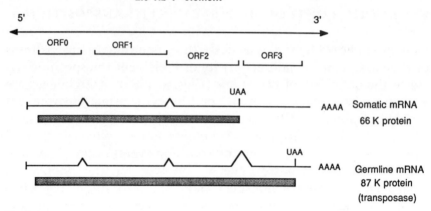

Fig. 1.

be present not only in the germline but also in the soma to prevent 'somatic dysgenesis'.

Biochemical analyses have been carried out using P element-specific antibodies to identify P element-encoded proteins that were expressed in cultured *Drosophila* cells (Rio *et al.*, 1986). These studies identified two proteins: a protein of 87 kD made from an mRNA in which all three introns are removed and a smaller protein of 66 kD made from the somatic mRNA that retains the ORF2-3 intron (Fig. 1). The 87 kD protein encoded by ORF 0, 1, 2 and 3 was predicted from genetic experiments (Karess & Rubin, 1984) to be the transposase and, using a transient assay for P element excision, was shown to be the biologically active transposase polypeptide (Rio *et al.*, 1986). The existence of the 66 kD protein encoded by ORF 0, 1 and 2 has led to the postulation that this polypeptide might be responsible for negatively regulating transposition. In fact, genetic data from Engels & Robertson (personal communication) suggest that truncated forms of the 87 kD transposase, as well as the 66 kD protein which carries 15 C-terminal amino acids not found in transposase, can function as 'repressors' in genetic tests. Current studies are directed toward the purification and biochemical characterisation of the 66 kD and 87 kD proteins in order to understand how one protein might negatively regulate transposition while the other one catalyses it. It is conceivable that if both proteins bind to identical or similar DNA sites then the 66 kD could compete for binding sites with transposase. Alternatively, since both proteins are very homologous except for the extended length of the 87 kD protein and the C-terminal 15 amino acids of the 66 kD protein it is possible that the 66 kD protein could interact with the 87 kD transposase to form inactive heteroprotomers. Hopefully, these possibilities can be addressed with monoclonal antibodies that recognise either one or both proteins. These biochemical probes should allow detection of protein–protein or protein–DNA interactions as well as determination of the relative stoichiometries of the two proteins in the *Drosophila* germline.

The P element termini are required for P elements to undergo transposition and excision. It is known that, in addition to the 31 bp inverted repeats which are absolutely required for transposition (Rubin & Spradling, 1983; Karess & Rubin, 1984), adjacent unique DNA sequences are also required (Rubin & Spradling, 1983). Furthermore, P element termini carry 11 bp inverted repeats which differ slightly (by 9 bp) in their location at either end (K. Moses, unpublished observations). Mullins, Rio & Rubin (unpublished results) have analysed the *cis*-acting DNA sequences required for P element transposition at the 3′ P element end. Using an *in vivo* assay for transposition involving a transposon with duplicated 3′ ends, it was shown that over 150 bp of DNA are required at the 3′ end for wildtype levels of transposition. This includes the 11 bp inverted repeat and unique DNA sequences between the 11 bp and 31 bp inverted repeats, although it has not been determined what role, if any, these unique DNA segments play in transposition. In addition, DNA sequences upstream from the 11 bp inverted repeat, although not essential for transposition, appear to significantly enhance the levels of transposition. Thus, the picture that emerges is one in which a large region of DNA is required at the 3′ end for wildtype levels of transposition. The fact that different unique DNA sequences are located between the 31 bp and 11 bp inverted repeats, as well as the different locations of the 11 bp inverted repeats at either end suggested that the 5′ and 3′ ends are non-equivalent. This idea was tested directly by Mullins, Rio & Rubin (unpublished results) and it was shown that a 5′ end will not substitute for a 3′ end in transposition. This result may not be surprising because the 5′ and 3′ ends are necessarily different since the 5′ end also functions as the promoter sequence to initiate transcription of P element mRNA and so must contain DNA sequence signals for this process. Current studies are directed toward determining whether the 66 and 87 kD P element proteins might bind to the P element termini and if other non-P element-encoded proteins, presumably accessory proteins involved in P element transposition, might recognise the terminal sequences. These studies should provide insight into the molecular mechanisms involved in P element transposition and its regulation.

TISSUE SPECIFICITY OF P ELEMENT TRANSPOSITION

Normally, P element transposition occurs only in the germline of *Drosophila* (Engels, 1983). This tissue-specificity has been studied by using modified P element derivatives, reintroducing them into the *Drosophila* genome and genetically determining if transposition occurs in the germline, somatically or both (Laski *et al.*, 1986). Transcription of P element coding sequences was placed under the control of the inducible *hsp*70 (70 kD heat shock protein) promoter, which is known to be active in somatic tissues. This *hsp*70-P element fusion gene was cloned into a transformation vector, introduced into *Drosophila* and then assayed for its ability to produce P element transcripts and to undergo transposition either somatically or in the germline (see below). Although transcription of this gene fusion occurred at high levels in somatic tissues upon induction, no somatic transposition occurred suggesting a post-transcriptional block to transposase activity.

The detailed structure of the somatic P element mRNA was determined using RNase protection mapping, cDNA cloning and direct RNA sequencing (Laski *et al.*, 1986). These results indicated that in the soma a transcript was synthesised that had two intervening sequences removed such that the first three open reading frames (ORF 0, 1, and 2) were joined in the same translational frame (Fig. 1). There was no splicing event that joined ORF2 to ORF3. This observation was inconsistent with the previous genetic observations of Karess & Rubin (1984) who found that all four ORFs were needed to encode transposase and should be linked in frame to encode a single polypeptide. This led to the hypothesis that there was a third intervening sequence that joined ORF2 to ORF3 but that this intron was removed only in the germline. Thus, the somatic mRNA would encode a truncated, inactive form of transposase. This idea was tested using oligonucleotide-directed mutagenesis (Laski *et al.*, 1986). Putative 5' and 3' consensus splice site sequences (Mount, 1982) were located in ORF2 and ORF3 respectively, such that when this putative intron was removed the two open reading frames would be joined translationally. Single base pair changes were made in each of the conserved dinucleotides at the 5' (GT) and 3' (AG) splice sites. These mutations were known from other work to inactivate splice site utilisation (Padgett *et al.*, 1986; Green, 1986). In addition, a deletion was made that precisely removed the putative intron, thus joining ORF2 to ORF3 in the same translational frame. These mutant P elements were introduced into *Drosophila* and then assayed for their ability to produce transposase activity in genetic tests. Both the 5' and 3' splice site mutations failed to give transposase activity in the sn^w test suggesting that these single base changes prevented splicing of the putative third P element intron. Moreover, using the sn^w test, the element carrying the deletion of the third intron (called Δ2–3) exhibited transposase activity not only in the germline but also in somatic tissues as observed by bristle mosaics in which sn^w, sn^+, sn^e bristles were observed in the same individual. This somatic transposase activity was visualised independently using P element transformants carrying the cell autonomous w^+ eye colour gene (Hazelrigg *et al.*, 1984), thus allowing detection of somatic transposase-mediated events as patches of wild-type or mutant eye tissue. Genetic tests using $P[w^+]$ transposons indicated that both somatic excision and transposition could occur in strains carrying the Δ2–3 transposon (Laski *et al.*, 1986).

These experiments illustrated three important points. First, the fact that single base changes in the 5' and 3' splice site consensus sequences abolished the ability of those elements to encode transposase suggested that there was an intron joining ORF2 to ORF3. Secondly, removal of the ORF2–ORF3 intron appears to be the sole determinant of transposase tissue specificity since the Δ2–3 P element is capable of producing somatic transposase activity. Thirdly, no germline-limited *Drosophila* proteins are required for P element excision and transposition since both activities are observed somatically with Δ2–3. However, the fact that transposase expression in somatic tissues allows transposition does not mean that there are not *Drosophila*-specific accessory proteins required for P element transposition. Thus, expression of transposase in other organisms may not be sufficient to allow transposition.

The regulation of P element germline-specific splicing seems likely to involve a positive factor present in the germline, since even with the *hsp*70-P element fusion construct where P element RNA is expressed at about 10–20 fold higher levels no somatic transposase activity is observed. It is possible, however, that a negative factor in the soma could be present in vast excess over the P element mRNA. Candidates for a positive factor in the germline involved in splicing might be a germline-specific small nuclear ribonucleoprotein particles (snRNPs) which are known to participate in mRNA splicing (Maniatis & Reed, 1987) or perhaps an RNA binding protein (specific or non-specific) that alters the P element RNA structure such that it can now be recognised and spliced by the normal cellular splicing machinery. These two models make specific and testable predictions. For instance, a germline-specific snRNP or specific RNA binding protein would function in trans and act at defined sites at or near the intron whereas a specific RNA structure might be perturbed by *cis*-acting mutations (not necessarily near the intron) that changed the RNA–RNA base pairing or tertiary structural interactions. Genetic and biochemical studies aimed at addressing these possibilities are currently under way. Removal of the third P element intron is unique among examples of alternative mRNA splicing in higher eukaryotes because, rather than using an alternate combination of donor and/or acceptor splice sites, the intron is either removed in the germline or not removed in the soma. This fact should provide a powerful biochemical assay for trying to define germline-specific factors involved in splicing the ORF2–ORF3 intron.

P ELEMENTS AS GENETIC TOOLS

The molecular analysis of P transposable elements led to the development of P elements as vectors for the efficient germline transformation of *Drosophila* (Spradling & Rubin, 1982; Rubin & Spradling, 1982). P element-mediated transformation involves cloning a DNA segment of interest between the two ends of a non-autonomous P element. This DNA and a helper plasmid that expresses transposase are co-injected into the posterior pole of preblastoderm embryos. Transposase then catalyses transposition of the non-autonomous P element from the plasmid into the chromosomes in the germ cells of the injected embryos. The flies that develop from the injected embryos are mated, allowing transformed gametes to develop in the next generation. Usually, the transposon of interest carries a phenotypic marker to allow identification of transformed individuals.

It has been shown using the rosy (ry) gene (the structural gene for xanthine dehydrogenase) that the quantitative expression of transformed genes is normal at a wide variety of genomic sites

(Spradling & Rubin, 1983). Furthermore, a number of genes, such as alcohol dehydrogenase (Goldberg *et al.*, 1983), white (Hazelrigg *et al.*, 1984) and dopa decarboxylase (Scholnick *et al.*, 1983), show correct tissue-specificity and timing of expression following transformation. These initial experiments have led to the use of P element vectors to analyse the *cis*-acting transcriptional regulatory sequences involved in the spatial and temporal expression of several *Drosophila* genes, such as alcohol dehydrogenase (Adh; Fischer & Maniatis, 1986), dopa decarboxylase (Ddc; Scholnick *et al.*, 1986), white (w; Levis *et al.*, 1985; Pirotta *et al.*, 1985), the yolk protein genes YP1 and YP2 (Garabedian *et al.*, 1985), the Rh1 rhodopsin gene (Mismer & Rubin, 1987), the fushi tarazu (*ftz*) gene (Hiromi *et al.*, 1985), the salivary gland glue protein *sgs*3 (Bourouis & Richards, 1985) and *sgs*4 genes (McNabb & Beckendorf, 1986), and the chorion gene amplification control elements (deCicco & Spradling, 1984; Orr-Weaver & Spradling, 1986). These studies have been extremely informative in identifying key regulatory elements in these genes. P element vectors have also been used to transform cloned genes to rescue a mutant phenotype to prove that a DNA segment carries the corresponding gene. This has been done for white (Hazelrigg *et al.*, 1984), nina C (Montell *et al.*, 1985), Rh1 (Zuker, unpublished), transformer (Butler *et al.*, 1986), extra sex combs (Frei *et al.*, 1985), *ftz* (Hiromi *et al.*, 1985) and sevenless (Hafen *et al.*, 1987). P elements can also be used to alter the tissue specificity of genes by incorporating other tissue-specific regulatory elements (McNabb & Beckendorf, 1986).

In addition to their use as vectors for transformation, P elements can be used to induce mutations in genes during dysgenic crosses (Kidwell, 1986). This allows P elements to be used in transposon tagging experiments (Levis *et al.*, 1981) as a starting point for isolating molecular clones of the corresponding gene. This has been successfully applied to the gene for the α-amanitin resistant subunit of RNA polymerase II (Searles *et al.*, 1982) as well as a number of other genes. P elements can also function, in theory, as cell autonomous markers using transposons carrying the white (w) gene or other genes and inducing somatic excision events with the Δ2-3 transposon (Laski *et al.*, 1986). It is also possible to mobilise P elements to new locations by micro-injection of embryos with plasmid DNA that encodes transposase (Levis *et al.*, 1985*b*) or genetically using chromosomal copies of the transposase gene (Steller & Pirotta, 1986; F. Spencer, unpublished; W. Engels, unpublished). This type of mobilisation has been used to study genomic position effects on gene expression (Levis *et al.*, 1985*b*) and in P element mutagenesis experiments (A. Spradling, unpublished).

A number of studies have been aimed at using P elements as genetic tools in other organisms. Although P elements have been shown to function in other *Drosophila* species (Scavarda & Hartl, 1984), all attempts to detect transposase activity in other organisms have been unsuccessful. The early studies involved mammalian tissue culture cells (Clough *et al.*, 1985) or transgenic mice (Khillan *et al.*, 1985) and used genomic P elements that carried the ORF2–ORF3 intron. Recent work using the modified Δ2-3 transposase gene has allowed detection of transposase excision and nuclease activities in mammalian cells and yeast, respectively (Rio *et al.*, 1988). However, no evidence for forward transposition in these or other organisms has been obtained. It is possible that other *Drosophila*-specific proteins, not found in other organisms, are required for P element transposition. Hopefully, as more is learned about the P element transposition reaction, we will be in a position to perform better experiments to use P element transposition as a genetic tool in other species.

ACKNOWLEDGEMENTS

I would like to thank F. Laski, M. Mullins and G. Rubin for their interest in and enthusiasm for P elements. I would especially like to thank F. Laski for many stimulating and lively discussions about P element regulation. I wish to thank G. Barnes, F. Laski, M. Michael, M. Mullins, T. Orr-Weaver and K. Rendahl for critical reading of this manuscript. D.C.R. is supported in part by the Lucille P. Markey Charitable Trust and by the National Institutes of Health.

REFERENCES

BINGHAM, P. M., KIDWELL, M. G. & RUBIN, G. M. (1982). *Cell*, **29**, 995–1004.
BINGHAM, P. M., LEVIS, R. & RUBIN, G. M. (1981). *Cell*, **25**, 693–704.
BLACKMAN, R. K., GRIMAILA, R., KOEHLER, M. M. D. & GELBERT, W. M. (1987). *Cell*, **49**, 497–505.
BOUROUIS, M. & RICHARDS, S. G. (1985). *Cell*, **40**, 349–57.
BREGLIANO, J. C. & KIDWELL, M. G. (1983). *Mobile Genetic Elements*, ed. J. A. Shapiro, pp. 363–410, New York, Academic Press.
BREGLIANO, J. C., PICARD, G., BUCHETON, A., PELISSON, A., LAVIGE, J. M. & L'HERITIER, P. (1980). *Science*, **207**, 606–11.
BUTLER, B., PIROTTA, V., IRMINGER-FINGER, I. & NOTHIGER, R. (1986). The *EMBO Journal*, **5**, 3607–13.

CLOUGH, D. W., LEPINSKI, H. M., DAVIDSON, R. L. & STORTI, R. V. (1985). *Molecular and Cellular Biology*, **5**, 898–901.

DANIELS, S. B., MCCARREN, M., LOWE, C. & CHOVNICK, A. (1985). *Genetics*, **109**, 95–117.

DECICCO, D. & SPRADLING, A. C. (1984). *Cell*, **38**, 45–54.

ENGELS, W. R. (1979*a*). *Genetics Research Cambridge*, **33**, 219–23.

ENGELS, W. R. (1979*b*). *Proceedings of the National Academy of Sciences, USA*, **76**, 4011–15.

ENGELS, W. R. (1981). *Cold Spring Harbor Symposium of Quantitative Biology*, **46**, 561–5.

ENGELS, W. R. (1983). *Annual Review of Genetics*, **17**, 315–44.

ENGELS, W. R. (1984). *Science*, **226**, 1194–6.

FISCHER, J. A. & MANIATIS, T. (1986). The *EMBO Journal*, **5**, 1275–89.

FREI, E., BAUMGARTNER, S., EDSTROM, J.-E. & NOLL, M. (1985). The *EMBO Journal*, **4**, 979–87.

GARABEDIAN, M. J., HUNG, M. C. & WENSINK, P. C. (1985). *Proceedings of the National Academy of Sciences, USA*, **82**, 1396–1400.

GOLDBERG, D., POSAKONY, J. & MANIATIS, T. (1983). *Cell*, **34**, 59–73.

GREEN, M. R. (1986). *Annual Review of Genetics*, **20**, 671–708.

HAFEN, E., BASLER, K., EDSTROM, J.-E. & RUBIN, G. M. (1987). *Science*, **236**, 55–63.

HAZELRIGG, T., LEVIS, R. & RUBIN, G. M. (1984). *Cell*, **36**, 469–81.

HIROMI, Y., KUROIWA, A. & GEHRING, W. (1985). *Cell*, **43**, 603–13.

KARESS, R. E. & RUBIN, G. M. (1984). *Cell*, **38**, 135–46.

KHILLAN, J. S., OVERBREAK, P. A. & WESTPHAL, H. (1985). *Developmental Biology*, **109**, 247–50.

KIDWELL, M. G. (1981). *Genetics*, **98**, 275–90.

KIDWELL, M. G. (1986). In *Drosophila: a practical approach*. 1st edn, ed. D. B. Roberts, pp. 59–81. Washington, DC: IRL Press.

KIDWELL, M. G., KIDWELL, J. F. & SVED, J. A. (1977). *Genetics*, **86**, 813–33.

LASKI, F. A., RIO, D. C. & RUBIN, G. M. (1986). *Cell*, **44**, 7–19.

LEVIS, R., HAZELRIGG, T. & RUBIN, G. M. (1985*a*). The *EMBO Journal*, **4**, 3489–99.

LEVIS, R., HAZELRIGG, T. & RUBIN, G. M. (1985*b*). *Science*, **229**, 558–61.

MANIATIS, T. & REED, R. (1987). *Nature*, **325**, 673–8.

MCNABB, S. L. & BECKENDORF, S. K. (1986). The *EMBO Journal*, **5**, 2331–40.

MISMER, D. & RUBIN, G. M. (1987). *Genetics*, in press.

MONTELL, C., JONES, K., HAFEN, E. & RUBIN, G. M. (1985). *Science*, **230**, 1040–3.

MOUNT, S. M. (1982). *Nucleic Acids Research*, **10**, 459–72.

O'HARE, K. (1987). *Trends in Genetics*, **1**, 250–4.

O'HARE, K. & RUBIN, G. M. (1983). *Cell*, **34**, 25–35.

ORR-WEAVER, T. L. & SPRADLING, A. C. (1986). *Molecular Cell. Biology*, **6**, 4624–33.

PADGETT, R. A., GRABOWSKI, P. J., KONARSKA, M. M., SEILER, S. & SHARP, P. A. (1986). *Annual Review of Biochemistry*, **55**, 1119–50.

PIROTTA, V., STELLER, H. & BOZETTI, M. P. (1985). The *EMBO Journal*, **4**, 3501–8.

RIO, D. C., BARNES, G., LASKI, F. A., RINE, J. & RUBIN, G. M. (1988). *Journal of Molecular Biology*, in press.

RIO, D. C., LASKI, F. A. & RUBIN, G. M. (1986). *Cell*, **44**, 21–32.

ROIHA, H., RUBIN, G. M. & O'HARE, K. (1988). *Genetics*, **118**, in press.

RUBIN, G. M. & SPRADLING, A. C. (1982). *Science*, **218**, 348–53.

RUBIN, G. M. & SPRADLING, A. C. (1983). *Nucleic Acids Research*, **11**, 6341–51.

RUBIN, G. M. (1983). *Mobile Genetic Elements*, J. A. Shapiro, ed. (New York: Academic Press), pp. 329–62.

RUBIN, G. M., KIDWELL, M. G. & BINGHAM, P. M. (1982). *Cell*, **29**, 987–94.

SCAVARDA, N. & HARTL, D. L. (1984) *Proceedings of the National Academy of Sciences, USA*, **81**, 7515–19.

SCHOLNICK, S. B., BRAY, S. J., MORGAN, B. A., McCORMICK, C. A. & HIRSH, J. (1986). *Science*, **234**, 998–1002.

SCHOLNICK, S. B., MORGAN, B. A. & HIRSH, J. (1983). *Cell*, **34**, 37–45.

SEARLES, L. L., JOKERST, R. S., BINGHAM, P. M., VOELKER, R. A. & GREENLEAF, A. L. (1982). *Cell*, **31**, 585–92.

SIMMONS, M. J. & BUCHOLZ, L. M. (1985). *Proceedings of the National Academy of Sciences, USA*, **82**, 8119–23.

SPRADLING, A. C. & RUBIN, G. M. (1982). *Science*, **218**, 341–7.

SPRADLING, A. C. & RUBIN, G. M. (1983). *Cell*, **34**, 47–57.

STELLER, H. & PIROTTA, V. (1986). *Molecular and Cellular Biology*, **6**, 1640–9.

SVED, J. A. (1979). *BioScience*, **29**, 659–64.

ZUKER, C. S., COWMAN, A. C. & RUBIN, G. M. (1985). *Cell*, **40**, 851–8.

DNA RECOMBINATIONS AND TRANSPOSITION IN TRYPANOSOMES

ETIENNE PAYS

Department of Molecular Biology, Free University of Brussels, 67 rue des Chevaux, B1640 Rhode St Genèse, Belgium

DNA REARRANGEMENTS AND ANTIGENIC VARIATION

Outlines of antigenic variation at the gene level

The observations summarised in this section are discussed in several recent reviews (Boothroyd, 1985; Donelson & Rice-Ficht, 1985; Steinert & Pays, 1986; Pays, 1986; Borst, 1986).

The parasite life-cycle

African trypanosomes are unicellular protozoa that inhabit the bloodstream of mammalian hosts, following inoculation by infected *Glossina* (tsetse) flies (Fig. 1). They can multiply for long periods in the blood, and thus maintain prolonged chronic infections. Trypanosomiasis is characterised by the irregular succession of parasite populations, each exhibiting a different antigenic specificity. This antigenic variation allows the parasite to escape the immune response of the host, by preventing at any time the complete destruction of the trypanosome population. Many antigen types can appear successively in the blood, even starting from an inoculum of a single trypanosome. Some antigen types are expressed early, while others are only found late in infection.

In each parasitaemic wave, the trypanosomes undergo morphological changes from rapidly dividing long slender forms, into nondividing stumpy forms, probably depending on changes in the concentration of growth factors in the blood. The stumpy forms may be destroyed by the antibody response, or alternatively, transmitted to a tsetse fly following a blood meal by the insect. In the mid-gut of the fly, the trypanosomes differentiate into procyclic forms, which, among other changes, lose their surface antigen. At the end of their development in the fly, the trypanosomes prepare themselves for a future invasion of the blood. In the fly salivary glands, they transform into metacyclics, where the surface coat reappears. Upon the bite of a mammal by the fly, the metacyclic trypanosomes are injected into the host, invade the blood and differentiate into bloodstream

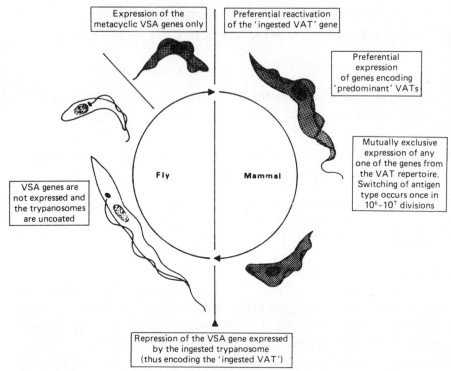

Fig. 1. Surface antigen gene expression during the life-cycle of *Trypanosoma brucei*. The developmental stages illustrated are, clockwise from top, the long slender form and the stumpy form (in the mammalian host), the procyclic, the epimastigote and the metacyclic trypanosomes (in the fly). Reprinted, with permission, from Steinert & Pays, 1986.

forms again. This is accompanied by changes at the cell surface; the metacyclic surface antigens are rapidly replaced by early variants, after a transitory re-expression of the antigen type present before uptake by the fly.

The repertoire of antigen genes

A large fraction of the trypanosome genome is devoted to antigenic variation. Titrations of the total number of antigen genes have given an estimate of about 1000 sequences per genome, although some may be either incomplete, or present in several identical copies. The majority of these genes are clustered in tandem arrays, but many are also found at chromosome ends (telomeres), generally with their 3' extremity oriented towards the DNA terminus.

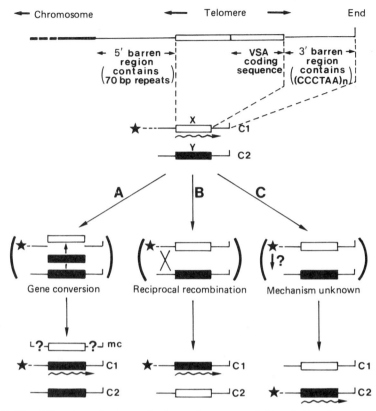

Fig. 2. The different mechanisms allowing antigen type switching in bloodstream trypanosomes. Before antigenic variation, the two antigen genes involved, X and Y, are located at the end of chromosomes C1 and C2, respectively. The telomere of chromosome C1 is activated, as represented by the star, leading to transcription of gene X (wavy arrow). The factors conditioning the selective telomere activation are still conjectural. Antigenic variation can be achieved in at least three ways, which can alternatively apply for the same telomeric gene. In (A), a variably sized gene conversion replaces the previously expressed gene (X) by the copy of the ensuing gene, which may be telomeric as represented here (Y), or chromosome-internal. In some instances, when the conversion involves extensive stretches of DNA encompassing elements allowing autonomous replication, the target (sequence X) could possibly be conserved in the form of a minichromosome (mc), although this is only speculative (see text). In (B), exchange of two telomeric genes occurs, leading to the expression of gene Y in the active telomere of chromosome C1, along with the inactivation of gene X. In (C), the telomere of chromosome C1 is deactivated and the resident antigen gene (X) silenced, while the telomere of chromosome C2 becomes the expression site of gene Y. The mechanism involved is totally unknown. VSA = variant-specific antigen. Reprinted, with permission, from Pays (1986).

Antigen genes in telomeres

The trypanosome genome contains many chromosomes, more particularly over a hundred 50–150 kb minichromosomes, hence many telomeres. As shown in the upper scheme of Fig. 2, the typical structure of a telomere carrying an antigen gene is, from 5' to 3', a variably sized 'barren' region made up of imperfect 70 bp repeats, then the antigen gene, generally preceded by a non-coding region of a size approximately equal to that of the gene, then a terminal 'barren' region which is virtually impossible to clone, but is known to contain CCCTAA repeats. At each cell division, this terminal portion is extended by several nucleotides, probably due to the particular mechanism of DNA replication at the end of the DNA molecule. This regular 'growth' is corrected by occasional large deletions.

Selective expression of the antigen gene

As a rule, only one antigen gene is transcribed at any one time. For some unknown reason, this transcription must take place in a telomere. Among other possibilities, this requirement may be linked to peculiarities of the necessary RNA polymerase, or to specific features of the telomeric DNA. The RNA polymerase transcribing the antigen gene has been reported to resist inhibition by the drug a-amanitin, and to be preferentially inactivated when trypanosomes are isolated from the blood. These two characteristics are not associated with the transcription of housekeeping genes, which are not telomeric. On the other hand, telomeric DNA is a specific target for an unusual form of DNA modification, which prevents the complete cleavage of some sites by restriction endonucleases. This particular modification seems restricted to telomeric genes, although, strikingly, it does not operate on the selectively activated telomere.

The telomeric antigen gene expression site

The telomere used for expression of the antigen gene is organised in much the same way as other telomeres. Some features, such as an 'open' chromatin configuration and absence of telomeric DNA modification, allow this telomere to be distinguished from others, but these characteristics most probably reflect rather than cause transcription. Of greater significance may be the presence of a set of genes called ESAG, for Expression Site Associated Genes, upstream of the 5' barren region. This set of genes may belong to the antigen gene expression site, as shown by their transcription characteristics (see next section).

The polycistronic antigen gene transcription unit

The identification of a trypanosome transcription promoter is difficult and must rely on indirect observations, because no assay for the initiation of transcription is presently available for this organism.

Several features linked to the transcription of the antigen gene suggest that its promoter may be located very far upstream:

(i) The antigen gene, as well as the immediate upstream sequences and ESAG genes, are all subject to the same transcription control. They are only expressed in bloodstream forms, not in cultured procyclics. All three sets of sequences are transcribed on the same strand, in the same direction. If these sequences are repeated in the genome, only the family member present in the expression site is transcribed. The same RNA polymerase is involved in each case, as judged by its particular behaviour towards a-amanitin inhibition (Cully *et al.*, 1985; Bernards *et al.*, 1985; Kooter & Borst, 1984; Murphy *et al.*, 1987*a*; Van der Ploeg *et al.*, 1987).

(ii) Experiments involving protection of RNA transcripts against S1 or exoVII nucleases by hybridisation with cloned probes, have shown that the antigen gene and immediate upstream sequences are transcribed in large precursors, indicating a complex processing pattern (De Lange *et al.*, 1986).

(iii) By using an *in vitro* transcription system in isolated nuclei, it has been demonstrated that upstream sequences, including those in front of the 5' region, are all transcribed at about the same rate (Bernards *et al.*, 1985; De Lange *et al.*, 1986).

(iv) The chromatin configuration of the expression site is open far upstream of the antigen gene (Murphy *et al.*, 1987*a*).

(v) In some instances of DNA rearrangements leading to activation or inactivation of the expression site, the recombination site may be very far upstream of the gene. The simplest interpretation of this observation is that the transcription promoter resides upstream of the recombination site (see below for a more detailed discussion).

(vi) In one case, transcription of upstream sequence was found to be disconnected from that of the antigen gene (Cornelissen *et al.*, 1985*b*). The interpretation of these data again suggests that the transcription promoter must be located far upstream of the gene: the inhibition of antigen gene transcription appeared due to the insertion, largely upstream of the gene, of a DNA stretch which blocked the progression of the RNA polymerase towards the antigen gene.

(vii) In trypanosomes, other transcription units seem to be polycistronic, and a trans-splicing mode of RNA maturation may have been developed to rapidly process very large precursors (see a review by Van der Ploeg, 1986).

(viii) The strongest evidence comes from recent experiments (Johnson *et al.*, 1987) showing that the transcription unit for antigen genes exhibits a sensitivity to UV irradiation compatible with a size of about 60 kb.

Control of antigen gene expression in the fly

Passive inhibition in procyclics? When taken up by the fly, the trypanosomes immediately stop synthesis of variant-specific antigens. The same phenomenon occurs when the parasites are cultivated *in vitro* at low temperature (27 °C). The transcription of the antigen gene is completely switched off, and the degradation of the residual antigen mRNA is accelerated (Ehlers *et al.*, 1987). *In vitro*, the process of gene inactivation seems to occur in two phases, since during an initial period it can be reversed by switching back to bloodstream conditions, before it subsequently becomes irreversible (Ehlers *et al.*, 1987). *In vivo*, there is evidence that inactivation of the gene never reaches the irreversible phase, since the antigen gene silenced during the transformation to procyclics in the fly is readily reactivated when the parasites are retransmitted to the mammal (Delauw *et al.*, 1985; unpublished results by D. Aerts strongly suggest that this reactivation may systematically accompany the differentiation into bloodstream forms).

The rapidity of the gene inactivation, as well as the genes' ability to be immediately re-expressed in the blood, suggest that silencing of the antigen gene in the fly is not due to an active process of gene repression, but rather to a depletion of a factor necessary for gene expression. Experimental data strengthen this conclusion. When inhibitors of protein synthesis are administered to bloodstream forms *in vitro*, the synthesis of the surface antigen is very rapidly and selectively inhibited (Ehlers *et al.*, 1987). This result suggests that bloodstream forms continuously synthesise a short-lived factor necessary for antigen gene expression. The synthesis or activity of this factor might be prevented during transformation to procyclic forms.

Active induction in metacyclics? At the metacyclic stage, antigen genes are re-expressed, but under a system of gene activation control totally different from that operating in the blood. The best evidence for the dual control of antigen gene expression is the observation that the 'bloodstream' expression site, which is about to be reactivated in the blood, is kept silent at this stage (Delauw *et al.*, 1985). Instead, the activity of several other expression sites is triggered in all trypanosomes simultaneously. Only a specific and very limited fraction (around 1%) of the total antigen repertoire is expressed. This particular activation can be mimicked *in vitro*, since the metacyclic transformation can be induced by salivary gland extracts (Cunningham, 1977), and indeed by tissue explants from other Diptera (Cunningham & Kaminski, 1986). The expression of metacyclic antigen genes is unstable. When transmitted to the host, there is a rapid interconversion to other metacyclic antigen types, before switching to the expression of bloodstream antigens (Barry *et al.*, 1985). Thus, at the metacyclic stage, the expression of antigen genes appears to be triggered by a specific inducer, present in the salivary glands of the fly. The progressive loss of this factor would account for the instability of metacyclic gene expression in the blood.

The susceptibility of only a limited number of antigen genes to activation by this factor suggests that specific features must characterise the metacyclic expression sites. Such a characteristic might be the exclusive localisation of these sites at the ends of large chromosomes (Cornelissen *et al.*, 1985a; Lenardo *et al.*, 1986; Delauw *et al.*, 1987). However, some antigen genes present in telomeres of large chromosomes do not encode metacyclic antigens (for instance, AnTat 1.1 and 1.10: Pays *et al.*, 1985a). Moreover, some bloodstream antigen genes may become metacyclic in only a single antigenic switch, without apparent major DNA rearrangement (D. Aerts & E. Pays, unpublished observations). Finally, the metacyclic expression sites may also be used for the transcription of antigen genes in the blood (Delauw *et al.*, 1987). The metacyclic marker, probably a receptor for the fly inducer, would thus be present in several, but not all, telomeres of large chromosomes. This marker could perhaps be transferable, and would not prevent the transcription of the antigen gene under bloodstream conditions.

Switches of gene expression in the blood
Several telomeres may be used as sites for the expression of antigen genes in the blood. Interestingly, there appears to be a timing in the use of these sites during chronic infection (Liu *et al.*, 1985; Timmers *et al.*, 1987; Delauw *et al.*, 1987). In particular, we have found that, in the strain EATRO 1125 of *Trypanosoma brucei*, the telomere of a 200 kb chromosome is systematically preferred for the expression of late antigenic types (Delauw *et al.*, 1987). The telomeres of the 50–150 kb minichromosome class, although carrying antigen genes and despite their high number, do not seem to be competent for

transcription. It is tempting to relate this observation to the fact that these telomeres are devoid of ESAGs (Murphy *et al.*, 1987*a*). If only telomeres carrying ESAGs can be activated, the number of antigen gene expression sites would be around 20, which is in accordance with the available evidence (Longacre *et al.*, 1983; Pays *et al.*, 1983*b*; Myler *et al.*, 1984*a*; Timmers *et al.*, 1987; Delauw *et al.*, 1987).

The mechanism governing the switches in the use of the alternative expression sites is totally unknown (Fig. 2, pathway C). No apparent DNA rearrangement seems linked to this process, but discrete alterations in the vicinity of the transcription promoter, far ahead of the antigen gene, cannot be excluded. However, models involving selective telomere activation exclusively by the translocation of a unique transcription inducer can be ruled out. Two expression sites can be active simultaneously, although this seems unstable or exceptional (Cornelissen *et al.*, 1985*b*; Esser & Schoenbechler, 1985; Baltz *et al.*, 1986). Thus, at least two transcription promoters may co-exist, one of which is rapidly silenced. The telomere activation may be linked to positional effects, such as a specific, but variable anchorage of telomeres in the nuclear matrix; but no evidence substantiates this hypothesis. A clue may perhaps exist in the intriguing characteristics of the telomeric DNA modification system, whose activity, restricted to silent telomeres, seems to increase near the chromosome tip and progressively, but irregularly, decreases towards the chromosome centre (Bernards *et al.*, 1984*b*; Pays *et al.*, 1984). The relative activity of the modification system may be interpreted in two ways: either by the variable association of the telomere with a structure-bound enzyme, or by the progressive inactivation of an enzyme running on the DNA from the chromosome end. The irregular level of DNA modification, even for sites in close proximity, seems best explained by the first model, although other explanations cannot be ruled out. That telomeres may interact with structural elements, such as the nuclear envelope or other telomeres, has been directly observed at meiosis in plant, fungi and animal cells (Lima-de-Faria, 1983). According to these views, the inactive telomeres may associate with structural elements, and the selectively activated telomere would be the only one free from that constraint. This situation would be unstable, leading to switches of expression site.

Once activated, a telomere may be used as expression site for several antigen genes successively, thus achieving antigenic variation. This occurs through partial or total gene replacements, by either gene conversion or reciprocal recombination (Fig. 2, pathways A and B).

On the DNA rearrangements in antigen genes

Gene conversion

Gene conversion is a non-reciprocal recombination event, which involves the replacement of a sequence by the copy of another. It was first described when analysing the products of meiosis in fungi. In some instances the genetic information from one allele turned out to have changed into that coded for by the other allele, without loss of the latter. The straightforward interpretation was that alleles can dictate the replacement of other alleles by their own copy. Subsequently, this mode of recombination has been invoked to explain the evolution of multigene families in mammals (Sligthon *et al.*, 1980; Baltimore, 1981; Ollo & Rougeon, 1983; Weiss *et al.*, 1983), or the mating type interconversion in yeast (Strathern *et al.*, 1982).

Similar events occur in trypanosomes. Antigenic variation is often achieved by the replacement of the active antigen gene by the copy of a silent one. A model for gene conversion, applied to the trypanosome antigenic variation system, is presented in Fig. 3. This model postulates that gene conversion occurs through recombinations in areas of sequence homology flanking the region of the gene encoding the surface-exposed antigenic determinants.

Usual 5' homology region

The sequence homology shared by antigen genes upstream of the coding region is an extensive and variably sized array of imperfect repeats of about 70 bp. This sequence starts approximately 1.5 kb ahead of the antigen gene, and usually extends over several kilobases. Two main features characterise the repeats: a conserved stretch of alternating TG dimers and a variable AT-rich region (Liu *et al.*, 1983; Aline *et al.*, 1985; Florent *et al.*, 1987). The TG oligomers belong to the 'CACA' family of sequences, which includes the enhancer 'core' element, the 'chi' sequence of bacteriophage lambda, and the hypervariable minisatellites of mammalian genomes (Rogers, 1983). These sequences are hotspots for homologous recombination (Steinmetz *et al.*, 1987), perhaps due to their capacity to adopt a left-handed 'Z' DNA configuration (Stollar, 1986). Such sequences can force homologous duplexes to interact, and may be the target of recombinases activities, leading to strand scission and

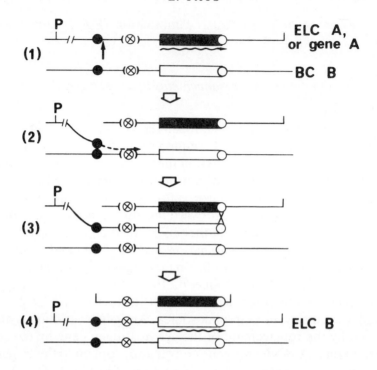

Fig. 3. Hypothetical scheme for the gene conversion leading to the switch from A to B antigen gene expression. This scheme is a modified version of one of the gene conversion models by Strathern *et al.* (1982). Each line represents a double-stranded DNA sequence. The dark and open boxes represent the antigen-specific A and B sequences, flanked by 5′ (●) and 3′ (○) homology blocks whose location respective to the gene may vary considerably. The length of the different elements is not drawn to scale. The wavy arrow represents the transcription, and P the transcription promoter. In (1), the actively transcribed sequence A could be either an expression-linked copy (ELC), or a telomeric gene activated without duplication; gene B could also be telomeric. The possible presence of an element allowing autonomous replication is indicated by (⊗). A double-strand cut (↑) would trigger the gene conversion process. It could occur in, or near, 'CACA'-type stretches located upstream from the gene and downstream from a homology block. The latter can be the 5′ repeat, located generally about 1.5 kb upstream from the gene, but sometimes more than 40 kb upstream; it could also be within the gene, if the target and the donor share sequence similarities. In (2), one of the free DNA ends of the target invades the donor, perhaps after some exonucleolytic digestion. In (3), the donor is copied by the invading strands, up to the 3′ repeat; if the donor is telomeric, the invasion could extend further downstream in the telomere, possibly up to the DNA end, although this supposition is only speculative. In this figure, recombination is shown to take place between the copy and the target sequence in the 3′ repeat. Alternatively, the converted stretch of the target could be first completely digested by an exonuclease, and the second free DNA end of the target could invade the donor in the 3′ region of homology. In (4), sequence A is destroyed or diluted during the cell divisions. We speculate that, in some instances, it could be conserved as a minichromosome, provided that a replication origin is located between the gene and the 5′ repeat. However, this remains to be demonstrated. Reprinted, with permission, from Pays (1985).

exchange (Kmiec & Holloman, 1984). Similarly, that AT-rich region may contain sites sensitive to nuclease activities (Michiels et al., 1985). These combined features make it likely that the 5' region of homology may be used for the initiation of conversion in antigen genes. Actual 5' endpoints have been mapped in various locations in the 70 bp repeats, in or near the AT-rich region (Liu et al., 1983; Campbell et al., 1984; De Lange et al., 1986; Timmers et al., 1987; Florent et al., 1987).

Usual 3' homology region

The analysis of several 3' conversion endpoints has revealed that recombination usually occurs in a conserved region at the end of the antigen gene, or in the adjacent downstream sequences (Michels et al., 1983). In this area, homology is patchy, and does not exhibit obvious characteristics, except the presence of a perfectly conserved 14 bp stretch. The poor overall sequence conservation downstream of the gene, as opposed to the high level of upstream sequence homology, has led to the suggestion that gene conversion is not initiated, but is terminated, in the 3' homology region (Pays, 1985).

Unusual 5' homology

The normal extent of the converted domain between the 5' and 3' recombination endpoints is about 3.5 kb. This includes the gene and upstream 'co-transposed' region of about the same length as the gene. However, cases of longer or smaller conversions are not exceptional. Very large conversions have been noticed on several occasions: between 29 and 42 kb (Bernards et al., 1986a), more than 40 kb (Pays et al., 1983a; Laurent et al., 1984b), or even 90 kb (Van der Ploeg et al., 1987). These observations show that the 5' conversion endpoint may be far upstream of the antigen gene and ESAG region, and thus that chromosomes may share homologies over considerable distances. Extended regions of repeated sequences have been observed within a chromosome, and may relate to this point (Van der Ploeg et al., 1987). In two instances, large conversion events have been found linked to preferential switching of antigen types. In each of these cases, the donor to the conversion process was a recently inactivated expression site (Pays et al., 1983a; Laurent et al., 1984b). Although these data cannot be easily interpreted, they suggest that recombinations in regions of homology unusually far upstream may occur with high frequency.

Unusual 3' homology

The 3' usual region of homology can also be bypassed on some occasions (De Lange *et al.*, 1983). It has even been speculated that the conversion may extend up to the end of the telomere, but this has not been proven. An obvious region where the 3' cross-over could occur is the telomere tip, thought to consist of extensive arrays of short repeats (Van der Ploeg *et al.*, 1984*c*). Telomeric sequences are indeed believed to show a high degree of homology, not only between chromosomes of the same species, but also between species widely divergent in evolution (reviewed in Blackburn & Szostak, 1984). Several considerations indicate that telomeres may associate and interact. Cytological examinations have revealed a linkage between chromosome ends in different plant and animal species in early meiosis ('bouquet' stage) and in spermatozoa (Lima-de-Faria, 1983). Moreover, the alignment of telomeres would readily explain the puzzling observation that the AnTat 1.1 antigen gene of the *T. brucei* stock EATRO 1125 is only expressed late in chronic infection (Hajduk & Vickerman, 1981), despite that fact that this gene normally possesses the usual flanking regions of homology (A. Van der Werf & E. Pays, unpublished results). Indeed, the AnTat 1.1 gene is the only known example of a telomeric gene with the 'wrong' polarity with respect to the chromosome end. The only simple explanation for the low frequency of interaction of this gene with the expression site would be that the recombining sequences first align themselves through a pairing between telomeres. Such a pairing would place the AnTat 1.1 gene in the reverse polarity with respect to that present in the expression site, preventing rapid recombination. The size of the terminal barren region, although variable, is very often around the same value (6 kb) in keeping with a possible role in the alignment of the partner sequences in the conversion event.

Large conversions and the generation of minichromosomes

The abundance of minichromosomes varies considerably between strains and species of trypanosomes (Van der Ploeg *et al.*, 1984*a*), and is noticeably depleted in *T. gambiense* (Dero *et al.*, 1987). The minichromosomes therefore, do not seem to play an essential role in the physiology of trypanosomes. Indeed, they appear to be made up exclusively of barren regions, with the exception of telomeric antigen genes (Williams *et al.*, 1982; Pays *et al.*, 1983*d*; Van der Ploeg *et al.*, 1984*b*). An obvious way to account for the generation of such structures would be the conservation of target sequences of very large antigen gene conversion events, as suggested in Fig. 3. Indeed, some gene conversion events can affect sizes (50–90 kb) which are in the mean size range of the minichromosomes. The conservation of these targets could occur provided that they contain the necessary elements for their autonomous replication, and that they acquire telomeric ends. It is known that telomeres may contain replication origins, and telomeric ends can be generated by a terminal transferase-like activity (Blackburn & Szostak, 1984). However, this hypothesis, which is not substantiated by any direct evidence, does not account for the absence of ESAGs from minichromosomes (Murphy *et al.*, 1987*a*).

Intragenic homology
Usually, the recombining antigen genes share homology only in their 3′ terminal region. However, most antigen genes belong to families of related sequences, and in one case it could be proven that two different members of the same family can encode different antigen types (Pays *et al.*, 1983*c*). Successive interconversions between these two variants occur through partial gene conversions, where the recombination endpoints are outside of the usual regions of homology, due to the presence of extensive homology all along the recombining sequences. In these instances, recombination is found to occur within the coding sequence or in close proximity (Pays *et al.*, 1985*b*). A striking example of partial gene conversion is the AnTat 1.1B gene which, together with its 5′ cotransposed region, has been constructed from four segments, each involving a different conversion donor. The coding region is made up of three sections, the smallest one being about 150 bp, each encoded by different members of the same multigene family. Two main conclusions can be drawn from such analyses. Firstly, the conversion of the region encoding the N-terminal domain of the antigen is sufficient to achieve antigenic variation, showing that the surface-exposed antigenic determinants must lie in that domain. This observation confirms serological data on trypanosome variant surface glycoproteins (Miller *et al.*, 1984; Gomes *et al.*, 1986). Secondly, a clear relationship is found between the extent of the partial conversions and the coding specificity of the DNA. More precisely, the limits of conversion events ending within the gene are found clustered in the region encoding the 'hinge' between the N- and C-terminal domains (Pays *et al.*, 1985*b*). This

suggests that only some recombinational events can be selected, as discussed below.

Extent of homology and gene expression frequency

In chronic infection, the repertoire of antigen genes is not expressed at random. Some antigen types appear frequently at the beginning of infection, while others are only observed exceptionally. The loosely ordered appearance of antigens is beneficial to the parasite, because it avoids the rapid exhaustion of its defences against the host, and leaves rare antigens to maintain prolonged survival. This phenomenon is due to an unequal probability of expression for the different antigen genes of the repertoire. As a working hypothesis, we have suggested that this probability may be determined by the extent of homology between the gene and actual expression site (Laurent et al., 1984b). It has been mentioned above that the choice of the expression site seems programmed during infection; for instance, a 200 kb chromosome is preferred for expression of late variants in a given *T. brucei* strain (Delauw et al., 1987). Depending on the site used, antigen genes will be expressed with variable frequency. Since telomeric genes are usually flanked by large barren regions, like the telomeric expression site, their probability of being expressed may be increased. Indeed early variants are virtually always encoded by telomeric genes (Laurent et al., 1983; Liu et al., 1985; Myler et al., 1984b; Aline et al., 1985b; but see also Longacre & Eisen, 1986). In this respect, it is highly significant to note that a very late-expressed gene may become one of the most predominant only if translocated to a telomere (Laurent et al., 1984b). The nature of the gene itself is thus without influence, contrary to speculations about the role played in the timing of antigen expression by the intrinsic growth rate of the variants (Seed et al., 1984). Along the same line, we have noticed on several occasions that inactivated expression sites have a strong tendency to be re-used, allowing a rapid reappearance of former variants (Pays et al., 1983a; Laurent et al., 1984a, b). This tendency, which is obscured in the host due to the antibody response, certainly reflects the very high probability of activation of *bona fide* telomeric expression sites. It also helps to explain why *in situ* activation of several telomeres, particularly from large chromosomes, is frequent at the beginning of infection (Laurent et al., 1984a; Liu et al., 1985).

Chromosome-internal antigen genes, which cannot be transcribed *in situ*, have to recombine with the telomeric expression sites in order to be expressed. Their activation thus completely depends on the nature and extent of homology with the expression site. Since the size of the barren regions flanking chromosome-internal genes is variable (Aline *et al.*, 1985*a*), the probability of the interaction of these genes with the expression site may vary accordingly. This is supported by our observation that sequences preceding the gene of a very late variant are strongly depleted in 5′ repeats (A. Van der Werf & E. Pays, unpublished results). Going a step further, one may imagine that genes may totally lack the usual homology regions necessary for the recognition of the expression site. This is actually suggested by data showing that the usual homology blocks are not always found in pairs in the genome (Van der Ploeg *et al.*, 1982). The expression of such genes would necessarily be late, because it would depend on the presence, in the expression site, of genes sharing with them unusual homology, for instance near or in the coding region. The resulting conversions are expected to give rise to chimaeric constructs, made up of only a fraction of the incoming sequence together with the residual portion of the previously expressed gene. This explanation satisfactorily accounts for the generation of hybrid genes when late variants are to be expressed (Laurent *et al.*, 1984*b*; Pays *et al.*, 1985*b*; Longacre & Eisen, 1986; Delauw *et al.*, 1987). In addition, this mechanism would be the only one allowing incomplete genes, or pseudogenes, to be expressed (Roth *et al.*, 1986). Therefore, late variants may be encoded by complex recombinations with a low probability of success, but which would expand the variation potential after the expression of a large fraction of the antigen repertoire.

Recombinations outside of the usual regions of homology would also explain cases of preferential switching of variants to a limited number of other antigen types, such as the interconversion between AnTat 1.1 and 1.10, two members of the same multigene family (Pays *et al.*, 1985*b*). The risk of switching on the same set of genes repeatedly during chronic infection is avoided by the selection pressure imposed by the antibody response of the host.

Reciprocal recombination
Gene conversion is by far the most common recombination mechanism used to achieve antigenic variation. In particular, it is the only one allowing chromosome-genes access to a telomeric expression site. However, telomeric antigen genes may use other pathways, such

as *in situ* activation. Moreover, in a single instance, reciprocal recombination has been found linked to the concomitant activation/inactivation of telomeric antigen genes, leading to antigenic switching (Pays *et al.*, 1985a). This event was due to an asymmetric cross-over taking place in the 5' barren region. So far, this report is unique, apart from another case where it has been speculated that reciprocal exchange has occurred, but only linked to the inactivation of a gene (Shea *et al.*, 1986). Moreover, we have not been able to detect subsequent reversal of the successful reciprocal exchange (E. Pays & D. Aerts, unpublished observations). Thus, the exchange of telomeres as a means of achieving antigenic variation appears to be used only rarely by the parasite. Since the telomeric antigen genes are generally amply provided with the barren regions necessary for the cross-over, it is difficult to explain why reciprocal recombination does not occur more frequently.

Telomere translocation?
Antigenic variation may be linked to complex DNA rearrangements including telomere translocations (Van der Ploeg & Cornelissen, 1984; Shea *et al.*, 1986). Whether telomere translocations may actually be used to bring about antigenic switching, however remains to be demonstrated.

The translocation of telomeric genes to the ends of some chromosomes, especially large chromosomes, would easily account for the internalisation, as well as the clustering of antigen genes, if the recipient telomeres also carry antigen genes. This hypothesis finds some support in the observation that the chromosome-internal genes are flanked by barren regions typical of telomeres (Aline *et al.*, 1985a; Blackburn & Challoner, 1984).

Alternate use of different activation mechanisms, and evolution of the antigen gene families
The successive use of different gene activation procedures, one of which involves the generation of a new gene copy, may lead to the rapid evolution of the antigen repertoire. Indeed, the activation of a new telomeric expression site leaves the previous one intact, with a silenced antigen gene, which is generally a gene copy, inserted in the expression site by gene conversion. Thus, *in situ* activation of telomeres allows the parasite to keep additional gene copies, and therefore to expand gene families (Pays *et al.*, 1983a; Laurent *et*

al., 1984*a*). In contrast, gene conversion operating on a freshly inaugurated expression site (a telomere activated *in situ*), leads to the replacement of the antigen gene residing in that site. This mechanism may account for the loss of antigen genes, as demonstrated in several instances (Pays *et al.*, 1983*a*; Laurent *et al.*, 1984*a*).

The gain and loss of genes might thus be dictated by the way *in situ* activation and gene conversion alternate to achieve antigenic variation. Since telomeric genes can be activated by either of these procedures (Pays *et al.*, 1983*a*; Young *et al.*, 1983; Bernards *et al.*, 1984*a*; Myler *et al.*, 1984*b*), genes may be lost in some strains or duplicated in others, depending on the choice and order of succession of the activation mechanisms. The duplicated gene copies are likely to undergo rapid changes, since, in their telomeric location, they may be prone to mutations (Frasch *et al.*, 1982) or rearrangements (Pays *et al.*, 1985*b*; Longacre & Eisen, 1986). Moreover, frequent sister chromated exchanges, as demonstrated for chromosome-internal genes, may also lead to deletions or duplications of antigen genes (Bernards *et al.*, 1986*b*). These phenomena probably account for the rapid diversification of antigen repertoires, even in closely related strains (Paindavoine *et al.*, 1986).

DNA rearrangements and selection

Are the rearrangements programmed?

Antigenic variation does not seem to be induced by external stimuli, such as antibodies (Myler *et al.*, 1985). The frequency of switching (10-5 for early variants; 10-6 to 10-7 during chronic infection: Lamont *et al.*, 1986) is within the range of the frequency of spontaneous DNA rearrangements in mitotic cells (Jackson & Fink, 1981; Liskay & Stachelek, 1983). Furthermore, the analysis of several breakpoints of recombination events did not reveal any conspicuous conserved sequence, which might be classed as a target for a specialised site-specific recombinase. Finally, the expression site, being in a telomere, may be inherently prone to recombination, as telomeres are known to be recombination hotspots in yeast (Horowitz *et al.*, 1984). All these considerations suggest that antigenic variation could be due to spontaneous DNA rearrangements, effected at random by the general recombination machinery in sites particularly susceptible to rearrangement.

This conclusion is supported by the high variability of the DNA rearrangements underlying antigenic variation. These rearrangements are of a diverse nature, and can affect DNA stretches differing widely in size, from several tens of bases to several tens of kilobases. Of particular interest is that the very same gene can be expressed following different activation procedures. The telomeric gene AnTat 1.10 has been found to be activated by either *in situ* activation, different types of segmental gene conversion, or reciprocal gene exchange (Pays *et al.*, 1985*a*). Even when a parasite population looks antigenically homogeneous, as in parasitaemic peaks where the same antigen type is massively expressed, it is actually composed of different genetic sub-populations having switched to the same gene expression by different means (Pays *et al.*, 1983*a*; Timmers *et al.*, 1987).

Evidence for a selection between many types of rearrangements
Despite the lack of any rule or specificity in the activation mechanism, a close examination of some instances of segmental gene conversions have led to the conclusion that DNA rearrangements do not occur totally at random. A clear relationship has been found between the extent of conversion and the coding specificity of the DNA. More precisely, different partial gene conversions operating between the same pair of genes, led each time to the same exchange between the two domains of the antigens (Pays *et al.*, 1985*b*; Gomes *et al.*, 1986). These observations suggest that antigenic variation results from the continuous selection of new variants, emerging from many trials of DNA reassortments. According to this view, the frequency of recombination in the expression site might be considerably higher than the actual switching rate. Only a few combinations might be successful, since, for example, rearrangements occurring within the N-terminal domain could not give rise to functional products, perhaps due to inadequate structure or cooperation between domains at the cell surface. Evidence for interactions between domains of the antigen has actually been presented (Gomes *et al.*, 1986). The final selection may further depend on several other parameters, such as the presence of antibodies in the blood, the intrinsic growth rate of the new variants, the nature of the other competing variants or even the compatibility between the new antigen and the old one during coat renewal.

Nature of the selection mechanism
Antigenic variation seems to occur at a similar rate *in vivo* as *in vitro*, in the presence or absence of antibodies (Doyle *et al.*, 1980). The immune system is thus not the main factor involved in the post-

recombinational selection of viable variants. The surface coat probably has other functions as well as eluding the immune response. At the cell surface, the variant-specific antigens are not arranged independently of each other. They interact to form oligomers (Auffret & Turner, 1981; Strickler & Patton, 1982), which in turn compose a matrix tightly wrapping the cell body. If the coat is defective, the trypanosomes are lysed, even in non-immune serum (Cross, 1978). It is clear therefore that the coat is necessary to enable the parasite to survive in the mammalian serum. Since the building of the coat required co-operation between antigen oligomers, mixed coats may be less ordered, and indeed they appear to be looser than homogeneous coats (Esser & Schoenbechler, 1985). This is supported by the observations made when studying *in vitro* interactions between antigens; oligomers do not form properly between different antigens (Cardoso de Almeida *et al.*, 1984). Thus, mixed coats may be less efficient to protect the trypanosomes against the serum.

Another speculation pertains to the possible involvement of the surface antigens in the interaction with growth factors from the serum. Trypanosomes need factors from the blood to divide. Depending on the relative depletion of these factors, they differentiate into non-dividing stumpy forms (Black *et al.*, 1983). Since variant-specific antigens cover the whole surface of the cell, it is tempting to hypothesise that they carry the receptors which interact with these growth factors. Two kinds of argument support this hypothesis. First, it is clear that the rate of trypanosome growth is influenced by the antigen type (Seed *et al.*, 1984). Secondly, trypanosomes harbouring different antigens may behave differently under the same growth conditions. This is particularly evident at the time the major part of a trypanosome population expressing a main antigen type (homotype) differentiates into stumpy forms. At that stage, trypanosomes expressing a different antigen type (heterotypes) actively proliferate as slender forms. Since the only difference between cells differentiating in either stumpy or slender forms is their antigenic specificity, and since this differentiation appears mainly due to interactions with growth factors, one may conclude that the putative receptors for growth factors are linked to the nature of the surface coat. Again, this coat function may require homogeneous antigen networks.

Mixed coats may thus be selected against, even *in vitro*, in the absence of an immune response. In fact, trypanosomes with mixed coats can be observed *in vitro*, although transiently (Baltz *et al.*,

1986). The equal tendency of these 'zebras' to rapidly switch either to the new or to the old antigenic type, strengthens the hypothesis that a selection, less stringent *in vitro* than *in vivo*, dictates the expression of an homogeneous coat, whatever the variant.

Linkage to gene expression

The active antigen gene may be preferred target for recombinases
The DNA rearrangements discussed above have been selected on the grounds that they alter the antigenic specificity. However, in some instances, rearrangements apparently unlinked to gene expression have been detected (Van der Ploeg & Cornelissen, 1984; Buck *et al.*, 1984; Liu *et al.*, 1985). These observations cannot be taken as indications that all telomeric antigen genes are rearranged with equal likelihood, independently of selective pressures. It is conceivable that antigenic switches may occur through short-lived intermediate variants. In such an event, the DNA rearrangements generated in the transiently activated gene might be conserved in the final variant, but could not be associated with the antigenic switch and would appear unlinked to gene expression, due to the loss of the short-lived intermediate. This interpretation probably accounts for the segmental gene conversion observed in the gene AnTat 1.10 in variant AnTat 1.1C (Pays *et al.*, 1983a). The switch from AnTat 1.16 to 1.1C has most probably involved a short-lived AnTat 1.10 intermediate, where the rearrangement in the expressed AnTat 1.10 gene may have taken place to achieve the antigenic switch to AnTat 1.1C (E. Pays, unpublished data). Therefore, it remains to be proven that silent antigen genes recombine at the same rate as expressed sequences.

On the contrary, it seems very likely that transcribed genes may behave as preferred targets for recombination. Actively transcribed antigen genes exhibit an 'open' chromatin configuration, easily accessible to nucleases such as DNAase I (Pays *et al.*, 1981). This particular configuration extends as far as the telomere end, since the tip of the active telomere is preferentially degraded by endogenous nucleases. As compared with the other, silent, telomeres, the expression site undergoes much more frequent deletions in its terminal region (Pays *et al.*, 1983b). These deletions, which reduce the size of the chromosomes to compensate for their constant growth, only occur occasionally in silent telomeres, and consist of the excision of large stretches of DNA. In the expression site, these deletions are less extensive, but occur continuously. This leads to a typical size heterogeneity of the expressed telomere, visible as a smear in Southern blots. Similarly, the domain of hypersensitivity to DNAase I extends far upstream of the antigen gene, in front of the 5′ barren region (Murphy *et al.*, 1987*a*). It seems therefore reasonable to assume that the active expression site can be preferentially attacked by the nucleases which trigger recombination.

Comparison with other systems

That recombination may preferentially affect transcribed sequences is also suggested by the analysis of other biological systems, where DNA rearrangements alter the expression of genes (for a detailed review on such systems, see Borst & Greaves, 1987).

In the bacterium *Neisseria gonorrhoeae*, antigenic variation proceeds in much the same way as in trypanosomes. Variation in the antigenicity of pilin is due to a transposition of silent gene copies into an active expression site. In this system, it seems that the gene conversion events are oriented, being targeted towards the active locus only. The donor sequences all lack the 5′ terminal section of the only functional gene, which suggests that gene conversions between these sequences do not take place in all orientations equally (Haas & Meyer, 1986).

In the yeast *Saccharomyces cerevisiae*, the switch of mating type depends on the transposition of either copy of two silent genes (HML and HMR) into a unique expression site (MAT). In this case the silent genes are complete, but are under active repression by transcriptional silencer sequences located in their vicinity. This repression depends on the full expression of four genes (SIR) (Nasmyth, 1982). Usually the HML and HMR genes only act as donors for the gene conversion event, which is triggered by a very specific double-strand DNA cut in the MAT locus. In mutants in which the normally silent HML and HMR genes are expressed, these loci start behaving as targets for gene conversion, exactly as it occurs normally at the MAT locus (Klar *et al.*, 1981). This observation clearly indicates that the chromatin configuration dictates the role of the interacting genes in the conversion process; target when transcribed, or donor when silent.

In B and T lymphocytes of mammals, the immunoglobulin and T-cell receptor genes, respectively, are rearranged so as to create the enormous diversity required to achieve an effective immune response. Two observations suggest the involvement of the chromatin structure in the triggering of these rearrangements. Firstly, transcription, though at a low level, precedes the rearrangement of immunoglobulin genes in pre-B cells (Yancopoulos & Alt, 1985). Secondly, in the same cells, T-cell receptor genes appear to be prevented from undergoing rearrangement only because they are packaged in an inactive chromatin configuration (Yancopoulos *et al.*, 1986).

In the chicken, generation of the somatic genes for immunoglobulins follows a different route than in mammals, and more closely resembles antigenic variation in *Neisseria* or *Trypanosoma*. The immunological diversity is achieved by multiple segmental conversions where only pseudogenes act as donors. Once more, the target appears to be restricted to the transcribed gene (Reynaud *et al.*, 1987).

Finally, the conversion events in genes of the major histocompatibility complex of mammals appear, as in trypanosomes, to be under functional constraints as to the nature of the product (Geliebter & Nathenson, 1987), which again argues that the rearrangements occur within expressed sequences.

One can conclude that, as it is suggested for DNA repair (Mellon *et al.*, 1986), DNA rearrangements require that the chromatin is in an accessible configuration for transcription.

Impact on the evolution of the antigen repertoire
In trypanosomes, the silent genes can be divided into two classes: the potentially active genes and the pseudogenes. The former are those few telomeric genes of the competent expression sites. How they are kept silent is not known. The pseudogenes are either incomplete genes, chromosome-internal genes or genes from non-transcribable telomeres, such as those of the minichromosomes. Why genes from these latter categories cannot be transcribed is also unknown. The 'pseudogene' category thus accounts for the vast majority of genes in the total repertoire. According to the views expressed above, these genes can only act as donors in the gene conversion process. Such a bias in the orientation of rearrangements between members of multigene families is sufficient to explain how they can diversify. Computer modelling has shown that the presence of preferred targets in a collection of interacting sequences leads to the generation of polymorphism among them (Brégégère, 1983; Otha, 1984).

DNA rearrangements and promoter activation

In some programmed gene rearrangements, the alteration of gene expression is linked to the activation of a transcription promoter, located within the rearranged sequence (see, for instance, the MAT locus in yeast, reviewed in Borst & Greaves, 1987). This mode of activation can be excluded in some cases of antigenic switching in trypanosomes, at least when *in situ* activation is not involved. When the extent of gene conversion is small, as in the AnTat 1.1B variant (Pays *et al.*, 1983c), it can virtually be ruled out that the transcription promoter is in the rearranged gene segment. Moreover, in at least two instances of 'canonical' gene conversion (about 3 kb), it has been shown that transcription is initiated upstream from the 5' conversion endpoint (Pays *et al.*, 1982; De Lange *et al.*, 1986). Finally, in the only clear example of antigenic switching linked to an exchange of telomeres, the simultaneous activation/inactivation of the exchanging genes can only be easily explained if it is assumed that the active promoter is not affected by the rearrangement, being located upstream of the cross-over (Pays *et al.*, 1985a). The occasional very large gene conversion events linked to antigenic variation are more difficult to interpret in this way, since this would imply that the promoter lies more than 50 kb, or even more than 90 kb upstream of the antigen gene. However, the available evidence tends to support this assumption (see above).

These observations suggest that the transcription promoter is permanently active ahead of rearranging antigen genes. However, they do not preclude the possibility of promoter activation. At least two transcription promoters may be active simultaneously on different telomeres (Cornelissen *et al.*, 1985b). Indirect evidence gives a number of about 20 potentially activatable telomeres per genone, hence 20 putative promoters. Thus, the *in situ* activation of a telomeric gene, which creates a new expression site, may be due to the derepression of a silent promoter. Whether this derepression involves a DNA rearrangement is not known.

TRANSPOSONS IN TRYPANOSOMES?

Evidences for trypanosomal LINEs

Characteristics of mammalian LINEs

Long interspersed nuclear elements (LINEs) are retroposons present in mammalian genomes in very large numbers (for a review, see

Singer & Skowronski, 1985). With retrotransposons, as with retroviruses, yeast Ty elements, *Drosophila* copia-like elements or mouse IAP elements, they share a coding region for reverse transcriptase, as well as a DNA-binding domain (Fanning & Singer, 1987). Furthermore, both kinds of transposons generate the duplication of a short target site. However, the mode of transposition of retroviruses involves the generation of long terminal direct repeats (LTRs), which are noticeably absent from LINEs. It would seem that LINEs, like processed pseudogenes or Alu repeats, transpose through retrotranscription starting at the 3' end of mRNAs, since both kinds of elements carry a 3' poly(A) tail. A clear distinction between LINEs and other retroposons is that the latter do not encode the enzymatic activity required for their transposition.

Among the very large collection of LINE copies, probably only a few are fully functional, since the majority are either truncated at their 5' end, or contain disrupted reading frames. The available evidence suggests that some elements may carry a unique uninterrupted open reading frame, transcribed in a large RNA probably used for the amplification of the family. In all mammalian species studied, this amplification is thought to have occurred very recently in evolution (Deininger & Daniels, 1986). However, direct evidence for amplification and transposition of LINEs is still lacking. Interestingly, transcription of LINEs is markedly elevated in undifferentiated cells (Weiner *et al.*, 1986).

LINEs in trypanosomes

In parallel with another group, we have recently characterised a 5.2 kb repeated sequence from *T. brucei*, which shares all the features described above for mammalian LINEs (Kimmel *et al.*, 1987; Murphy *et al.*, 1987*b*). This element is present at 400 copies per genome, which gives a density of one copy per 175 kb, very close to the overall density of mammalian LINEs, which is one per 150 kb. The trypanosome LINEs, named 'ingi' (Kimmel *et al.*, 1987) or TRS-1 (for Trypanosome Repeated Sequence: Murphy *et al.*, 1987*b*) are found in several *Trypanosoma* species, all closely related to *T. brucei*. These elements can probably undergo transposition, since they are dispersed in all chromosome size classes and are bordered by direct repeats of about 4 bp. The few copies sequenced each contain several large open reading frames, but the situation found in one case (TRS-1.6: Murphy *et al.*, 1987*b*) strongly suggested that some elements may carry a unique coding frame over their entire length (1651 codons). TRS-1.6 is indeed interrupted by only one stop codon, which does not change the downstream reading frame, and which is replaced by a sense codon in other TRS-1 copies.

The putative polypeptide encoded by TRS-1 shows homology with the conserved sequences of reverse transcriptase, which indicates that these elements may retrotranspose. This hypothesis is strengthened by the presence of a poly(A) tail at the 3' end of the repeat. Another domain of the polypeptide shows characteristic features of DNA-binding fingers, namely five putative zinc-binding domains homologous to those of the transcription factor TFIIIA of *Xenopus laevis* (Pays & Murphy, 1987).

Association with the transposable RIME elements

The RIME transposon

A transposable element of 511 bp, arranged as a dimer, has been found inserted in ribosomal genes of *T. brucei*, where it generated a 7 bp direct duplication (Hasan *et al.*, 1984). This element, named RIME for Ribosomal Mobile Element, is supposed to have transposed through reverse transcription, since the 3' end of each monomer was found polyadenylated (see the upper scheme in Fig. 4). Each monomer exhibits an open reading frame encoding the same 160 amino acid protein. Hybridisation studies suggested that this element was widespread in the genome of several trypanosome species, mainly as a monomer.

RIME halves usually flank the trypanosome LINEs

Surprisingly, it was found that the trypanosome LINEs are generally flanked by the separate different halves of RIME (Kimmel *et al.*, 1987; Murphy *et al.*, 1987*b*). The central 4.7 kb portion of TRS-1 appears inserted in the middle of a RIME monomer (see bottom scheme in Fig. 4). The open reading frame starting in the first RIME half is conserved through the insertion point, and as far as the second RIME half. Interestingly, the conservation of the RIME reading frame in TRS-1 is due to an additional G at nucleotide 34 of the first RIME half, as compared to the published sequence of RIME (Hasan *et al.*, 1984).

Usually TRS-1 is flanked by a single copy of each of the two RIME halves, as determined both by sequencing of cloned copies and by hybridisation of Southern blots. However, some copies differ from this general scheme. The structure of TRS-1.6 (Fig. 4) shows

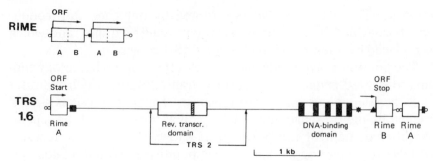

Fig. 4. Structure of two trypanosome retroposons. *Top*: structure of the RIME element, as found in ribosomal genes (Hasan *et al.*, 1984). A 511 bp transposon, terminated by a poly(A) tail (∿) is duplicated in tandem and flanked by 7 bp direct repeats (○). Between the two monomers is a 6 bp sequence interpreted as a residual target (●). Each of the monomers carries an open reading frame (ORF). The subdivision into A and B halves (dotted lines) is only for the purpose of comparison with the structure of TRS-1.6. *Bottom*: structure of the TRS-1.6 element (Murphy *et al.*, 1987*b*). A 4.7 kb sequence appears inserted between the two RIME halves. A 3′ poly(A) tail (∿) and direct repeats (∞) are reminiscent of the RIME structure. The whole 5.2 kb element may be translatable into one large polypeptide. The latter exhibits homology with reverse transcriptase (open box in the middle; the hatched strip corresponds to the well conserved YXDD box: Yuki *et al.*, 1986) and with DNA-binding 'fingers' (box near the right end; the black bands correspond to the five 'fingers': Pays & Murphy, 1987). Additional features include a putative tRNA binding site (■; this site is interrupted at the very 3′ end of the whole structure), a poly(PuPy) Stretch (*) and a polypurine track (▲). Usually, TRS-1 elements do not carry the second RIME A half downstream from the poly(A) tail. The extent of a related sequence (TRS-2) is indicated below. In both schemes, the size of the segment between RIME B and A is not drawn to scale.

the beginning of a dimeric organisation, reminiscent of that of the original RIME dimer. The 3′ end of the complete element is immediately followed by the beginning of a second, starting with the first RIME half and ending after an insertion of only 62 bp. This particular arrangement could result from a large deletion in the second member of a TRS-1 dimer, although other interpretations are possible (see below). The variant copy TRS-2 exhibits a rather different picture; it contains only a 1.45 kb fragment from the central TRS-1 portion, and lacks RIME (Fig. 4, bottom scheme). Again, a possible interpretation is that extensive deletion has occurred, but this time at both ends of a TRS-1 monomer. Other possibilities are discussed below.

Evolutionary relationship between RIME and LINEs
The association of RIME with TRS-1 raises the question of their respective origin. RIME may have arisen by deletion of the 4.7 kb central region of the complete element, or the latter may have arisen by insertion of the central sequences into RIME. The monomeric RIME sequence appears to be a typical retroposon, devoid of the

coding capacity for its transposition. It is possible then that monomeric RIME elements are processed retrotranscripts of TRS-1. However, there are no canonical splice junction sequences between the RIME halves and no sequence homologies that could easily account for a looping out of the intervening 4.7 kb sequence. Also there is an additional G residue in the first RIME half of TRS-1, as compared with the monomeric RIME, arguing that the latter may not be directly derived from TRS-1. The alternative model, that the complete element is derived by insertion into RIME, would be in accordance with the observation that retroposons frequently insert within each others (Rogers, 1985). However, no feature at the junctions of the 4.7 kb central portion with either of the RIME halves indicates that this portion alone is transposable. The only salient feature so far is that these junctions are identical in all 'ingi'/TRS-1 elements sequenced, suggesting that RIME and TRS-1 diverged after a single insertion or deletion event.

Possible interplay between RIME and LINEs
The simultaneous presence of two kinds of related transposable elements in the trypanosome genome suggests a possible functional relationship between them. As in Maize and Drosophila, for example, short transposable elements may be mobilised by the expression of complete elements, which encode the transposase activity (reviewed by Finnegan, 1985). Thus, RIME may carry the terminal signals necessary for the transposition, but would depend on TRS-1 reverse transcriptase for transposition. Alternatively, the association of TRS-1 with RIME might have produced a more efficient mobile element, in the same way that some LINEs are found associated with Alu retroposons, which carry the internal RNA polymerase III promoters necessary for retrotranscription. In relation to this, we found trypanosome Alu repeats in the close vicinity of a TRS-1 copy (TRS-1.2: N. Murphy & E. Pays, unpublished observations). However, these repeats do not appear to be transcribed (Sloof et al., 1983). On the other hand, RIME elements do not seem to carry internal transcription promoters, although this conclusion is far from proven (see next section).

Evolutionary origin of the trypanosome LINEs
The narrow species specificity of TRS-1, together with its abundance in *T. brucei*, suggest a recent origin and amplification. The same conclusion has been drawn for mammalian LINEs, leading to the

paradoxical observation that a sequence that has appeared recently in evolution is found in two categories of organisms as widely separated as protozoa and mammals. One may argue that LINEs are distributed among more eukaryotic genomes than has so far been established, which would then imply that LINEs may have originated very early in evolution, in contradiction to the proposal cited above. An alternative explanation would be that the trypanosome LINEs have been acquired from the genome of their mammalian host.

In this context, the particular organisation of TRS-2 deserves some comment. TRS-2 only contains a central 1.45 kb fragment of TRS-1, which roughly corresponds to the reverse transcriptase domain (Fig. 4). This element may be the relic of an incomplete excision event. However, a more interesting possibility is that TRS-2 is an antecedent or parent gene, from which TRS-1 is derived. There are two lines of evidence that point to this possibility. First, TRS-2 is linked to a second gene with an open reading frame of one kb, and this structural arrangement is conserved in different subspecies of *T. brucei* (Fig. 5). This conservation would not be expected if TRS-2 were the result of an imperfect excision event. Secondly, this linked gene is transcribed in *T. brucei*, albeit at a very low level, and generates large transcripts (about 8 kb), which may include the TRS-2 sequence. Recent models of the origin of repetitive DNA families in eukaryotes implicate a parent gene that, through structural alterations, generates an active progenitor leading to amplification of this progenitor sequence (Deininger & Daniels, 1986; Weiner *et al.*, 1986). According to these views, the origin of the reverse transcriptase gene and, subsequently, of TRS-1, would be found in a rapid amplification within the trypanosome genome, rather than from external sources. This interpretation gains tentative support from the fact that the homologies between 'ingi'/TRS-1 and the mammalian LINEs seem restricted to the reverse transcriptase domain (Kimmel *et al.*, 1987).

Transcription control

General characteristics (Murphy *et al.*, 1987*b*)

The transcription of RIME and TRS-1 exhibit the same general characteristics. In both cases, heterogeneous transcripts are produced on both strands, with a size generally larger than the element itself. The transcription proceeds much more often in the direction of the coding message, and the transcription level is far higher in this direc-

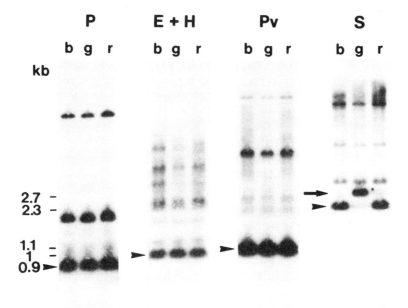

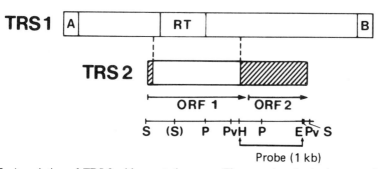

Fig. 5. Association of TRS-2 with a putative gene. The open box in the bottom scheme shows the extent of the TRS-2 element, and its relative position with respect to TRS-1 (A and B = RIME halves; RT = reverse transcriptase domain). The hatched boxes flanking TRS-2 are not homologous to TRS-1, but belong to open reading frames located in or near TRS-2 (ORF 1 and 2, respectively). Both ORF's are in the same orientation, but in different phases. The restriction map allows to see the extent of fragments which hybridise to the 1 kb HindIII-EcoRI probe, in the Southern blots of genomic digests shown above. All digests are designed to generate fragments spanning both TRS-2 and ORF 2, in order to see if the association between TRS-2 and ORF 2 is conserved in different trypanosome strains. These fragments are marked by arrow heads. In SalI digests, the fragment marked by an arrow (DNA in second lane) is probably extending from a site 0.4 kb upstream from the one (bracketed) present in the two other DNAs (and in the sequenced copy of TRS-2). In this DNA, the total replacement of the original fragment by its polymorphic version suggests that the number of TRS-2 copies is limited. The DNAs analysed in each panel are from *T. b. brucei* AnTat 1.3B (b), *T. b. gambiense* LiTat 1.6 (g) and *T. b. rhodesiense* AnTat 12.1 (r) clones, from Uganda, Ivory Coast and Rwanda, respectively. Abbreviation for restriction endonuclease sites are: E = EcoRI; H = HindIII; P = PstI; Pv = PvuII; S = SalI.

tion in bloodstream forms than in procyclics cultivated *in vitro*. The RNA polymerase transcribing this DNA exhibits the same sensitivity to a-amanitin as the trypanosome homologue to eukaryotic RNA polymerase II.

Where is the transcription promoter?

Despite their high copy number, RIME and TRS-1 elements thus appear mainly to be transcribed under specific conditions. This apparent paradox can be resolved in two ways. Either the RIME sequence contains an inducible transcription promoter, or some elements are extensively transcribed in a controlled way. In both cases, a background of non-specific transcription would occur on both strands. Clearly, the controlled transcription takes place in bloodstream forms, over a non-specific background also found in procyclics, although it should be noted that even in procyclics, transcription in the sense direction is favoured over the anti-sense.

The fact that the transcripts are larger than the transposable elements may suggest that the latter do not carry transcription promoters, but depend on those provided by neighbouring sequences. Indeed, the heterogeneity of the transcripts does not seem to be linked to a variation of the 3' terminal size, but rather to differences in the initiation of transcription. This is suggested by three lines of evidence.

(i) It is probable that the sequence at the end of RIME carries the proper termination signal, since this sequence is polyadenylated at a specific site.

(ii) Priming from external promoters would account for the symmetrical transcription on both DNA strands, since the two extremities of TRS-1 do not share the conserved features to be expected if they both carried transcription promoters.

(iii) The 5' end of trypanosome mRNAs is usually capped by a short sequence, a spliced leader often improperly named 'miniexon', which is added to the RNA by trans-splicing (see Borst, 1986). This spliced leader is clearly absent from TRS-1, despite the fact that the RNA polymerase transcribing TRS-1 is the same as that transcribing the bulk of the mRNAs. This observation also suggests that TRS-1 transcription initiates upstream of the transposable section. Indeed, it is known that reverse transcription of the retrotransposons is remarkably accurate, so that the first nucleotide can be copied, even when it is capped (Arkhipova *et al.*, 1986). One may thus have expected that the spliced leader would be copied during reverse transcription of TRS-1, if the mRNA 5' limit was at the beginning of the element.

In conclusion, it would appear that in the blood, the main transcription of TRS-1 starts at variable loci upstream of the element. However, retrotranscription and/or transposition would only affect the conserved 3' portion of these transcripts. An analysis of cloned cDNAs of TRS-1, as well as intermediates in the retrotransposition process, should help to clarify this question.

Nature of the transcription control

Wherever the transcription promoter for TRS-1 may be, it is clearly induced in bloodstream forms. Since the latter grow at a higher temperature than procyclics (37 °C versus 27 °C), it is tempting to ascribe the transcription control to a heat-shock factor. In this respect, no significant counterpart to the consensus sequence for heat-shock promoters has been found within RIME or TRS-1. However, only a partial equivalent to that consensus has been found ahead of true heat-shock induced genes of *T. brucei* (Glass *et al.*, 1986), which leaves open the possibility that a relaxed version of the heat-shock consensus sequence may operate in TRS-1.

In mammalian LINEs, transcription also appears to be controlled. Although both DNA strands can be used as templates, one strand may be preferred, and its transcription is much higher in undifferentiated cells (Weiner *et al.*, 1986). The recent demonstration by Chowdhury *et al.* (1987) that repeated sequences encoding DNA-binding 'fingers' are preferentially transcribed in undifferentiated cells may be related to this point. According to this criterion, trypanosome bloodstream forms could thus be compared to undifferentiated mammalian cells, in contrast to procyclic forms *in vitro*, in which transcription of TRS-1 is suppressed.

Possible modes of amplification and dispersion

Evidence for transposition

Although the TRS-1 elements are numerous and dispersed, no direct evidence for their transposition has been obtained so far. The pattern of size distribution of restriction fragments was found to be invariant when comparing related trypanosome clones or strains. This observation also holds for mammalian LINEs. However, several considerations leave no doubt that these elements may transpose. First, RIME has been found inserted in ribosomal genes, with the duplication of a short stretch of the target, a clear indication of transposition. Secondly, we found that a TRS-1 copy (TRS-1.2) had been inserted in the middle of a known gene (ESAG 1: N. Murphy & E. Pays, unpublished data). Finally, TRS-1 harbours distinctive features of retroposons, such as a poly(A) tail and a gene for reverse transcriptase. Since TRS-1 can be transcribed, its retrotransposition is theoretically possible, according to the recent demonstration that any mRNA can be inserted as cDNA into the genome, if reverse transcriptase is active in a cell (Linial, 1987).

Mode of transposition

Even if retrotransposition of TRS-1 is likely, the mechanism involved is far from clear. Retrotransposons cannot be taken as models, since a key feature in their transposition process is the generation of long terminal direct repeats (LTRs), and the latter are absent from TRS-1. One may argue that the RIME A half (together with the first 62 bp of the 4.7 kb central portion) is such a repeat in the case of TRS-1.6 (Fig. 4). However, the TRS-1.6 arrangement is exceptional, and cannot account for the transposition of the majority of TRS-1 copies.

Despite the absence of LTRs, TRS-1 exhibits two essential features of retrotransposons: a putative tRNA binding site and polypurine stretch, at the 5' and 3' ends of the central 4.7 kb region, respectively (Fig. 4). In retrotransposons, these two sequences are used as primer sites for the synthesis of either cDNA strands (see Arkhipova *et al.*, 1986). Some support for the possible involvement of these sequences in the retrotransposition of TRS-1 has been provided by the analysis of the TRS-1.6 copy. Besides the full TRS-1 element, an apparently abortive TRS-1 copy is interrupted precisely within the putative tRNA homology region (Fig. 4). Regarding the polypurine stretch, this kind of sequence could also be used for LINE amplification or recombination by means different from that used in retrotransposons. Due to its particular base stacking, it may stop the synthesis of DNA (d'Ambrosio & Furano, 1987), perhaps leading to amplification in 'onion-skin' structures (Schimke, 1984), or it may promote the recombination in A-rich regions located downstream, by unpairing the DNA in non-contiguous sequences (Kohwi-

Shigematsu & Kohwi, 1985). However, these hypotheses do not easily account for the presence of the second RIME half downstream from the polypurine stretch.

Speculations about the role of these elements

Function of the trypanosome LINEs
As for other retroposons, the intrinsic function of RIME and TRS-1 is not obvious. They probably belong to the category of 'selfish' DNA sequences, able to invade the genome by virtue of their own enzymatic machinery for amplification and dispersion (Doolittle & Sapienza, 1980; Orgel & Crick, 1980). The extent of this invasion seems more or less controlled, since the density of LINE elements appears constant whatever the genome size, about one copy per chromatin domain (one per 150 kb).

One of the main functions of TRS-1 would be to encode reverse transcriptase activity. Interestingly, the reverse transcriptase domain is totally conserved in the severely truncated TRS-2 copy (see Figs 4 and 5), suggesting that the latter may still be under selective pressure. In contrast, the putative protein encoded by RIME bears no resemblance to reverse transcriptase, and its function is totally unknown.

The TRS-1 elements also encode a DNA-binding domain, as also found in retrotransposons (Berg, 1986). This domain may be necessary to package the cDNA in a karyophilic particle, capable of being reinserted into the nucleus.

Apart from these two regions, the other putative functions of the protein encoded by TRS-1 are unknown. Since the homologies with other retrotransposons and mammalian LINEs are restricted to the reverse transcriptase gene and DNA-binding fingers, it seems possible that additional functions, if any exist, are not essential for the retrotransposition process.

It should be stressed that only a few TRS-1 copies may be truly functional. It is likely that the majority of these elements cannot code for the complete protein, because of interruptions in the reading frame. However, if the large polypeptide is a polyprotein precursor, it is possible that a complementation could occur between imperfect translation products. Transcription studies have indicated that some elements may be preferentially expressed, under the control of developmentally regulated factors (see above). It is then possible that the function of TRS-1 can only be expressed at certain stages of the parasite life-cycle. Whether these functions directly relate to the trypanosome life-cycle is not clear, but the absence of TRS-1 from related kinetoplastida such as *T. congolense* would tend to rule this out.

Impact on the trypanosome genome

The number and dispersion of TRS-1 elements make it unlikely that they are without influence on the organisation of the genome. They contribute up to 2–3% of the nuclear DNA, and therefore appear as one of the most abundant dispersed repetitive sequences in the *T. brucei* genome. One obvious inference would be to provide many dispersed sites of homology, where pairing and recombination can occur. One may expect translocations or chromosome fusions to occur as a result of recombinations between elements located on different chromosomes. Clear examples of such situations can be found in *Drosophila* (in recombinations between the LINE-like 'Doc' elements: Schneuwly *et al.*, 1987) or in yeast (in recombinations between the retrotransposons Ty: Roeder & Fink, 1983). Chromosome rearrangements have been amply documented in *T. brucei* (Van der Ploeg & Cornelissen, 1984).

Impact on gene expression

In two cases, RIME or TRS-1 elements have been found inserted in known genes. Of particular interest is the observation of a TRS-1 copy (TRS-1.2) inserted in the middle of an ESAG-1 copy. Although that copy does not belong to the functional antigen gene expression site, this finding suggests that the insertion of transposable elements may switch off the activity of such sites. In accordance with this hypothesis, the insertion of a 30 kb DNA stretch upstream of the antigen gene has actually been found linked to the inactivation of an expression site (Cornelissen *et al.*, 1985*b*).

Going a step further, one may ask if the specific transcription of TRS-1 in bloodstream forms may not be linked to the expression of the antigen gene transcription unit. However, it seems likely that the promoter for transcription of the antigen gene is not the only one to be specifically induced in the bloodstream form. It has been shown in several instances, in particular in oncogenic transformation, that the insertion of a retrotransposon may also influence the expression of neighbouring genes in a positive manner, due to the transcription promoter and enhancer sequences present in the LTRs (see Finnegan, 1985). In this context, it should be noted that, if TRS-1 does not carry a transcription promoter, it contains near the 3' end a poly(PuPy) sequence which, in line with the considerations discussed above, may perhaps behave as an enhancer for transcription promoters located in the vicinity (Fig. 4).

NOTES ADDED IN PROOF

1. A putative pol I-like transcription promoter has been found about 5 kb upstream from an antigen gene (Shea et al., 1987, Cell, 50, 603–12). This promoter would only allow re-initiation of transcription, since upstream sequences also seem to belong to the antigen gene transcription unit, as determined both by their transcription characteristics and by comparison with other antigen gene expression sites. In the particular case studied, the promoter appears to have been activated by transposition. The basis for this activation was tentatively ascribed to a positioning of the telomere in the nucleolus.
2. A new family of retroposons, distinct from TRS-1 and RIME, has been found associated with the 'mini-exon' genes in Trypanosoma brucei (Aksoy et al., 1987, The Embo Journal, 6, 3819–26).

ACKNOWLEDGEMENTS

I wish to thank Dr David Jefferies and Professor M. Steinert for comments on the manuscript. Work reported herein from my laboratory has been supported by research contracts between the University of Brussels and the Commission of the European Communities (TDS-M-023B), and between the University and Solvay & Cie, SA (Brussels). It was also funded by the Agreement for Collaborative Research on African Trypanosomiasis between ILRAD (Nairobi) and Belgian Research Centres.

REFERENCES

ALINE, R. F. J., MAC DONALD, G., BROWN, E., ALLISON, J., MYLER, P., ROTHWELL, V. & STUART, K. (1985a). (TAA)n within sequences flanking several intrachromosomal variant surface glycoprotein genes in Trypanosoma brucei. Nucleic Acids Research, 13, 3161–77.

ALINE, R. F. J., SCHOLLER, J. K., NELSON, R. G., AGABIAN, N. & STUART, K. (1985b). Preferential activation of telomeric variant surface glycoprotein in Trypanosoma brucei. Molecular and Biochemical Parasitology, 17, 311–20.

ARKHIPOVA, I. R., MAZO, A. M., CHERKASOVA, V. A., GORELOVA, T. V., SCHUPPE, N. G. & ILYIN, Y. V. (1986). The steps of reverse transcription of Drosophila mobile dispersed genetic elements and U3-R-U5 structure of their LTRs. Cell, 44, 555–63.

AUFFRET, C. A. & TURNER, M. J. (1981). Variant specific antigens of Trypanosoma brucei exist in solution as glycoprotein dimers. Biochemical Journal, 193, 647–50.

BALTIMORE, D. (1981). Gene conversion: some implications for immunoglobulin genes. Cell, 24, 592–4.

BALTZ, T., GIROUND, C., BALTZ, D., ROTH, A., RAIBAUD, A. & EISEN, H. (1986). Stable expression of two variable surface glycoproteins by cloned *Trypanosoma equiperdum*. *Nature*, **319**, 602–4.

BARRY, J. D., CROWE, J. S. & VICKERMAN, K. L. (1985). Neutralization of individual variable antigen types in metacyclic populations of *Trypanosoma brucei* does not prevent their subsequent expression in mice. *Parasitology*, **90**, 79–88.

BERG, J. M. (1986). Potential metal-binding domains in nucleic acid binding proteins. *Science*, **232**, 485–7.

BERNARDS, A., DE LANGE, T., MICHELS, P. A. M., HUISMAN, M. J. & BORST, P. (1984a). Two modes of activation of a single surface antigen gene of *Trypanosoma brucei*. *Cell*, **35**, 163–70.

BERNARDS, A., VAN HARTEN-LOOSBROEK, N. & BORST, P. (1984b). Modification of telomeric DNA in *Trypanosoma brucei*, a role in antigenic variation? *Nucleic Acids Research*, **12**, 4153–70.

BERNARDS, A., KOOTER, J. M. & BORST, P. (1985). Structure and transcription of a telomeric surface antigen gene of *Trypanosoma brucei*. *Molecular and Cellular Biology*, **5**, 545–3.

BERNARDS, A., KOOTER, J. M., MICHELS, P. A. M., MOBERTS, R. M. P. & BORST, P. (1986a). Pulsed field gradient electrophoresis of DNA digested in agarose allows the sizing of the large duplication unit of a surface antigen gene in trypanosomes. *Gene*, **42**, 313–22.

BERNARDS, A., VAN DER PLOEG, L. H. T., GIBSON, W. C., LEEGWATER, P., EIJGENRAAM, F., DE LANGE, T., WEIJERS, P., CALAFAT, J. & BORST, P. (1986b). Rapid change of the repertoire of variant surface glycoprotein genes in trypanosomes by gene duplication and deletion. *Journal of Molecular Biology*, **190**, 1–10.

BLACK, S. J., JACK, R. M. & MORRISON, W. I. (1983). Host–parasite interactions which influence the virulence of *Trypanosoma brucei* organisms. *Acta Tropica* (Basel) **40**, 11–18.

BLACKBURN, E. H. & CHALLONER, P. B. (1984). Identification of a telomeric DNA sequence in *Trypanosoma brucei*. *Cell*, **36**, 447–57.

BLACKBURN, E. H. & SZOSTAK, J. W. (1984). The molecular structure of centromeres and telomeres. *Annual Review of Biochemistry*, **53**, 163–94.

BOOTHROYD, J. C. (1985). Antigenic variation in African trypanosomes. *Annual Review of Microbiology*, **39**, 475–502.

BORST, P. (1986). Discontinuous transcription and antigenic variation in trypanosomes. *Annual Review of Biochemistry*, **55**, 701–32.

BORST, P. & GREAVES, D. R. (1987). Programmed gene rearrangements altering gene expression. *Science*, **235**, 658–67.

BRÉGÉGÈRE, F. (1983). A directional process of gene conversion is expected to yield dynamic polymorphism associated with stability of alternative alleles in class I histocompatibility antigens gene family. *Biochimie*, **65**, 229–37.

BUCK, G. A., JACQUEMOT, C., BALTZ, T. & EISEN, H. (1984). Re-expression of an inactivated variable surface glycoprotein gene in *Trypanosoma equiperdum*. *Gene*, **32**, 329–36.

CAMPBELL, D. A., VAN BREE, M. P. & BOOTHROYD, J. C. (1984). The 5′ limit of transposition and upstream barren region of a trypanosome VSG gene: tandem 76 base-pairs repeats flanking (TAA) 90. *Nucleic Acids Research*, **12**, 2759–74.

CARDOSO DE ALMEIDA, M. L., ALLAN, L. M. & TURNER, M. J. (1984). Purification and properties of the membrane form of VSGs from *Trypanosoma brucei*. *Journal of Protozoology*, **31**, 53–60.

CHOWDHURY, K., DEUTSCH, U. & GRUSS, P. (1987). A multigene family encoding several 'finger' structures is present and differentially active in mammalian genomes. *Cell*, **48**, 771–8.

CORNELISSEN, A. W. C. A., BAKKEREN, J. A. M., BARRY, J. D., MICHELS, P. A. M. & BORST, P. (1985a). Characteristics of trypanosome antigen genes active in the fly. *Nucleic Acids Research*, **13**, 4661–76.
CORNELISSEN, A. W. C. A., JOHNSON, P. J., KOOTER, J. M., VAN DER PLOEG, L. H. T. & BORST, P. (1985b). Two simultaneously active VSG gene transcription units in a single *Trypanosoma brucei* variant. *Cell*, **41**, 825–32.
CROSS, G. A. M. (1978). Antigenic variation in trypanosomes. *Proceedings of the Royal Society of London B*, **202**, 55–72.
CULLY, D. F., IP, H. S. & CROSS, G. A. M. (1985). Coordinate transcription of variant surface glycoprotein genes and an expression site associated gene family in *Trypanosoma brucei*. *Cell*, **42**, 173–82.
CUNNINGHAM, I. (1977). New culture medium for maintenance of tsetse tissues and growth of trypanosomatids. *Journal of Protozoology*, **24**, 325–9.
CUNNINGHAM, I. & KAMINSKI, R. (1986). Development of metacyclic forms of *Trypanosoma brucei* spp. in cultures containing explants of *Phormia regina* Meigen. *Journal of Parasitology*, **72**, 944–8.
D'AMBROSIO, E. & FURANO, A. V. (1987). DNA synthesis arrest sites at the right terminus of rat long interspersed repeated (LINE or L1Rn) DNA family members. *Nucleic Acids Research*, **15**, 3155–75.
DEININGER, P. L. & DANIELS, G. R. (1986). The recent evolution of mammalian repetitive DNA elements. *Trends in Genetics*, **2**, 76–80.
DE LANGE, T., KOOTER, J. M., MICHELS, P. A. M. & BORST, P. (1983). Telomere conversion in trypanosomes. *Nucleic Acids Research*, **11**, 8149–65.
DE LANGE, T., KOOTER, J. M., LUIRINK, J. & BORST, P. (1986). Transcription of transposed trypanosome surface antigen gene starts upstream of the transposed segment. *The EMBO Journal*, **4**, 3299–306.
DELAUW, M. F., PAYS, E., STEINERT, M., AERTS, D., VAN MEIRVENNE, N. & LE RAY, D. (1985). Inactivation and reactivation of a variant-specific antigen gene in cyclically transmitted *Trypanosoma brucei*. *The EMBO Journal*, **4**, 989–93.
DELAUW, M. F., LAURENT, M., PAINDAVOINE, P., AERTS, D., PAYS, E., LE RAY, D. & STEINERT, M. (1987). Characterization of genes coding for two major metacyclic surface antigens in *Trypanosoma brucei*. *Molecular and Biochemical Parasitology*, **23**, 9–17.
DERO, B., ZAMPETTI-BOSSELER, F., PAYS, E. & STEINERT, M. (1987). The genome and the antigen gene repertoire of *Trypanosoma b. gambiense* are smaller than those of *T. b. brucei*. *Molecular and Biochemical Parasitology*, **26**, 247–56.
DONELSON, J. E. & RICE-FICHT, A. C. (1985). Molecular biology of trypanosome antigenic variation. *Microbiological Review*, **49**, 107–25.
DOOLITTLE, W. F. & SAPIENZA, C. (1980). Selfish genes, the phenotype paradigm and genome evolution. *Nature*, **284**, 601–3.
DOYLE, J. J., HIRUMI, H., HIRUMI, K., LUPTON, E. N. & CROSS, G. A. M. (1980). Antigenic variation in clones of animal-infective *Trypanosoma brucei* derived and maintained *in vitro*. *Parasitology*, **80**, 359–69.
EHLERS, B., CZICHOS, J. & OVERATH, P. (1987). RNA turnover in *Trypanosoma brucei*. *Molecular and Cellular Biology*, **7**, 1242–9.
ESSER, K. M. & SCHOENBECHLER, M. (1985). Expression of two variant surface glycoproteins on individual African trypanosomes during antigen switching. *Science*, **229**, 190–3.
FANNING, T. & SINGER, M. (1987). The LINE-1 DNA sequences in four mammalian orders predict proteins that conserve homologies to retrovirus proteins. *Nucleic Acids Research*, **15**, 2251–60.
FINNEGAN, D. J. (1985). Transposable elements in eukaryotes. *International Review of Cytology*, **93**, 281–326.

FLORENT, I., BALTZ, T., RAIBAUD, A. & EISEN, H. (1987). On the role of repeated sequences 5' to variant surface glycoprotein genes in African trypanosomes. *Gene*, **53**, 55–62.

FRASCH, A. C. C., BORST, P. & VAN DEN BURG, J. (1982). Rapid evolution of genes coding for variant surface glycoproteins in trypanosomes. *Gene*, **17**, 197–211.

GELIEBTER, J. & NATHENSON, S. G. (1987). Recombination and the concerted evolution of the murine MHC. *Trends in Genetics*, **3**, 107–12.

GLASS, D. J., POLVERE, R. I. & VAN DER PLOEG, L. H. T. (1986). Conserved sequences and transcription of the *hsp*70 gene family in *Trypanosoma brucei. Molecular and Cellular Biology*, **6**, 4657–66.

GOMES, V., HUET-DUVILLIER, G., AUBERT, J. P., DIRAT, I., TETAERT, D., MONCANY, M. L. J., RICHET, C., VERVOORT, T., PAYS, E. & DEGAND, P. (1986). Physical and immunological analysis of the two domains isolated from a variant surface glycoprotein of *Trypanosoma brucei. Archives in Biochemistry and Biophysics*, **249**, 427–36.

HAAS, R. & MEYER, T. F. (1986). The repertoire of silent pilus genes in *Neisseria gonorrhoeae*: evidence for gene conversion. *Cell*, **44**, 107–15.

HAJDUK, S. L. & VICKERMAN, K. (1981). Antigenic variation in cyclically transmitted *Trypanosoma brucei*. Variable antigen type composition of the first parasitaemia in mice bitten by trypanosome-infected *Glossina morsitans. Parasitology*, **83**, 609–21.

HASAN, G., TURNER, M. J. & CORDINGLEY, J. S. (1984). Complete nucleotide sequence of an unusual mobile element from *Trypanosoma brucei. Cell*, **37**, 333–41.

HOROWITZ, H., THORBURN, P. & HABER, J. E. (1984). Rearrangements of highly polymorphic regions near telomeres of *Saccharomyces cerevisiae. Molecular and Cellular Biology*, **4**, 2509–17.

JACKSON, J. A. & FINK, G. R. (1981). Gene conversion between duplicated genetic elements in yeast. *Nature*, **292**, 306–11.

JOHNSON, P. J., KOOTER, J. M. & BORST, P. (1987). Inactivation of transcription by UV irradiation in *Trypanosoma brucei* provides evidence for a multicistronic transcription unit that includes a variant specific glycoprotein gene. *Cell*, **51**, 273–81.

KIMMEL, B. E., OLE-MOIYOI, O. K. & YOUNG, J. R. (1987). Ingi, a 5.2 kb dispersed sequence element from *Trypanosoma brucei* that carries half of a smaller mobile element at either end and has homology with mammalian LINEs. *Molecular and Cellular Biology*, **7**, 1465–75.

KLAR, A. J. S., STRATHERN, J. N. & HICKS, J. B. (1981). A position-effect control for gene transposition: state of expression of yeast mating type genes affects their ability to switch. *Cell*, **25**, 517–24.

KMIEC, E. B. & HOLLOMAN, W. K. (1984). Synapsis promoted by *Ustilago* rec1 protein. *Cell*, **36**, 593–8.

KOHWI-SHIGEMATSU, T. & KOHWI, Y. (1985). Poly(dG)-poly(dC) sequences, under torsional stress, induce an altered DNA conformation upon neighboring DNA sequences. *Cell*, **43**, 199–206.

KOOTER, J. M. & BORST, P. (1984). Alpha-amanitin insensitive transcription of variant surface glycoprotein genes provides further evidence for discontinuous transcription in trypanosomes. *Nucleic Acids Research*, **12**, 9457–72.

LAMONT, G. S., TUCKER, R. S. & CROSS, G. A. M. (1986). Analysis of antigen switching rates in *Trypanosoma brucei. Parasitology*, **92**, 355–67.

LAURENT, M., PAYS, E., MAGNUS, E., VAN MEIRVENNE, N., MATTHYSSENS, G., WILLIAMS, R. O. & STEINERT, M. (1983). DNA rearrangements linked to expression of a predominant surface antigen gene of trypanosomes. *Nature*, **302**, 263–6.

LAURENT, M., PAYS, E., DELINTE, K., MAGNUS, E., VAN MEIRVENNE, N. & STEINERT, M. (1984a). Evolution of a trypanosome surface antigen gene repertoire linked to a non-duplicative gene activation. *Nature*, **308**, 370–3.

LAURENT, M., PAYS, E., VAN DER WERF, A., AERTS, D., MAGNUS, E., VAN MEIRVENNE, N. & STEINERT, M. (1984b). Translocation alters the activation rate of a trypanosome surface antigen gene. *Nucleic Acids Research*, **12**, 8319–28.

LENARDO, M. J., ESSER, K. M., MOON, A. M., VAN DER PLOEG, L. H. T. & DONELSON, J. E. (1986). Metacyclic variant surface glycoprotein genes of *Trypanosoma brucei* subsp. *rhodesiense* are activated *in situ*, and their expression is transcriptionally regulated. *Molecular and Cellular Biology*, **6**, 1991–7.

LIMA-DE-FARIA, A. (1983). Organization and function of telomeres. In: *Molecular evolution and organization of the chromosome*, pp. 701–21. Amsterdam, Elsevier Science Publishers.

LINIAL, M. (1987). Creation of a processed pseudogene by retroviral infection. *Cell*, **49**, 93–102.

LISKAY, R. M. & STACHELEK, J. L. (1983). Evidence for intrachromosomal gene conversion in cultured mouse cells. *Cell*, **35**, 157–65.

LIU, A. Y. C., VAN DER PLOEG, L. H. T., RIJSEWIJK, F. A. M. & BORST, P. (1983), The transcription unit of variant surface glycoprotein gene 118 of *Trypanosoma brucei*. Presence of repeated elements at its border and absence of promoter associated sequences. *Journal of Molecular Biology*, **167**, 57–75.

LIU, A. Y. C., MICHELS, P. A. M., BERNARDS, A. & BORST, P. (1985). Trypanosome variant surface glycoprotein genes expressed early in infection. *Journal of Molecular Biology*, **182**, 383–96.

LONGACRE, S., HIBNER, U., RAIBAUD, A., EISEN, H., BALTZ, T., GIROUD, C. & BALTZ, D. (1983). DNA rearrangements and antigenic variation in *Trypanosoma equiperdum*: multiple expression-linked sites in independent isolates of trypanosomes expressing the same antigen. *Molecular and Cellular Biology*, **3**, 399–409.

LONGACRE, S. & EISEN, H. (1986). Expression of whole and hybrid genes in *Trypanosoma equiperdum* antigenic variation. *The EMBO Journal*, **5**, 1057–63.

MELLON, I., BOHR, V. A., SMITH, C. A & HANAWALT, P. C. (1986). Preferential DNA repair of an active gene in human cells. *Proceedings of the National Academy of Science, USA*, **83**, 8878–82.

MICHELS, P. A. M., LIU, A. Y. C., BERNARDS, A., SLOOF, P., VAN DER BIJL, M. M. W., SCHINKEL, A. H., MENKE, H. H., BORST, P., VEENEMAN, G. H., TROMP, M. C. & VAN BOOM, J. H. (1983). Activation of the genes for variant surface glycoproteins 117 and 118 in *Trypanosoma brucei*. *Journal of Molecular Biology*, **166**, 537–56.

MICHIELS, F., MUYLDERMANS, S., HAMERS, R. & MATTHYSSENS, G. (1985). Putative regulatory sequences for the transcription of mini-exons in *Trypanosoma brucei* revealed by S1 sensitivity. *Gene*, **36**, 263–70.

MILLER, E. N., ALLAN, L. M. & TURNER, M. J. (1984). Topological analysis of antigenic determinants on a variant surface glycoprotein of *Trypanosoma brucei*. *Molecular and Biochemical Parasitology*, **13**, 67–81.

MURPHY, N. B., GUYAUX, M., PAYS, E. & STEINERT, M. (1987a). Analysis of VSG expression site sequences in *Trypanosoma brucei*. In: *Molecular strategies of parasitic invasion*, ed. N. Agabian, H. Goodman & N. Nogueira, pp. 449–69. New York, Alan R. Liss, Inc.

MURPHY, N. B., PAYS, A., TEBABI, P., COQUELET, H., GUYAUX, M., STEINERT, M. & PAYS, E. (1987b). A *Trypanosoma brucei* repeated element with unusual structural and transcriptional properties. *Journal of Molecular Biology*, **195**, 855–72.

340   E. PAYS

MYLER, P. J., ALLISON, J., AGABIAN, N. & STUART, K. (1984a). Antigenic variation in African trypanosomes by gene replacement or activation of alternate telomeres. *Cell*, **39**, 203–11.

MYLER, P., NELSON, R. G., AGABIAN, N. & STUART, K. (1984b). Two mechanisms of expression of a predominant variant antigen gene of *Trypanosoma brucei*. *Nature*, **309**, 292–294.

MYLER, P. J., ALLEN, A. L., AGABIAN, N. & STUART, K. (1985). Antigenic variation in clones of *Trypanosoma brucei* grown in immune-deficient mice. *Infection and Immunity*, **47**, 684–90.

NASMYTH, K. A. (1982). The regulation of yeast mating type chromatin structure by SIR: an action at a distance affecting both transcription and transposition. *Cell*, **30**, 567–78.

OLLO, R. & ROUGEON, F. (1983). Gene conversion and polymorphism: generation of mouse immunoglobulin 2a chain alleles by differential gene conversion by 2b chain gene. *Cell*, **32**, 515–23.

ORGEL, L. E. & CRICK, F. H. C. (1980). Selfish DNA: the ultimate parasite. *Nature*, **284**, 604–7.

OTHA, T. (1984). Some models of gene conversion for treating the evolution of multigene families. *Genetics*, **106**, 517–28.

PAINDAVOINE, P., PAYS, E., LAURENT, M., GELTMEYER, Y., LE RAY, D., MEHLITZ, D. & STEINERT, M. (1986). The use of DNA hybridization and numerical taxonomy in determining relationships between *Trypanosoma brucei* and subspecies. *Parasitology*, **92**, 31–50.

PAYS, E. (1985). Gene conversion in trypanosome antigenic variation. *Progress in Nucleic Acids Research in Molecular Biology*, **32**, 1–26.

PAYS, E. (1986). Variability of antigen genes in African trypanosomes. *Trends in Genetics*, **2**, 21–6.

PAYS, E., LHEUREUX, M. & STEINERT, M. (1981). The expression-linked copy of surface antigen gene in *Trypanosoma* is probably the one transcribed. *Nature*, **292**, 265–7.

PAYS, E., LHEUREUX, M. & STEINERT, M. (1982). Structure and expression of a *Trypansosma brucei gambiense* variant specific antigen gene. *Nucleic Acids Research*, **10**, 3149–63.

PAYS, E., DELAUW, M. F., VAN ASSEL, S., LAURENT, M., VERVOORT, T., VAN MEIRVENNE, N. & STEINERT, M. (1983a). Modifications of a *Trypanosoma brucei* antigen gene repertoire by different DNA recombinational mechanisms. *Cell*, **35**, 721–31.

PAYS, E., LAURENT, M., DELINTE, K., VAN MEIRVENNE, N. & STEINERT, M. (1983b). Differential size variations between transcriptionally active and inactive telomeres of *Trypanosoma brucei*. *Nucleic Acids Research*, **11**, 8137–47.

PAYS, E., VAN ASSEL, S., LAURENT, M., DARVILLE, M., VERVOORT, T., VAN MEIRVENNE, N. & STEINERT, M. (1983c). Gene conversion as a mechanism for antigenic variation in trypanosomes. *Cell*, **34**, 371–81.

PAYS, E., VAN ASSEL, S., LAURENT, M., DERO, B., MICHIELS, F., KRONENBERGER, P., MATTHYSSENS, G., VAN MEIRVENNE, N., LE RAY, D. & STEINERT, M. (1983d). At least two transposed sequences are associated in the expression site of a surface antigen gene in different trypanosome clones. *Cell*, **34**, 359–69.

PAYS, E., DELAUW, M. F., LAURENT, M. & STEINERT, M. (1984). Possible DNA modification in GC dinucleotides of *Trypanosoma brucei* telomeric sequences; relationship with antigen gene transcription. *Nucleic Acids Research*, **12**, 5235–47.

PAYS, E., GUYAUX, M., AERTS, D., VAN MEIRVENNE, N. & STEINERT, M. (1985a). Telomeric reciprocal recombination as a possible mechanism for antigenic variation in trypanosomes. *Nature*, **316**, 562–4.

PAYS, E., HOUARD, S., PAYS, A., VAN ASSEL, S., DUPONT, F., AERTS, D., HUET-DUVILLER, G., GOMES, V., RICHET, C., DEGAND, P., VAN MEIRVENNE, N. & STEINERT, M. (1985*b*). *Trypanosoma brucei*: the extent of conversion in antigen genes may be related to the DNA coding specificity. *Cell*, **43**, 821–9.

PAYS, E. & MURPHY, N. B. (1987). DNA-binding fingers encoded by a trypanosome retroposon. *Journal of Molecular Biology*, **197**, 147–8.

REYNAUD, C. A., ANQUEZ, V., GRIMAL, H. & WEILL, J. C. (1987). A hyperconversion mechanism generates the chicken light chain preimmune repertoire. *Cell*, **48**, 379–88.

ROEDER, G. S. & FINK, G. R. (1983). Transposable elements in yeast. In: *Mobile genetic elements*, ed. J. A. Shapiro, pp. 299–328. New York, Academic Press.

ROGERS, J. (1983). CACA sequences. The ends and the means? *Nature*, **305**, 101–2.

ROGERS, J. (1985). The origin and evolution of retroposons. *International Review of Cytology*, **93**, 187–279.

ROTH, C. W., LONGACRE, S., RAIBAUD, A., BALTZ, T. & EISEN, H. (1986). The use of incomplete genes for the construction of a *Trypanosoma equiperdum* variant surface glycoprotein gene. *The EMBO Journal*, **5**, 1965–70.

SCHIMKE, R. T. (1984). Gene amplification in cultured animal cells. *Cell*, **37**, 705–13.

SCHNEUWLY, S., KUROIWA, A. & GEHRING, W. J. (1987). Molecular analysis of the dominant homeotic Antennapedia phenotype. *The EMBO Journal*, **6**, 201–6.

SEED, J. R., EDWARDS, R. & SECHELSKI, J. (1984). The ecology of antigenic variation. *Journal of Protozoology*, **31**, 48–53.

SHEA, C., GLASS, D. J., PARANGI, S. & VAN DER PLOEG, L. H. T. (1986). Variant surface glycoprotein gene expression site switches in *Trypanosoma brucei*. *Journal of Biological Chemistry*, **261**, 6056–63.

SINGER, M. F. & SKOWRONSKI, J. (1985). Making sense out of LINEs: long interspersed repeat sequences in mammalian genomes. *Trends in Biochemical Science*, **10**, 119–22.

SLIGTHON, J. L., BLECHL, A. E. & SMITHIES, O. (1980). Human Fetal G and A globin genes: complete nucleotide sequences suggest that DNA can be exchanged between these duplicated genes. *Cell*, **21**, 627–38.

SLOOF, P., BOS., J. L., KONINGS, A. F. J. M., MENKE, H. H., BORST, P., GUTTERIDGE, W. E. & LEON, W. (1983). Characterization of satellite DNA in *Trypanosoma brucei* and *T. cruzi*. *Journal of Molecular Biology*, **167**, 1–21.

STEINERT, M. & PAYS, E. (1986). Selective expression of surface antigen genes in African trypanosomes. *Parasitology Today*, **2**, 15–19.

STEINMETZ, M., UEMATSU, Y. & LINDAHL, K. F. (1987). Hotspots of homologous recombination in mammalian genomes. *Trends in Genetics*, **3**, 7–10.

STOLLAR, B. D. (1986). Antibodies to DNA. *Critical Review of Cytology*, **20**, 1–36.

STRATHERN, J. N., KLAR, A. J. S., HICKS, J. B., ABRAHAM, J. A., IVY, J. M., NASMYTH, K. A. & McGILL, C. (1982). Homothallic switching of yeast mating type cassettes is initiated by a double-stranded cut in the MAT locus. *Cell*, **31**, 183–92.

STRICKLER, J. E. & PATTON, C. L. (1982). *Trypanosoma brucei*: effect of inhibition of N-linked glycosylation on the nearest neighbor analysis of the major variable surface coat glycoprotein. *Molecular and Biochemical Parasitology*, **5**, 117–31.

TIMMERS, H. T. M., DE LANGE. T., KOOTER, J. M. & BORST, P. (1987). Coincident multiple activations of the same surface antigen gene in *Trypanosoma brucei*. *Journal of Molecular Biology*, **194**, 81–90.

VAN DER PLOEG, L. H. T. (1986). Discontinuous transcription and splicing in trypanosomes. *Cell*, **47**, 479–80.

VAN DER PLOEG, L. H. T., VALERIO, D., DE LANGE, T., BERNARDS, A., BORST, P. & GROSVELD, F. G. (1982). An analysis of cosmid clones of nuclear DNA from *Trypanosoma brucei* shows that the genes for variant surface glycoproteins are clustered in the genome. *Nucleic Acids Research*, **10**, 5905–23.

VAN DER PLOEG, L. H. T. & CORNELISSEN, A. W. C. A. (1984). The contribution of chromosomal translocation to antigenic variation in *Trypanosoma brucei*. *Philosophical Transactions of the Royal Society of London*, B **307**, 13–26.

VAN DER PLOEG, L. H. T., CORNELISSEN, A. W. C. A., BARRY, J. D. & BORST, P. (1984*a*). Chromosomes of *Kinetoplastida*. *The EMBO Journal*, 3109–15.

VAN DER PLOEG, L. H. T., CORNELISSEN, A. W. C. A., MICHELS, P. A. M. & BORST, P. (1984*b*). Chromosome rearrangements in *Trypanosoma brucei*. *Cell*, **39**, 213–21.

VAN DER PLOEG, L. H. T., LIU, A. Y. C. & BORST, P. (1984*c*). Structure of the growing telomeres of trypanosomes. *Cell*, **36**, 459–68.

VAN DER PLOEG, L. H. T., SHEA, C., POLVERE, R. & LEE, G. S. M. (1987). Chromosomal rearrangements and VSG gene transcription in *Trypanosoma brucei*. In: *Molecular strategies of parasitic invasion*, ed. N. Agabian, H. Goodman & N. Nogueira, pp. 437–47. New York, Alan R. Liss, Inc.

WEINER, A. M., DEININGER, P. L. & EFSTRATIADIS, A. (1986). Non-viral retroposons: genes, pseudogenes, and transposable elements generated by the reverse flow of genetic information. *Annual Review of Biochemistry*, **55**, 631–61.

WEISS, E. H., MELLOR, A., GOLDEN, L., FAHRNER, K., SIMPSON, E., HURST, J. & FLAVELL, R. A. (1983). The structure of a mutant H2 gene suggests that the generation of polymorphism in H-2 genes may occur by gene conversion-like events. *Nature*, **301**, 671–4.

WILLIAMS, R. O., YOUNG, J. R. & MAJIWA, P. A. O. (1982). Genomic environment of *Trypanosoma brucei* VSG genes: presence of a minichromosome. *Nature*, **299**, 417–21.

YANCOPOULOS, G. D. & ALT, F. W. (1985). Developmentally controlled and tissue-specific expression of unrearranged VH gene segments. *Cell*, **40**, 271–81.

YANCOPOULOS, G. D., BLACKWELL, T. K., SUH, H., HOOD, L. & ALT, F. W. (1986). Introduced T-cell receptor variable region gene segments recombine in pre-B cells: evidence that B and T cells use a common recombinase. *Cell*, **44**, 251–9.

YOUNG, J. R., MILLER, N., WILLIAMS, R. O. & TURNER, M. J. (1983). Are there two classes of VSG genes in *Trypanosoma brucei*? *Nature*, **306**, 196–8.

YUKI, S., ISHIMARU, S., INOUYE, S. & SAIGO, K. (1986). Identification of genes for reverse transcriptase-like enzymes in two *Drosophila* retrotransposons, 412 and gypsy; a rapid detection method of reverse transcriptase genes using YXDD box probes. *Nucleic Acids Research*, **14**, 3017–30.

TRANSPOSONS AND RETROTRANSPOSONS IN PLANTS: ANALYSIS AND BIOLOGICAL RELEVANCE

ZSUZSANNA SCHWARZ-SOMMER AND HEINZ SAEDLER

Max-Planck Institut für Züchtungsforschung 5000 Köln, FRG

SOME ASPECTS OF TRANSPOSITION IN PLANTS

The mobile elements so far identified in bacteria, yeast, animals and plants possess common features such as the generation of somatic or germinal instability of genes affected by their insertion, the creation of target site duplication upon their integration and the structural organisation of their termini (for review see Schwarz-Sommer, 1987). In two aspects, however, plant transposons reveal unique properties not featured by elements in other organisms. First, the excision of plant transposons only rarely restores the wildtype gene sequence, although the gene function may be phenotypically restored (Saedler & Nevers, 1985; Schwarz-Sommer *et al.*, 1985a). Secondly, functionally defective elements generating a stable mutation in adjacent genes can still interact with 'signals' emitted in *trans* by functionally intact elements of the same family (so-called two-component systems, for review see Nevers, Shepherd & Saedler, 1986). The consequence of this interaction resembles generation of a novel regulatory unit which replaces the genuine control of genes (Schwarz-Sommer & Saedler, 1987). In this sense, plant transposable elements deserve the designation 'controlling element' (McClintock, 1956, 1984). Both processes result in rapid diversification of expressed genes and, as we shall outline below, may be indicative of the utility of plant transposons during evolution. In fact, molecular analysis of genetically defined plant transposons revealed the presence of multiple copies of silent elements in the plant genome. Most of these transposons were mobilised in the past by 'stress' or 'genomic shock' (for reviews see Fedoroff, 1983; Dellaporta & Chomet, 1985; Nevers *et al.*, 1986). The occurrence of silent elements with the potential to become active may indicate that functional transposable elements confer, under certain circumstances, selective advantage to the organism. This advantage could be the mutagenic effect of transposons by insertion

– and in plants also by excision – which generate diversity for subsequent selection during evolution. The mechanism of the reversible activation/inactivation of plant transposons is not clear. There is some evidence for the involvement of hypo- and hypermethylation for the induction and suppression of transposition (for review see Döring & Starlinger, 1986).

In contrast to the broadly documented transposition via a DNA intermediate in virtually all living organisms, RNA-mediated transposition seems to be restricted to higher eukaryotes (Rogers, 1985). In discussing the evolutionary relevance of plant transposons we also shall provide the first evidence for the existence of non-viral retrotransposition in plants which may also be involved in the evolution of plant genes.

PLANT TRANSPOSABLE ELEMENTS MAY GENERATE NOVEL REGULATORY UNITS

Insertion of active plant transposable elements, if it occurs into an expressed gene, generates an unstable mutation. If the element loses its capacity to excise from this location, the mutation becomes stable. Such stable mutations may either arise by internal deletions affecting the coding function of the element (Pohlman, Fedoroff & Messing, 1984; Pereira *et al.*, 1985) or by deletions affecting its termini serving as the substrate for proteins involved in excision (Hehl, Sommer & Saedler, 1987). Both types of deletions require proteins encoded by the element, thus many types of deletion derivatives may be produced within a single class of elements. Allelic series, so-called 'states', can be generated in this way, where every allele represents an independent deletion event affecting the same progenitor element at its original location. Although such series have been obtained with several transposon families, we shall restrict our discussion to the En(Spm) family (Peterson, 1953; McClintock, 1954), because genetic and molecular analyses have been most intensive here. Interest has mainly focused on mutable alleles of the *A1* gene of maize encoding an NADPH-dependent reductase and involved in anthocyanin biosynthesis of the plant (Coe & Neuffer, 1977; Rohde *et al.*, 1986).

Genetic analysis of states of the *a1-m1* and *a1-m2* alleles revealed that the phenotypic expression of colour in mutants varies within the allelic series (McClintock, 1965; Reddy & Peterson, 1984).

Through cloning and sequence analysis of various states of each allele, the relationship between the internal structure of the elements residing at identical location within the same gene and their effect on gene expression could be established (Schwarz-Sommer *et al.*, 1985*b*, 1987*a*; Tacke, Schwarz-Sommer *et al.*, 1986). Clearly, the extent of interference by the insert with transcription, splicing and translation of the affected gene depends on signals carried by the element. Deletion of such signals may result in expression of an altered but still functional gene product (the *a1-m1 5719A-1* allele) whereas their presence prohibits phenotypic gene expression (the *a1-m16078* allele, Tacke *et al.*, 1986 and see references quoted above). The immobile 'receptor' elements (termed Inhibitor or I) in these mutants *per se* exhibit properties of *cis*-acting modules on gene expression.

The presence of immobilised elements as modules within a gene can be detected in two ways. Either allelic variants of the gene exist which do or do not contain the insert (and may or may not be equally expressed) or the insert can be mobilised by a *trans*acting 'signal'. In the first case one deals with two wildtype genes in which an insert became an integral part of a transcription unit. In the second case the stability of the insert is transient and is due, for example, to lack of a functional transposase. Two-component systems in maize reveal this second property. Their detection is due to the observation that crosses between plants with a virtually stable mutation with plants carrying the functional ('autonomous' or 'regulator') element uncover instability at the given locus. The mobilisation of defective elements by gene product of a functional element is highly specific. By this means several families of plant transposons can be defined (Peterson, 1981). The interaction between the Spm (regulator) element and the I (receptor) element, however, is not restricted to mobilisation of the transposition-deficient receptor alone. The regulator in *trans* may also alter the receptor's influence on gene expression. This includes suppression or induction of gene expression as displayed by the mutable alleles *a1-m1 5719A-1* and *a1-m2 8004*, respectively (McClintock, 1965). Based on the analysis of transcripts of mutant alleles in the absence and in the presence of a regulator, we assume that *suppression* is the consequence of blocking transcriptional read-through after binding of a regulator-encoded protein to the termini of the receptor (Gierl, Schwarz-Sommer & Saedler, 1985; Schwarz-Sommer *et al.*, 1985*b*). Such an interaction will result in *negative control* of the gene if the *structure* (and position) of the particular receptor evokes little or no interference with the expression of the affected gene.

Protein signals emitted by Spm may also *induce* expression of an adjacent gene thereby conferring *positive control* on that gene. Molecular analysis of the *a1-m2 8004* allele which exhibits induction of gene expression by Spm activity indicates that in this case the particular *location* of the receptor element rather than its structure is crucial for the phenomenon. In this allele the receptor is located upstream of the CAAT and TATA boxes of the *A1* gene promoter thereby displacing upstream regulatory sequences (Schwarz-Sommer *et al.*, 1987a). *A1* gene expression in this mutant is abolished. From a formal point of view the induction of *A1* gene activity by Spm replaces the wildtype induction mechanism. Which Spm-encoded protein governs this induction and which DNA region within the receptor (or within the *A1* gene promoter) is involved in protein binding and initiation of *A1* gene transcription is not known. It is likely that the basic mechanism of suppression and induction by Spm of expression of adjacent genes affected by I inserts is similar (Schwarz-Sommer *et al.*, 1987a). Hence, whether negative or positive control is exhibited by the regulator element depends on the composition of the receptor insert and on its position with respect to the affected transcription unit.

If the gene control established by a transposable element became of value for evolution, both the regulator and the receptor component would have to be stabilised at their locations. As mentioned already, small deletions affecting the termini of elements perfectly immobilise them without affecting their coding function or their response to *trans*-acting signals. In theory, transposable elements possess the potential to become integral parts of genes as *cis*-acting modules which respond to *trans*-acting signals. Inserts of this kind may replace the genuine control of genes. *De facto*, none of the genetically identified transposable elements were found to be an integral part of a gene. McClintock's emphasis on the involvement of transposons as 'controlling elements' in gene regulation and plant development has therefore not yet received support at the molecular level. However, after gene control is established, the transposon-derived elements of control (receptor and regulator) may undergo mutational changes masking their original structure but not affecting basic properties needed for their regulatory interaction. Inserts altered in this way would no longer resemble transposons and may escape detection as such. Genetically unidentified inserts altering the structure of transcription units have recently been found by comparing alleles of wildtype plant genes (Doyle *et al.*, 1986; Sachs, Dennis *et al.*, 1986; Zack, Ferl & Hannah, 1986). In these cases it is not known what consequences these inserts have on the expression of the affected genes.

PLANT TRANSPOSABLE ELEMENTS AS TOOLS FOR PROTEIN EVOLUTION

DNA sequence analysis of several revertant genes arising from excision of a plant transposable element revealed that the excision process is rather imprecise but still follows certain rules (Saedler & Nevers, 1985; Coen, Carpenter & Martin, 1986). As a consequence, at the target of insertion altered sequences, 'footprints' are left behind which are only occasionally identical with the wildtype sequence. In most of the cases sequenced these alterations are not simple base substitutions but, rather, small insertions or deletions (Sachs, Peacock *et al.*, 1983; Pohlman, Fedoroff & Messing, 1984; Schwarz-Sommer *et al.*, 1985a). If the insertion of the transposable element occurs within a gene encoding a protein, its footprint generated upon excision will alter the structure of that protein in the resulting revertant (see citations above). In the cases mentioned above, the function of the wildtype gene was restored such that no difference between the phenotypic expression of wildtype and revertant could be detected. But the alterations may also affect the function of the protein, as documented by analysis of several excision products (Chen, Freeling & Merckelbach, 1986; Wessler, Baran *et al.*, 1986). These observations indicate that transposable elements can serve as generators of protein sequence diversity.

There is some indication that alterations which can be attributed to visitation of genes by transposable elements may have occurred during evolution. By comparing the sequence of alleles one can detect 'footprints' as perfect or imperfect duplications present in one allele but absent in the other. 'Footprints' most frequently occur within intron sequences which apparently tolerate various kinds of DNA alterations (for review see Schwarz-Sommer, 1987). But footprints can also be tolerated within exons (Rohde *et al.*, 1986; Schwarz-Sommer, 1987) resulting in allelic variants of wildtype proteins. One can therefore speculate that visitation of genes by transposable elements generates frequent mutations, not only by their integration but also if the element leaves that location. The consequence of such a visit is sequence diversity of the affected gene(s) within a population on which subsequently selection can operate.

RETROTRANSPOSITION IN PLANTS?

We argued in the previous sections that the DNA-mediated transposition of insertion elements in plants can become of value for the species during evolution. By generating novel regulatory units upon insertion or by generating sequence diversity within a gene upon excision plant transposons create a playground for subsequent selection. In the following section we wish to draw attention to mutagenesis by insertion of DNA via an RNA intermediate. Until now there is no evidence for activity of retrotransposons nor for genome-derived reverse transcriptase activity in plants except that some plant transposons resemble in structure viral retroposons possessing long terminal repeats (see compilation in Schwarz-Sommer, 1987). Recently we have discovered the first retrotransposon-like element in a plant (Schwarz-Sommer et al., 1987a and Schwarz-Sommer, Leclercq & Saedler, 1987b).

The Cin4 element in maize was detected as an insert altering the structure of the wildtype A1 transcription unit (Schwarz-Sommer et al., 1987a, Schwarz-Sommer, 1987b and unpublished results). Cin4-1 at the A1 locus is 1.1 kb long. It has no terminal structure but ends in 12 adenosine residues. One of the A residues can be part of a 7 bp long duplication flanking Cin4-1 and occurring only once in the other wildtype A1 allele. The copynumber of Cin4-1 related sequences is 50–100 per diploid maize genome, and the Cin4 family is composed of 5′-truncated family members. Cloning and sequencing of 5 independent truncated Cin4 elements revealed that the length of the polyA track (defining the 3′ end of the element) varies, and also the length of the direct duplication flanking the elements differs. The elements contain a long open reading frame (ORF). The length of the ORF in the longest (but still truncated) copy comprises 3198 amino acids (Schwarz-Sommer et al., 1987b). In all these features Cin4 resembles processed pseudogenes (Vanin, 1985).

Recently a class of elements, termed non-viral retrotransposons, have been detected in several organisms. The L1 family of mammals (Fanning & Singer, 1987), the I-element of Drosophila (Fawcett,

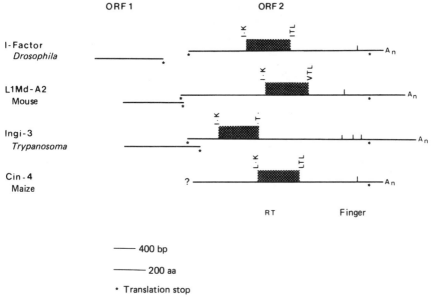

Fig. 1. Conserved structural features of non-viral retrotransposons. The published sequences of I (Fawcett *et al.*, 1986), L1Md-A2 (Loeb *et al.*, 1986) Ingi-3 (Kimmel *et al.*, 1987) and Cin4 elements (Schwarz-Sommer *et al.*, manuscript in preparation) are compiled in a linear form reflecting true size and distance relations. For comparison, the translation stop (*) within the longer ORF of the elements is aligned to an identical position. The conserved domain with homology to retroviral reverse transcriptase (RT, also see Fig. 2) is represented by boxes. Capital letters indicate the first and last amino acid of this domain. Horizontal lines show the position of putative DNA-binding 'fingers' in I and Cin4 elements with amino acid sequence homology to similar domains detected in the *gag* region of retroviruses (Covey, 1986), as discussed within the L1Md-A2 elements by Fanning & Singer (1987). We detected a similar region within the sequence of the Ingi-3 element revealing repeated homology to the zinc-binding domain of TFIIIA published by Miller, McLachlan & Klug (1985). The question mark indicates that the structure of the non-truncated Cin4 element is not yet known. In this figure, the longest analysed truncated Cin4 element is depicted.

Lister *et al.*, 1986), and the Ingi-element of *Trypanosoma* (Kimmel, Ole-Moiyoi & Young, 1987) share no extended sequence homologies at the nucleic acid level. Yet all these elements show striking similarities in their overall organisation (Fig. 1) and in the conservation of two regions at the amino acid level within their long ORFs. One of these regions is homologous to the conserved amino acid sequence within the *pol* region of retroviral reverse transcriptases (see citations above). The other conserved region is similar in structure to retroviral DNA-binding 'fingers', although non-viral retrotransposons differ with respect to position within the longer ORF of the putative 'finger' from that of viral retrotransposons (Fanning & Singer, 1987).

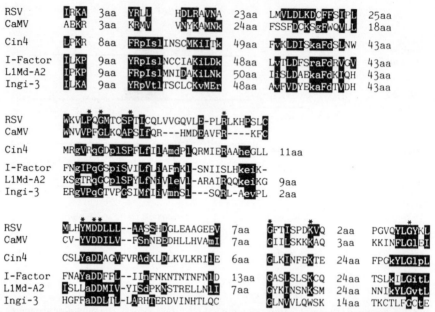

Fig. 2. Amino acid sequence homology within the conserved region of known and putative reverse transcriptases. The amino acid sequence within the conserved domain of Cin4 is aligned, together with the sequences of the conserved domains from other non-viral retrotransposons (Hattori, Kuhara *et al.*, 1986 for sequence references see Fig. 1), with the polymerase gene product of *Rous sarcoma* virus (RSV, Schwartz, Tizard & Gilbert, 1983) and with the putative polymerase gene product of cauliflower mosaic virus (CaMV, Gardner *et al.*, 1981). Conservative positions are indicated by dark boxes and correspond to the conservation first detected by Toh, Hayashida & Miyata (1983). The ten invariant amino acids found by these authors comparing the polymerase gene products of retroviruses and CaMV are indicated by asterisks. Positions conserved in at least three of the four non-viral retrotransposons are shown in dark boxes by small letters. For comparing sequences the same groups of residues have been used as by Fawcett *et al.* (1986: P,A,G,S (Neutral or weakly hydrophobic); Q,N,E,D (hydrophylic, acid amine); H,K,R (hydrophylic, basic); L,I,V,M (hydrophobic); F,Y,W (hydrophobic, aromatic); C (cross-link forming). Gaps were introduced to increase similarity and numbers give the distance between homology blocks.

Sequence analysis revealed that the Cin4 element of maize is homologous to non-viral retrotransposons, on the basis of those criteria mentioned above (Fig. 1). This homology includes the spacing between homology blocks within the conserved 'reverse transcriptase' region (Fig. 2) and also the position of the conserved domains within the long ORF compiled in Fig. 1.

There are several aspects to be discussed concerning the presence of a non-viral retrotransposon in plants. For example, the copy-number of Cin4 in maize is several orders of magnitude lower than the copynumber of L1 or Ingi elements. This could mean that the mobility of the element is suppressed and that Cin4 has to be supplied with an 'activator' during crosses, in analogy to the activation of I elements in *Drosophila* during hybrid dysgenesis (Fawcett *et al.*, 1986). Alternatively, the low copynumber can also reflect that Cin4 was only active for a short period of time and that the element is a remnant of an 'infection'. In fact, we did not find evidence for Cin4 activity in maize, except the variability in the chromosomal location of the element by comparing different maize lines in Southern blot experiments (Schwarz-Sommer *et al.*, 1987*b*). Northern blot analysis with mRNAs from different plant tissues (including leaves, roots, tassels, fertilised and unfertilised ears and somatic embryos) reveals a smear if the 3' region of Cin4 was used as a probe (data not shown). Also cDNA cloning gave no indication for the presence of long Cin4 transcripts synthesised from the authentic Cin4 copy. Although many cDNA clones were found by hybridisation to Cin4, no two of them carried identical sequences and all seem to be transcribed from external promoters (Schwarz-Sommer *et al.*, manuscript in preparation). Further, none of the cDNAs contain sequences homologous to upstream regions of the longest truncated Cin4 copy. These observations argue rather in favour of an early dysgenic cross which mobilised Cin4 or in favour of a past infection by Cin4 leaving the question open which organism was the infectious agent. In this light the differences in the chromosomal locations of Cin4 copies could be the consequence of large-scale rearrangements and might not be due to transposition of elements.

Our failure to detect the full-size Cin4 transcript as an indication for Cin4 activity may reside in not having looked at the proper tissue or not at the appropriate time. The functional relevance of Cin4 for maize (and perhaps the relevance of similar elements in other plants) may be revealed by its homology to non-viral retrotransposons in other organisms. L1, l, Ingi and Cin4 elements differ in nucleotide sequence. However, certain regions are conserved in all elements at the amino acid level suggesting that functional constraint conserves the integrity of these regions. Whether this homology indicates the common functional relevance of all non-viral retrotransposons for the organisms or whether it indicates their common origin during evolution remain open questions. Nevertheless, detection of Cin4 as an integral part of some wildtype *A1* genes indicates that mobility of retrotransposons adds to the variability of genes and not only serves to reorganise genomes by large-scale rearrangements.

REFERENCES

CHEN, C. H., FREELING, M. & MERCKELBACH, A. (1986). Enzymatic and morphological consequences of Ds excisions from maize *Adh1*. *Maydica*, XXXI, 93–108.

COE, E. H. & NEUFFER, M. G. (1977). The genetics of corn. In *Corn and corn improvement*, ed. Sprague, G. F., pp. 11–223. Madison, WI, USA, American Society of Agronomy Inc.

COEN, E. S., CARPENTER, R. & MARTIN, C. (1986). Transposable elements generate novel spatial patterns of gene expression in *Antirrhinum majus*. *Cell*, **47**, 285–96.

COVEY, S. (1986). Amino acid sequence homology in *gag* region of reverse transcribing elements and the coat protein gene of cauliflower mosaic virus. *Nucleic Acids Research*, **14**, 623–33.

DELLAPORTA, S. L. & CHOMET, P. S. (1985). The activation of maize controlling elements. In *Genetic flux in plants*, ed. Hohn, B. & Dennis, E. S., pp. 169–216. Wien–New York, Springer Verlag.

DÖRING, H.-P. & STARLINGER, P. (1986). Molecular genetics of transposable elements in plants. *Annual Review of Genetics*, **120**, 175–200.

DOYLE, J. J., SCHULER, M. A., GODETTE, W. D., ZENGER, V. & BEACHY, R. N. (1986). The glycosylated seed storage proteins of *Glycine max* and *Phaseolus vulgaris*: Structural homologies of genes and proteins. *The Journal of Biological Chemistry*, **261**, 9228–38.

FANNING, T. & SINGER, M. (1987). The LINE-1 DNA sequences in four mammalian orders predict proteins that conserve homologies to retrovirus proteins. *Nucleic Acids Research*, **15**, 2251–60.

FAWCETT, D. H., LISTER, C. K., KELLETT, E. & FINNEGAN, D. J. (1986). Transposable elements controlling I–R hybrid dysgenesis in *D. melanogaster* are similar to mammalian LINEs. *Cell*, **47**, 1007–15.

FEDOROFF, N. V. (1983). Controlling elements in maize. In *Mobile genetic elements* ed. Shapiro, J. A., pp. 1–63. London—New York, Academic Press.

GARDNER, R. C., HOWARTH, A. J., HAHN, P., BROWN-LUEDI, M., SHEPHERD, R. J. & MESSING, J. (1981). The complete nucleotide sequence of an infections clone of cauliflower mosaic virus by M13mp7 shotgun sequencing. *Nucleic Acids Research*, **9**, 2871–87.

GIERL, A., SCHWARZ-SOMMER, ZS. & SAEDLER, H. (1985). Molecular interactions between the components of the En-I transposable elements in *Zea mays*. *The EMBO Journal*, **4**, 579–83.

HATTORI, M., KUHARA, S., TAKENAKA, O. & SAKAKI, Y. (1986). L1 family of repetitive DNA sequences in primates may be derived from a sequence encoding a reverse transcriptase-related protein. *Nature*, **321**, 625–8.

HEHL, R., SOMMER, H. & SAEDLER, H. (1987). Interaction between the Tam1 and Tam2 transposable elements of *Antirrhinum majus*. *Molecular and General Genetics*, in press.

KIMMEL, B. E., OLE-MOIYOI, O. K. & YOUNG, J. R. (1987). Ingi, a 5.2 kb dispersed sequence element from *Trypanosoma* that carries half of a smaller mobile element at either end and has homology with mammalian LINEs. *Molecular and Cellular Biology*, **7**, 1465–75.

LOEB, D. D., PADGETT, R. W., HARDIES, S. C., SHEHEE, W. R., COMER, M. B., EDGELL, M. B. & HUTCHISON, C. A. (1986). The sequence of a large L1Md element reveals tandemly repeated 5' end and several features found in retrotransposons. *Molecular and Cellular Biology*, **6**, 168–82.

McCLINTOCK, B. (1954). Mutations in maize and chromosomal aberrations in *Neurospora*. *Carnegie Institute Washington Yearbook*, **53**, 254–60.

McCLINTOCK, B. (1956). Controlling elements and the gene. *Cold Spring Harbor Symposia on Quantitative Biology*, **21**, 197–216.

McCLINTOCK, B. (1965). The control of gene action in maize. In *Genetic control of differentiation, Brookhaven Emposia in Biology*, **18**, 162–84.

McCLINTOCK, B. (1984). The significance of responses of the genome to challenge. *Science*, **226**, 792–801.

MILLER, J., McLACHLAN & KLUG, A. (1985). Repetitive zinc-binding domains in the protein transcription factor IIIA from *Xenopus* oocytes. *The EMBO Journal*, **6**, 1609–14.

NEVERS, P., SHEPHERD, N. S. & SAEDLER, H. (1986). Plant transposable elements. *Advances in Botanical Research*, **12**, 103–203.

PEREIRA, A., SCHWARZ-SOMMER, Zs., GIERL, A., BERTRAM, I., PETERSON, P. A. & SAEDLER, H. (1985). Genetic and molecular analysis of the enhancer (En) transposable element system of *Zea mays*. *The EMBO Journal*, **4**, 17–23.

PETERSON, P. A. (1953). A mutable pale green locus in maize. *Genetics*, **45**, 682–3.

PETERSON, P. A. (1981). Instability among the components of a regulatory element transposon in maize. *Cold Spring Harbor Symposia on Quantitative Biology*, XLV, 447–54.

POHLMAN, R. F., FEDOROFF, N. & MESSING, J. (1984). The nucleotide sequence of the maize controlling element activator. *Cell*, **37**, 635–43.

REDDY, L. V. & PETERSON, P. A. (1984). Enhancer transposable element induced changes at the *A* locus in maize: The *a-m-1 6078* allele. *Molecular and General Genetics*, **194**, 124–37.

ROGERS, J. H. (1985). The origin and evolution of retrotransposons. *International Review of Cytology*, **93**, 187–279.

ROHDE, W., BARZEN, E., MAROCCO, A., SCHWARZ-SOMMER, Zs., SAEDLER, H. & SALAMINI, F. (1986). Isolation of genes that could serve as traps for transposable elements in *Hordeum vulgare*. *Barley Genetics*, V, in press.

SACHS, M. M., PEACOCK, W. J., DENNIS, E. S. & GERLACH, W. L. (1983). Maize Ac/Ds controlling elements: a molecular viewpoint. *Maydica*, XXVIII, 289–302.

SACHS, M. M., DENNIS, E. S., GERLACH, W. L. & PEACOCK, W. J. (1986). Two alleles of maize *alcohol dehydrogenase 1* have 3' structural and poly(A) addition polymorphisms. *Genetics*, **113**, 449–67.

SAEDLER, H. & NEVERS, P. (1985). Transposition in plants: a molecular model. *The EMBO Journal*, **4**, 585–90.

SCHWARTZ, D. E., TIZARD, R. & GILBERT, W. (1983). Nucleotide sequence of Rous sarcoma virus. *Cell*, **32**, 853–69.

SCHWARZ-SOMMER, Zs., GIERL, A., CUYPERS, H., PETERSON, P. A. & SAEDLER, H. (1985a). Plant transposable elements generate the DNA sequence diversity needed in evolution. *The EMBO Journal*, **4**, 591–7.

SCHWARZ-SOMMER, Zs., GIERL, A., BERNDTGEN, R. & SAEDLER, H. (1985b). Sequence comparison of 'states' of *a-ml* suggests a model of Spm(En) action. *The EMBO Journal*, **4**, 2439–43.

SCHWARZ-SOMMER, Zs. (1987). The significance of plant transposable elements in biological processes. In *Results and problems in cell differentiation 14, Structure and function of eukaryotic chromosomes*, ed. W. Hennig, in press. Berlin–Heidelberg, Springer Verlag.

SCHWARZ-SOMMER, Zs. & SAEDLER, H. (1987). Can plant transposable elements generate novel regulatory units? *Molecular and General Genetics*, **209**, 213–21.

SCHWARZ-SOMMER, Zs., SHEPHERD, N., TACKE, E., GIERL, A., ROHDE, W., LECLERCQ, L., MATTES, M., BERNDTGEN, R., PETERSON, P. A. & SAEDLER, H. (1987a). Influence of transposable elements on the structure and function of the *A1* gene of *Zea mays*. *The EMBO Journal*, **6**, 287–94.

SCHWARZ-SOMMER, Zs., LECLERCQ, L. & SAEDLER, H. (1987b). Cin4, a retrotransposon-like element in *Zea mays*. In *Plant Molecular Biology, NATO Advanced*

Study Institute, eds von Wettstein, D. & Chua, N.-H., New York: Plenum Publishing Corp., 191–8.

TACKE, E., SCHWARZ-SOMMER, ZS., PETERSON, P. A. & SAEDLER, H. (1986). Molecular analysis of 'states' of the *A1* locus of *Zea mays*. *Maydica*, XXXI, 83–92.

TOH, H., HAYASHIDA, H. & MIYATA, T. (1983). Sequence homology between retroviral reverse transcriptase and putative polymerases of hepatitis B virus and cauliflower mosaic virus. *Nature*, **305**, 827–9.

VANIN, E. F. (1985). Processed pseudogenes: characteristics and evolution. *Annual Review of Genetics*, **19**, 253–72.

WESSLER, S. R., BARAN, G., VARAGONA, M. & DELLAPORTA, S. L. (1986). Excision of Ds produces *waxy* proteins with a range of enzymatic activities. *The EMBO Journal*, **5**, 2427–32.

ZACK, C. D., FERL, R. J. & HANNAH, L. C. (1986). DNA sequence of a *shrunken* allele of maize: evidence for vistation by insertional sequences. *Maydica*, XXXI, 5–16.

TRANSFER OF T-DNA FROM AGROBACTERIUM INTO PLANTS

J. SCHELL

Max-Planck Institut für Züchtungsforschung 5000 Köln 30, West Germany

INTRODUCTION

The soil phytopathogen, *Agrobacterium tumefaciens*, is a sophisticated parasite which uses genetic engineering processes to force infected plant cells to divert some of their organic carbon and nitrogen supplies for the synthesis of nutrients (called opines) which the infecting bacteria can specifically catabolise (Tempe & Schell, 1977; Tempe *et al.*, 1978; Schell *et al.*, 1979). The genetically engineered plant cells are also stimulated to proliferate and thus form typical tumour tissues or 'crown-galls'.

A detailed genetic and molecular analysis of this phenomenon became possible when it was discovered (Zaenen *et al.*, 1974; Van Larebecke *et al.*, 1974, 1975; Watson *et al.*, 1975; Bomhoff *et al.*, 1976) that large extrachromosomal plasmids in agrobacteria carried the genes responsible for both tumorous growth and opine synthesis of crown-gall tissues as well as for opine catabolism by agrobacteria, thus providing genetic evidence pointing to specific transfer of genes from the Ti-plasmids in agrobacteria to the genome of plant cells.

Transposon mutagenesis of Ti-plasmids (Hernalsteens *et al.*, 1978; Holsters *et al.*, 1980; Garfinkel & Nester, 1980; Ooms *et al.*, 1980; Garfinkel *et al.*, 1981; De Greve *et al.*, 1981; Leemans *et al.*, 1982; Joos *et al.*, 1983) indicated that two segments of these plasmids were involved in oncogenicity and therefore presumably in transfer of DNA from bacteria to plant cells. One segment, called the *vir* region, contains genes the inactivation of which lead to a loss of tumour-inducing capacity by the resultant *Agrobacterium* mutant strains. Inserts in the other segment, called the T-region, produced mutant strains still capable of DNA transfer but producing transformed plant cells either lacking the capacity to synthesise opines or growing as tissues with aberrant morphologies (e.g. shoot-forming teratomas or root-forming calli).

One of the most stimulating observations made with these transposon insertions was the demonstration that a transposon located within the T-region of a Ti-plasmid was physically integrated into the genome of transformed plant cells (Hernalsteens *et al.*, 1980), thus confirming that the T-region of Ti-plasmids was physically transferred to plant cells and that Ti-plasmids can be used as a host vector system for the introduction of foreign DNA into plant cells.

Probing with smaller DNA fragments derived from Ti-plasmids demonstrated that a well-defined segment, (called T-DNA) derived from the T-region of these plasmids, was covalently integrated into the plant nuclear genome (Chilton *et al.*, 1977; Lemmers *et al.*, 1980; Thomashow *et al.*, 1980; Willmitzer *et al.*, 1980; Chilton *et al.*, 1980). RNA/DNA hybridisations demonstrated that the T-DNA in plant cells was transcribed into a number of well defined polyadenylated transcripts. Some of these transcripts clearly correlated with the T-DNA functions genetically identified by transposon mutagenesis of the T-region of Ti-plasmids (Willmitzer *et al.*, 1982, 1983).

I. CHEMICAL SIGNALS ATTRACT AND ACTIVATE AGROBACTERIA

The initial phenomena leading to the transfer of T-DNA from *Agrobacterium* to plant cells, involve chemoattraction of the bacteria to the susceptible host plants, followed by attachment of the bacteria to the plant cells and the induction via plant signal molecules, of the Ti-plasmid *vir* genes.

Actively dividing plant cells are known to excrete molecules some of which are toxic and have a general role in protecting plant tissues against phytopathogens.

Agrobacteria have apparently developed the ability to recognise some of these exudates and to use them as signal molecules to guide them to susceptible host tissues. Agrobacteria were indeed shown to be able to move up a concentration gradient of compounds such as vanillyl alcohol and acetosyringone, which are excreted by wounded plant tissues (Shaw *et al.*, 1986; Ashby *et al.*, 1987). The actual attachment of the bacteria to the plant cell wall is a two-step process. In the first step the bacteria bind to receptor sites on the plant cell surface. Binding assays suggest that lipopolysaccharides present in the bacterial cell envelope are involved in this initial attachment step. In a second step this binding is made largely irreversible through the synthesis by the bacteria of an extensive cellulose fibril network, affixing the bacteria to the plant cells (for a review see Mathysse, 1986).

Two chromosomal genes of *Agrobacterium* (*chvA* and *chvB*) are essential for the formation of a stable complex (Douglas *et al.*, 1985). It is likely that the products of these constitutive genes play a role in the constitution of bacterial cell walls. Thus it was shown that *chvB* encodes a protein involved in the biosynthesis of β-2-glucan (Puvanesarajah *et al.*, 1985). Although the exact nature of the plant cell wall receptor is not known, competition experiments with isolated plant cell wall components indicate that polygalacturonic acid and demethylated forms of pectin are involved in the binding of *Agrobacterium* (Pueppke & Benny, 1981, 1983).

Signal molecules released from the plant (such as acetosyringone) also play a role in the specific activation of the Ti-plasmid *vir* genes (Stachel *et al.*, 1985). By transposon insertion mutagenesis several non-virulent mutants of *Agrobacterium* were isolated most of which turned out to be located in the *vir* region of the Ti-plasmid (Holsters *et al.*, 1980; Garfinkel & Nester, 1980; De Greve *et al.*, 1981). Complementation tests with these Vir-mutants led to the identification of six *vir* loci, designated *A*, *B*, *C*, *D*, *E*, and *G* (Klee *et al.*, 1982; Stachel & Nester, 1986).

Using an ingenious transposon derivative (Tn*3*-HOHO1), Stachel & Nester (1986) constructed β-galactosidase gene fusions to all of the *vir* genes. These were used to study the influence of environmental factors on the expression of the Ti-plasmid *vir* genes. It turned out that only *virA* and *virG* are expressed by bacteria in the absence of plant cells. Filtrates of plant cell cultures however induced the expression of *virB*, *virC*, *virD*, *virE* and *virG*. Purification and structural analysis of the inducing factor of plant cell filtrates, led to the identification of the plant signal molecules acetosyringone and alpha-hydroxy-acetosyringone (Stachel *et al.*, 1985).

Interestingly the same phenolic compound (acetosyringone) was thus found to be involved in both chemoattraction of agrobacteria (at nM concentrations) and in the transcriptional activation (at mM concentrations) of the virulence functions of *Agrobacterium tumefaciens*. *VirA* and *virG* functions appear to be required for the transcriptional activation of the other *vir* genes. Stachel & Zambryski (1986) have presented evidence to indicate that *virA* might code for a permease for the efficient uptake of the plant signal molecule and that the *virG* function is activated by the signal molecule to stimulate transcription of *virG* itself and of all the other *vir* loci.

Sequence homology with other known genes indicates that *virG* could code for a DNA binding protein.

II. VIRULENCE FUNCTIONS INVOLVED IN T-DNA TRANSFER

One can distinguish two types of *vir* functions. Most of the *vir* functions, except for the one encoded by *virE*, appear to function within agrobacteria, e.g. by promoting the formation of T-DNA intermediates or their transfer to plant cells. *VirE*, however, synthesises a factor which has an effect on or in the target plant cell (Otten *et al.*, 1984), since *virE* mutants can be complemented simply by the co-infection of plant cells with a *virE* positive strain. The functions of *virB, C, D*, and *E* are only partially understood.

At least three abundant proteins are encoded by the *virB* locus. These proteins are specifically localised in the cell envelope of the induced agrobacteria. The size as well as the abundance of these proteins argue in favour of the hypothesis that they represent pilus proteins (Engström *et al.*, 1987).

The following observations indicate that the *virC* gene product is not essential for T-DNA transfer but enhances transformation efficiency. Strains with mutations in *virC* are still able to induce attenuated tumours on most plants. Biotype III agrobacteria, have a limited host range and do not have a *virC* gene. In these limited host range strains the *virC* gene product actually prevents tumour formation on the normal host (*Vitis vinifera*). This is probably due to the fact that these plants react to the bacteria expressing *virC* by a hypersensitive reaction (Janofsky *et al.*, 1985). The evidence demonstrating that two proteins coded by the *virD* locus (D_1 and D_2) are required for the processing of the T-region, prior to transfer to the plants, will be discussed in the next section.

The protein encoded by *virE* has a MW of 65Md (Hirooka & Kado, 1986; Engström *et al.*, 1987). It is conceivable that this protein functions extracellularly or in the plant cells. Two lines of evidence support this hypothesis, (1) *virE* mutant strains can be complemented by co-infection with *virE* positive strains and (2) the *virE* function, although essential for transformation involving the integrations of T-DNA in the host genome, appears not to be essential for the transfer of T-DNA regions that do not require integration. Indeed *virE⁻* strains, harbouring either viroid cDNA (Gardner & Knauf, 1986) or a tandem array of the cauliflower mosaic virus genome (Hille *et al.*, 1986), are capable of efficiently transferring these T-DNA's to plant cells, as detected by viroid or virus replication. It is therefore conceivable that the *virE* protein is playing a role in the integration of T-DNA in the plant genome.

III. A 25 BP 'BORDER' SEQUENCE IS THE TARGET FOR *vir*D FUNCTIONS AND DEFINES THE T-REGION

Genetic evidence demonstrated that most if not all genes of the T-region can be inactivated or deleted without affecting the T-DNA transfer-integration process (Leemans *et al.*, 1982; Zambryski *et al.*, 1983). The actual size of the T-region did not appear to play a major role either, since the insertion of relatively large DNA sequences, such as Tn7, within the T-region did not appear to affect T-DNA transfer and integration (Hernalsteens *et al.*, 1980). In contrast the DNA sequences flanking the T-region are essential for transfer-integration.

By cloning and sequencing T-DNA-plant DNA junction fragments from a number of independent transformed cell lines, it became obvious the process of T-DNA transfer and integration must be very precise. Especially the right T-DNA border was found to be very precise and to coincide with a 25 bp direct repeat which flanks all the known T-regions of Ti-plasmids (Zambryski *et al.*, 1982; Yadav *et al.*, 1982; Simpson *et al.*, 1982; Holsters *et al.*, 1983; Barker *et al.*, 1983).

In the following paragraphs the evidence will be reviewed demonstrating that these 25 bp border sequences are recognition signals which define in *cis* the region that will be transmitted to the plant cell and integrated in the plant genome. The *virD* locus, on the other hand, provides the *trans* functions that recognise the 25 bp recognition sequences to initiate T-DNA transfer. As a consequence it is understood why the virulence genes can be physically separated from the T-region (e.g. by inserting the 25 bp recognition sequences in an independent replicon) and still promote T-DNA transfer (Hoekema *et al.*, 1983; de Framond *et al.*, 1983).

By deleting a fragment containing the right 25 bp border sequence and by subsequently replacing this fragment by an oligonucleotide consisting of just the 25 bp border repeat sequence, Wang et al. (1987) demonstrated that the 25 bp recognition sequence is essential and sufficient for T-DNA transfer-integration. These authors also demonstrated that the 25 bp border sequence had a polar effect on T-DNA transfer-integration, depending on the orientation of the 25 bp border sequence. T-regions are flanked by a very similar 25 bp sequence at their left border. Apparently other sequences, adjacent to the left 25 bp repeat, somehow inhibit the recognition of this repeat (Wang et al., submitted). In this way the transfer of Ti-plasmid sequences to the left of the T-region is inhibited. Reciprocally, sequences flanking the right 25 bp border sequence appear to enhance T-DNA transfer (Peralta et al., 1986).

A detailed study of the early events occurring within agrobacteria as a result of the induction of the virulence regulon, allows the function of the 25 bp border sequences that flank the T-region of Ti-plasmids to be better understood. Koukolikova-Nicola et al. (1985) demonstrated that T-region rearrangements occur specially upon activation of vir functions. Circular T-region molecules were identified in which the 25 bp border repeats were fused, creating a composite sequence. At first sight these experiments seemed to indicate that such double stranded T-DNA circles could represent the actual T-DNA transfer intermediates. New evidence suggests that these T-DNA circles are in fact either a side product of the mechanism leading to the functional T-DNA transfer intermediate, or, since the T-DNA double stranded circles were isolated after transformation and replication in E. coli, they might also result from DNA repair of T-DNA transfer intermediate structures.

By SI nuclease mapping, Stachel et al. (1986) demonstrated that upon induction of the vir functions the right 25 bp border sequence became the target for an endonucleolytic cleavage of one of the strands (the so-called 'bottom' strand) of the T-region, just after the third or fourth left most base of the 25 bp border sequence. Coinciding with these site specific nicks, single stranded copies of the T-region, derived from the 'bottom' strand, were detected as free unipolar linear molecules in induced agrobacteria (so called T-strands). On average one copy of such a T-strand was detected per induced bacterial cell. A single T-strand is produced from a two border T-region. The single stranded nicks observed in the T-region flanking 25 bp direct repeat sequences therefore probably represent sites of initiation (right border) and termination (left border) of T-strand synthesis. The following evidence indicates that two proteins coded for by the *virD* locus are involved in the formation T-DNA intermediates (T-strands) and other rearrangements (double stranded circles). Alt-Moerbe *et al.* (1986) showed that $virD_1$ and $virD_2$ encode proteins of $+/- 15$ and 56 kD which control the production of T-DNA circles detected at low frequencies in induced agrobacteria. Recent work (Wang *et al.*, 1987; Timmerman *et al.*, in preparation) provides evidence that the double stranded T-DNA circles, observed in a small fraction of induced agrobacteria, result from a recombinational event between the repeat 25 bp border sequences. Yanofsky et al. (1986) sequenced the $virD_1$ and D_2 genes and found them to code for products of 16.2 and 47.4 kD. These authors also demonstrated that these proteins function as a single stranded endonuclease, specifically recognising the 25 bp border sequences. Finally Stachel *et al.* (1987) showed that $virD_1$ and D_2 are also essential for T-strand synthesis, thus supporting the evidence that border nicks serve as sites of initiation and termination of T-strand synthesis. It is supposed that bacterial helicase and polymerase are also involved in T-strand synthesis.

All of these observations were strikingly reminiscent of bacterial conjugation and led to the suggestion that T-DNA transfer to plants might in fact be a (very) special type of conjugation, involving bacterial and plant cells (Wang *et al.*, 1987; Stachel *et al.*, 1986).

That this suggestion is probably correct was elegantly demonstrated by Buchanan-Wollaston *et al.* (1987). Indeed, the single stranded DNA intermediates involved in bacterial conjugation are initiated by the introduction of a specific nick at *ori*T by so-called *mob* functions (Everett & Willets, 1980). The question therefore was whether the *virD*–T-DNA border sequences would in fact be the functional equivalent of the *mob-ori*T system of bacterial conjugation. In the case of the wide-host-range plasmid RSF1010 this appears to be the case, since this plasmid carrying a plant selectable marker gene was shown to be transferred and integrated to plant cells. *Vir* functions (other than *virD*) as well as *mob* and *ori*T of RSF1010 were essential for this transfer to take place. It remains to be seen whether integration in plant DNA took place through the *ori*T sequence. These results also show that it is possible that a variety of wide-host-range plasmid can be transferred to plants via transition in agrobacteria and that therefore plants might have access to the gene pool of Gram-negative bacteria (Buchanan-Wollaston *et al.*, 1987).

ACKNOWLEDGEMENT

The author wishes to express his gratitude to his former PhD student Dr B. Timmerman and to his colleagues Dr P. Zambryski and Dr S. Stachel for their help and their profound contribution to the understanding of the remarkable *Agrobacterium*–plant cell conjugation system.

REFERENCES

ALT-MOERBE, J., RAK, B. & SCHRÖDER, J. (1986). A 3.6-kbp segment from the *vir* region of Ti plasmids contains genes responsible for border sequence-directed production of T-region circles in *E. coli*. *The EMBO Journal*, **5**, 1129–35.

ASHBY, A. M., WATSON, M. D. & SHAW, C. H. (1987). Chemotactical attraction of *A. tumefaciens* by plant signal molecules. *FEMS Microbiological Letters*, in press.

BARKER, R. F., IDLER, K. B., THOMPSON, D. V. & KEMP, J. D. (1983). Nucleotide sequence of the T-DNA region from the *Agrobacterium tumefaciens* octopine Ti plasmid pTi15955. *Plant Molecular Biology*, **2**, 335–50.

BOMHOFF, G., KLAPWIJK, P., KESTER, H. C. M., SCHILPEROORT, R. A., HERNALSTEENS, J. P. & SCHELL, J. (1976). Octopine and nopaline synthesis and breakdown genetically controlled by a plasmid of *Agrobacterium tumefaciens*. *Molecular & General Genetics*, **145**, 177.

BUCHANAN-WOLLASTON, V., PASSIATORE, J. E. & CANNON, F. (1987). The *mob* and *ori*T mobilization functions of a bacterial plasmid promote its transfer to plants. *Nature*, **328**, 172–5.

CHILTON, M.-D., DRUMMOND, M. D., MERLO, D. J., SCIAKY, D., MONTOYA, A. L., GORDON, M. P. & NESTER, E. W. (1977). Stable incorporation of plasmid DNA into higher plant cells: the molecular basis of crown gall tumorigenesis. *Cell*, **11**, 263.

CHILTON, M.-D., SAIKI, R. K., YADAV, N., GORDON, M. P. & QUETIER, F. (1980). T-DNA from *Agrobacterium* Ti plasmid is in the nuclear DNA fraction of crown gall tumor cells. *Proceedings of National Academy of Sciences, U.S.A.*, **77**, 4060–4.

DE FRAMOND, A. J., BARTON, K. A. & CHILTON, M.-D. (1983). Mini-Ti: a new vector strategy for plant genetic engineering. *Biotechnology*, **1**, 262–9.

DE GREVE, H., DECRAEMER, H., SEURINCK, J., VAN MONTAGU, M. & SCHELL, J. (1981). The functional organization of the octopine *Agrobacterium tumefaciens* plasmid pTiB6S3. *Plasmid*, **6**, 235–48.

DOUGLAS, C., HALPERIN, W., GORDON, M. & NESTER, E. (1985). Specific attachment of *Agrobacterium tumefaciens* to bamboo cells in suspension culture. *Journal of Bacteriology*, **171**, 764–6.

ENGSTRÖM, P., ZAMBRYSKI, P., VAN MONTAGU, M. & STACHEL, S. (1987). Characterization of *A. tumefaciens* virulence proteins induced by the plant factor acetosyringone. *Journal of Molecular Biology* (submitted).

EVERETT, R. & WILLETTS, N. (1980). Characterization of an *in vivo* system for nicking at the origin of conjugal DNA transfer of the sex factor F. *Journal of Molecular Biology*, **136**, 129–50.

GARDNER, R. C. & KNAUF, V. C. (1986). Transfer of *Agrobacterium* DNA to plants requires a T-DNA border but not the *vir*E locus. *Science*, **231**, 725–7.

GARFINKEL, D. J. & NESTER, E. W. (1980). *Agrobacterium tumefaciens* mutants affected in crown gall tumorigenesis and octopine catabolism. *Journal of Bacteriology*, **144**, 732–43.

GARFINKEL, D. J., SIMPSON, R. B., REAM, L. W., WHITE, F. F., GORDON, M. P. & NESTER, E. W. (1981). Genetic analysis of crown gall: fine structure map of the T-DNA by site-directed mutagenesis. *Cell*, **27**, 143.

HEERNALSTEENS, J.-P., DE GREVE, H., VAN MONTAGU, M. & SCHELL, J. (1978). Mutagenesis by insertion of the drug resistance transposon Tn7 applied to the Ti plasmid of *Agrobacterium tumefaciens*. *Plasmid*, **1**, 218–25.

HEERNALSTEENS, J. P., VAN VLIET, F., DE BEUCKELEER, M., DEPICKER, A., ENGLER, G., LEMMERS, M., HOLSTERS, M., VAN MONTAGU, M. & SCHELL, J. (1980). The *Agrobacterium tumefaciens* Ti plasmid as a host vector system for introducting foreign DNA in plant cells. *Nature*, **287**, 654–6.

HILLE, J., DEKKER, M., LUTTIGHUIS, H. O., VAN KAMMEN, A. & ZABEL, P. (1986). Detection of T-DNA transfer of plant cells by *A. tumefaciens* virulence mutants using agroinfection. *Molecular & General Genetics*, **205**, 411–16.

HIROOKA, T. & KADO, C. (1986). Location of the right boundary of the virulence region on *Agrobacterium tumefaciens* plasmid pTiC58 and a host-specific gene next to the boundary. *Journal of Bacteriology*, **168**, 237–43.

HOEKEMA, A., HIRSCH, P. R., HOOYKAAS, P. J. J. & SCHILPEROORT, R. A. (1983). A binary plant vector strategy based on separation of *vir-* and T-region of the *Agrobacterium tumefaciens* Ti plasmid. *Nature*, **303**, 179–81.

HOLSTERS, M., SILVA, B., VAN VLIET, F., GENETELLO, C., DE BLOCK, M., DHAESE, P., DEPICKER, A., INZE, D., ENGLER, G., VILLARROEL, R., VAN MONTAGU, M. & SCHELL, J. (1980). The function organization of the nopaline *A. tumefaciens* plasmid pTiC58. *Plasmid*, **3**, 212–30.

HOLSTERS, M., VILLARROEL, R., GIELEN, J. SEURINCK, J., DE GREVE, H., VAN MONTAGU, M. & SCHELL, J. (1983). An analysis of the boundaries of the octopine TL-DNA in tumors induced by *Agrobacterium tumefaciens*. *Molecular & General Genetics*, **190**, 35–41.

JOOS, H., INZE, D., CAPLAN, A., SORMANN, M., VAN MONTAGU, M. & SCHELL, J. (1983). Genetic analysis of T-DNA transcripts in nopaline crown galls. *Cell*, **32**, 1057–67.

KLEE, H. J., GORDON, M. P. & NESTER, E. W. (1982). Complementation analysis of *Agrobacterium tumefaciens* Ti plasmid mutations affecting oncogenicity. *Journal of Bacteriology*, **150**, 327–31.

KOUKOLIKOVA-NICOLA, Z., SHILLITO, R. D., HOHN, B., WANG, K., VAN MONTAGU, M. & ZAMBRYSKI, P. (1985). Involvement of circular intermediates in the transfer of T-DNA from *Agrobacterium tumefaciens* to plant cells. *Nature*, **313**, 191–6.

LEEMANS, J., DEBLAERE, R., WILLMITZER, L., DE GREVE, H. C., HERNALSTEENS, J.-P., VAN MONTAGU, M. & SCHELL, J. (1982). Genetic identification of functions of TL-DNA transcripts in octopine crown galls. *The EMBO Journal*, **1**, 147–52.

LEMMERS, M., DE BEUCKELEER, M., HOLSTERS, M., ZAMBRYSKI, P., DEPICKER, A., HERNALSTEENS, J. P., VAN MONTAGU, M. & SCHELL, J. (1980). Internal organization, boundaries and integration of Ti-plasmid DNA in nopaline crown gall tumours. *Journal of Molecular Biology*, **144**, 353–76.

MATTHYSSE, A. G. (1986). Initial interactions of *A. tumefaciens* with plant host cells. *Critical Reviews in Microbiology*, **13**, 281–307.

OOMS, G., KLAPWIJK, P., POULIS, J. & SCHILPEROORT, R. (1980). Characterization of Tn904 insertions in octopine Ti plasmid mutants of *Agrobacterium tumefaciens*. *Journal of Bacteriology*, **144**, 82–91.

OTTEN, L., DE GREVE, H., LEEMANS, J., HAIN, R., HOOYKAAS, P. J. J. & SCHELL, J. (1984). Restoration of virulence of Vir region mutants of *Agrobacterium tumefaciens* strain B6S3 by coinfection with normal and mutant *Agrobacterium* strains. *Molecular & General Genetics*, **195**, 159–63.

PERALTA, E. G., HELLMISS, R. & REAM, W. (1986). *Overdrive*, a T-DNA transmission enhancer on the *A. tumefaciens* tumour-inducing plasmid. *The EMBO Journal*, **5**, 1137–42.

PUEPPKE, S. G. & BENNY, U. K. (1981). Induction of tumors of *Solanum tuberosum* L. by *Agrobacterium*: quantitative analysis, inhibition by carbohydrates and virulence of selected strains. *Physiological Plant Pathology*, **18**, 169–79.

PUVANESARAJAH, V., SCHELL, F. M., STACEY, G., DOUGLAS, C. J. & NESTER E. W. (1985). A role for β-2-glucan in the virulence of *Agrobacterium tumefaciens*. *Journal of Bacteriology*, **164**, 102–6.

SHAW, C. H., ASHBY, A. M. & WATSON, M. D. (1986). Plant tumour induction. *Nature*, **324**, 415.

SCHELL, J., VAN MONTAGU, M., DE BEUCKELEER, M., DE BLOCK, M., DEPICKER, A., DE WILDE, M., ENGLER, G., GENETELLO, C., HERNALSTEENS, J. P., HOLSTERS, M., SEURINCK, J., SILVA, B., VAN VLIET, F. & VILLARROEL, R. (1979). Interactions and DNA transfer between *Agrobacterium tumefaciens*, the Ti-plasmid and the plant host. *Proceedings of the Royal Society of London B*, **204**, 251–66.

SIMPSON, R. B., O'HARA, P. J., KWOK, W., MONTOYA, A. L., LICHTENSTEIN, C., GORDON, M. P. & NESTER, E. W. (1982). DNA from the A6S/2 crown gall tumor contains scrambled Ti-plasmid sequences near its junctions with the plant DNA. *Cell*, **29**, 1005–14.

STACHEL, S. E., MESSENS, E., VAN MONTAGU, M. & ZAMBRYSKI, P. (1985). Identification of the signal molecule produced by wounded plant cells that activate T-DNA transfer in *Agrobacterium tumefaciens*. *Nature*, **318**, 624–9.

STACHEL, S. E. & NESTER, E. W. (1986). The genetic and transcriptional organization of the *vir* region of the A6 Ti plasmid of *Agrobacterium tumefaciens*. *The EMBO Journal*, **5**, 1445–54.

STACHEL, S. W., TIMMERMAN, B. & ZAMBRYSKI, P. (1986). Generation of single-stranded T-DNA molecules during the initial stages of T-DNA transfer from *Agrobacterium tumefaciens* to plant cells. *Nature*, **322**, 706–12.

STACHEL, S. E., TIMMERMAN, B. & ZAMBRYSKI, P. (1987). Activation of *A. tumefaciens vir* gene expression generates multiple single-stranded molecules from the pTiA6 T-region: requirement for 5' *virD* gene products. *The EMBO Journal*, in press.

STACHEL, S. E. & ZAMBRYSKI, P. C. (1986). *VirA* and *VirG* control the plant-induced activation of the T-DNA transfer process of *A. tumefaciens*. *Cell*, **46**, 325–33.

THOMASHOW, M. F., NUTTER, R., MONTOYA, A. L., GORDON, M. P. & NESTER, E. W. (1980). Integration and organization of Ti-plasmid sequences in crown gall tumors. *Cell*, **19**, 729–39.

TEMPE, J., ESTRADE, C. & PETIT, A. (1978). In: *Proceedings IVth International Conference on Plant Pathogenic Bacteria*, M. Ride (ed.) Angers, I.N.R.A., 153.

TEMPE, J. & SCHELL, J. (1977). In: *Translation of natural and synthetic polynucleotides*, AB Legocki (Ed.) Poznan, University of Agriculture, 416.

VAN LAREBEKE, N., ENGLER, G., HOLSTERS, M., VAN DEN ELSACKER, S., ZAENEN, J., SCHILPEROORT, R. A. & SCHELL, J. (1974). Large plasmid in *Agrobacterium tumefaciens* essential for crown gall-inducing ability. *Nature*, **252**, 169–70.

VAN LAREBEKE, N., GENETELLO, C., SCHELL, J., SCHILPEROORT, R. A., HERMANS, A. K., HERNALSTEENS, J. P. & VAN MONTAGU, M. (1975). Acquisition of tumour-inducing ability by non-oncogenic agrobacteria as a result of plasmid transfer. *Nature*, **255**, 742–3.

WANG, K., STACHEL, S., TIMMERMAN, B., VAN MONTAGU, M. & ZAMBRYSKI, P. (1987). Site-specific nick in the T-DNA border sequence as a result of *Agrobacterium vir* gene expression. *Science*, **235**, 587–91.

WATSON, B., CURRIER, T. C., GORDON, M. P., CHILTON, M.-D. & NESTER, E. W. (1975). Plasmid required for virulence of *Agrobacterium tumefaciens. Journal of Bacteriology*, **123**, 255.

WILLMITZER, L., DE BEUCKELEER, M., LEMMERS, M., VAN MONTAGU, M. & SCHELL, J. (1980). DNA from Ti-plasmid is present in the nucleus and absent from plastids of plant crown-gall cells. *Nature*, **287**, 359–61.

WILLMITZER, L., DHAESE, P., SCHREIER, P. H., SCHMALENBACH, W., VAN MONTAGU, M. & SCHELL, J. (1983). Size, location, and polarity of T-DNA-encoded transcripts in nopaline crown gall tumors; evidence for common transcripts present in both octopine and nopaline tumors. *Cell*, **32**, 1045–56.

WILLMITZER, L., SIMONS, G. & SCHELL, J. (1982). The TL-DNA in octopine crown gall tumours codes for seven well-defined polyadenylated transcripts. *The EMBO Journal*, **1**, 139–46.

YADAV, N. S., VANDERLEYDEN, J., BENNETT, D. R., BARNES, W. M. & CHILTON, M.-D. (1982). Short direct repeats flank the T-DNA on a nopaline Ti plasmid. *Proceedings of the National Academy of Sciences of the USA*, **79**, 6322–6.

YANOVSKY, M., LOWE, B., MONTOYA, A., RUBIN, R., KRUL, W., GORDON, M. & NESTER, E. (1985). Molecular and genetic analysis of factors controlling host range in *Agrobacterium tumefaciens. Molecular & General Genetics*, **201**, 237–46.

YANOFSKY, M. F., PORTER, S. G., YOUNG, C., ALBRIGHT, L. M., GORDON, M. P. & NESTER, E. W. (1986). The *virD* operon of *Agrobacterium tumefaciens* encodes a site-specific endonuclease. *Cell*, **47**, 471–7.

ZAENEN, I., VAN LAREBEKE, N., TEUCHY, H., VAN MONTAGU, M. & SCHELL, J. (1974). Supercoiled circular DNA in crown gall-inducing *Agrobacterium* strains. *Journal of Molecular Biology*, **86**, 109.

ZAMBRYSKI, P., DEPICKER, A., KRUGER, K. & GOODMAN, H. (1982). Tumor induction by *Agrobacterium tumefaciens*: analysis of the boundaries of T-DNA. *Journal of Molecular Applied Genetics*, **1**, 361–70.

ZAMBRYSKI, P., JOOS, H., GENETELLO, C., LEEMANS, J., VAN MONTAGU, M. & SCHELL, J. (1983). Ti plasmid vector for the introduction of DNA into plant cells without alternation of their normal regeneration capacity. *The EMBO Journal*, **2**, 2143.

INDEX